PHOTON COUNTING - FUNDAMENTALS AND APPLICATIONS

Edited by **Nikolay Britun**
and **Anton Nikiforov**

Photon Counting - Fundamentals and Applications
http://dx.doi.org/10.5772/intechopen.69183
Edited by Nikolay Britun and Anton Nikiforov

Contributors

Filippo Ambrosino, Franco Meddi, Edward M. D. Fisher, Keyu Xia, Robert Thomas Youker, Luca Caucci, Yijun Ding, Harrison Barrett, Jan Voráč, Pavel Dvořák, Martina Mrkvičková, Nicola D'Ascenzo, Qingguo Xie, Fabio Acerbi, Matteo Perenzoni, Frédéric Marchal, Jean-Pierre Gardou, N. Merbahi, Mohammed Yousfi, Neermalsing Sewraj, Tatsuo Shiina, Wilfried Uhring, Imane Malass, Jean-Pierre Le Normand, Norbert Dumas, Foudil Dadouche, Fardad Shakibaie, Laurence Walsh, Halina Rubinsztein-Dunlop, Laurent Lamard

Published by InTech
Janeza Trdine 9, 51000 Rijeka, Croatia

Publishing Process Manager Martina Usljebrka
Technical Editor SPi Global
Cover InTech Design team

First published March, 2018
Printed in Croatia
Legal deposit, Croatia: National and University Library in Zagreb

Additional hard copies can be obtained from orders@intechopen.com

Photon Counting - Fundamentals and Applications, Edited by Nikolay Britun and Anton Nikiforov
p. cm.
Print ISBN 978-953-51-3907-2
Online ISBN 978-953-51-3908-9

For EU product safety concerns:
IN TECH d.o.o., Prolaz Marije Krucifikse Kozulić 3, 51000 Rijeka, Croatia,
info@intechopen.com or visit our website at intechopen.com.

Meet the editors

Nikolay Britun was born in Kiev, Ukraine, in 1979. He received his bachelor's and master's degrees from the Radiophysics Department, Kiev National University, in 2000 and 2002, respectively, and PhD degree in spectroscopic diagnostics of the low-temperature plasma from the Sungkyunkwan University, Suwon, South Korea. He currently holds a postdoctoral position at the "Chimie des Interactions Plasma-Surface" (ChIPS) Laboratory, University of Mons, Belgium. His research interests are related to nonintrusive characterization of low-temperature and fusion plasmas by optical emission, absorption, laser-based, and related methods, including diagnostics of the magnetron, microwave, and various atmospheric discharges.

Anton Nikiforov received his master's degree in Chemistry in 2001 from the Ivanovo State University of Chemistry and Technology, Russia. In 2004, he obtained his PhD degree in Chemistry from the Institute of Solution Chemistry of the Russian Academy of Sciences. Since 2009, he is working at the Department of Applied Physics, Ghent University, Belgium. His main research activity is essentially experimental work that focuses on physics and applications of nonequilibrium and fusion-related plasmas and advanced plasma diagnostics. His current research interests include the study of plasma-surface interaction, plasma diagnostics and laser spectroscopy, and emission and absorption spectroscopy, as well as Doppler-free spectroscopy for electric field measurements.

Contents

Preface

Highly sensitive measurements are almost inevitable in the modern science and technology, particularly in the optical domain. There are numerous research applications involving optical imaging and optical spectroscopy, which deal with extremely low-photon fluxes. These applications are astronomical measurements, interferometry, fluorescence, medical imaging, and tomography, among them. For all of these applications, statistical detection of a single-photon arrival represents an important event.

Photon counting is a unified name for the techniques, which use single-photon detection for accumulative measurements of the light flux, normally occurring under extremely low-light conditions. Various types of single-photon detectors along with the corresponding electrical circuits are used in parallel in order to achieve this goal. Nowadays, photon counting can be applied to the wide variety of radiation wavelengths, starting from X-ray and deep ultraviolet transitions and ending with far-infrared part of the spectrum.

As a special tribute to the photon counting, we have to note that several important scientific advances, including the studies of cosmic microwave background radiation in astronomy, the experiments with muon detection, and the large-scale fundamental experiments on the nature of matter, are hardly imaginable without this technique.

Under the fast development of modern science and corresponding technological solutions, an extended overview combining the basic knowledge with the recent advances of a certain technique is always of a great interest. Thus, the goal of the book is to provide readers with an overview on the fundamentals of photon counting and to describe some state-of-the-art applications of this technique in the applied science and everyday life. The historical backgrounds and the important technical solutions, such as the development of key light-sensitive photodetectors, are among the main subjects.

The book consists of 12 chapters ranging from historical development of the fast photon detectors to the LIDAR and tomography applications with single-photon detection capability. It is subdivided into three main sections covering (I) the historical aspects of the low-light detection, (II) the fundamental aspects and applications of photon counting, and finally, (III) the additional technical applications of this technique, which may be demanded in the modern life.

We hope that the contributions from this book may serve as handbook materials not only for the experienced researchers but also for students and experimentalists in the areas of low-light detection and optical spectroscopy. We also hope that the described progress in the field should inspire some scientists with fruitful ideas in order to define new potential directions in the domain of single-photon and low-light detections. The editors would like to acknowledge their host institutions and the contributing authors for the support and fruitful interactivity during the preparation of this book.

Dr. Nikolay Britun
Plasma-Surface Interactions Chemistry Laboratory
University of Mons, Belgium

Dr. Anton Nikiforov
Research Unit of Plasma Technology
Ghent University, Belgium

Section 1

Overview and Historical Aspects

Chapter 1

Principles and Early Historical Development of Silicon Avalanche and Geiger-Mode Photodiodes

Edward M.D. Fisher

Additional information is available at the end of the chapter

http://dx.doi.org/10.5772/intechopen.72148

Abstract

The historical development of technology can inform future innovation, and while theses and review articles attempt to set technologies and methods in context, few can discuss the historical background of a scientific paradigm. In this chapter, the nature of the photon is discussed along with what physical mechanisms allow detection of single-photons using solid-state semiconductor-based technologies. By restricting the scope of this chapter to near-infrared, visible and near-ultraviolet detection we can focus upon the internal photoelectric effect. Likewise, by concentrating on single-photon semiconductor detectors, we can focus upon the carrier-multiplication gain that has allowed sensitivity to approach the single-photon level. This chapter and the references herein aim to provide a historical account and full literature review of key, early developments in the history of photodiodes (PDs), avalanche photodiodes (APDs), single-photon avalanche diodes (SPADs), other Geiger-mode avalanche photodiodes (GM-APDs) and silicon photo-multipliers (Si-PMs). As there are overlaps with the historical development of the transistor (1940s), we find that development of the p-n junction and the observation of noise from distinct crystal lattice or doping imperfections – called *"microplasmas"* – were catalysts for innovation. The study of microplasmas, and later dedicated structures acting as known-area, uniform-breakdown artificial microplasmas, allowed the avalanche gain mechanism to be observed, studied and utilised.

Keywords: single-photon avalanche diodes, SPAD, p-n junctions, photodiodes, avalanche, multiplication, internal photoelectric effect, photon counting, history, Geiger-mode, gain

1. Introduction

Optics has seen significant progress in the last 100 years. We now take as routine that we can detect single-photons, count them and time them in such an accurate manner that we can study phenomena that would have been treated as science fiction in the days of quantum pioneers such as Max Planck, Albert Einstein and Werner Heisenberg. We are able to use time of flight, at the

2.1. Photon or anti-photon: an elusive concept

In Loudon's treatise on the quantum theory of light [9], it is made clear from the outset that the word photon is somewhat of a vague, theory-loaded term [6] that can lead to confusion. The word was coined by G. Lewis in a 1926 Nature paper. This is surprising as many would cite Einstein's 1905 paper on the photoelectric effect [10] as a suitable definition of a 'photon' i.e. a single quantum of electromagnetic radiation [11]. The history of quantized optics is described by Lamb [12] and Loudon [9], with much modern theory and philosophical questions discussed in [13]. However, Lamb was highly critical of the term suggesting it should be used only by *"properly qualified people"*. Indeed, Einstein famously stated, *"All the fifty years of conscious brooding have brought me no closer to answering the question, "What are light quanta?" Of course, today every rascal thinks he knows the answer, but he is deluding himself."*

One may ask where Lamb's criticism of the word photon comes from. Firstly, Lewis when coining the term, suggested it as a real, physical particle (i.e. realism), as a method of explaining chemical valence. Specifically, he hypothesised the photon as a mediator of radiation from one atom to another, helping to explain how a molecule such as Hydrogen gas can be stabilised by two electrons that sometimes can have a strong attractive rather than repulsive force. However, he explicitly [12] denied it related to the quantum of light discussed by Planck and Einstein. Secondly, Lamb expounds the view that all uses of the term 'photon' can be more accurately thought of using quantization of wave interpretations, explicitly he states: *"With more complicated states, it is terribly difficult to talk meaningfully about 'photons' at all"*. He cites the work of Wentzel and Beck (1926) as they show that the photoelectric effect can be described by quantum theory, without the use of light quanta [12].

2.2. Photons: energy and statistics

As Loudon suggests, the impression the word gives is of an indistinct, fuzzily-bound globule of light that travels from point A to B through optical equipment or free space. Many would conceive photons as bullets travelling as a stream making up distinct rays. Despite that view, a photon can be more correctly thought of as an electromagnetic field within a cavity of length L. As with sound waves, there are an infinite set of spatial-modes discretized into integer divisions of the cavity spacing, i.e. $L/2$ etc. Extension to open cavities can be achieved by, considering an experiment of finite size but with no identifiable cavity or viewing the system as discrete travelling-wave modes [9]. This scenario can be described by a quantized harmonic-oscillator of angular frequency, ω. If we take a single spatial-mode, we can write its Hamiltonian in the form of Eq. (1) [9, 12]. This relates to a pseudo particle with an effective mass, μ, an angular frequency, ω, a position, x and a momentum, p. This a wave-mechanical description of a particle, where matrix mechanics (Heisenberg 1925–1926) can be used to form an analogy with the 4-dimensional (x, y, z, t) electric and magnetic fields within a cavity.

$$H = \frac{1}{2}\left(\frac{p^2}{\mu} + \mu\omega^2 x^2\right) \tag{1}$$

2.2.1. Energy

As Lamb suggests, rather than using x, y, z as spatial coordinate labels for a spatial-mode, the vector $\boldsymbol{k}$ can be used which is called the wave-vector. For a simple 1-D case, $k = |\boldsymbol{k}| = 1$ for the fundamental, i.e. a single wave period within the cavity, $k = 2$ for the second harmonic and $k = n$ for n periods of the wave in the cavity. To come back to 'photons' and their definition, if we assume we have a single mode, which has an associated 'number state', n or $|n\rangle$, we can give the energy of that state in the form of Eq. (2) [9, 12]. The state $|0\rangle$ is called the ground or vacuum state and still represents a finite energy level, although this cannot be detected without specialist experimentation and does not contribute to photon counting [9].

$$E_n^k = \left(n_k + \frac{1}{2}\right)\hbar\omega_k \tag{2}$$

The spatial-mode, k, therefore contains n_k 'photons', and the angular frequency is, $\omega = ck$. As we can consider photon 'creation' as being the increase in electromagnetic energy in a mode, while a decrease in energy represents photon 'destruction', the state can only be excited by integer multiples of $\hbar\omega$, where $\hbar$ is the reduced Planck constant, $h/2\pi$. The energy of a photon is defined using Eq. (3) [9], where c is the speed of light and λ is the wavelength.

$$E_p = \hbar\omega = h\nu = \frac{hc}{\lambda} \tag{3}$$

2.2.2. *Statistics*

One particular reason why the single packet of energy has become useful as a conceptual framework, is that the single-photon state can produce a single current-pulse in a photodetector that uses ionisation ([9] p2). This significant point allows us to use the word 'photon' from a detection viewpoint, removing some wave and probability distribution details. If we try to measure the number of photons within a mode, we would find a probability, $P(n)$, of finding a given number. This we will see leads to Poisson statistics. Photon creation, destruction and thermal variation will contribute to the fluctuation in photon number. In Eq. (4), $\langle n\rangle$ denotes the mean number of thermally excited photons in a field-mode at temperature, T, with Boltzmann constant k_B [9]. This is called the Planck thermal excitation function ([9] p10) and represents the probability that the mode is excited to $\langle n\rangle$ photons at thermal equilibrium.

$$\langle n\rangle = \frac{1}{exp\left(\frac{\hbar\omega}{k_B T}\right) - 1} \tag{4}$$

If we think of an ensemble of measurements (over time or separate identical systems), the probability, $P(n)$, of finding exactly n photons within an optical cavity with a mean photon number, $\langle n\rangle$, is given by the Geometric distribution of Eq. (5). This holds only if we consider a period, t, greater than the characteristic time scale of any fluctuations ([9] p14).

$$P(n) = \frac{\langle n\rangle^n}{(1 + \langle n\rangle)^{1+n}} \tag{5}$$

Lasers and the radiative transitions of excited atoms are often treated as ideal. Despite this, there are processes such as collision, power and Doppler broadening that modify the shape of a spectral emission line [9]. These allow the electric field and amplitude of a light beam to vary in a time dependent manner. The time scales of these fluctuations are inversely proportional to the differences in optical frequencies produced by line-broadening processes. Hence a fine line-width laser is more coherent than a chaotic source such as an incandescent lamp. If we consider a continuous wave emitted by an atom of a gas, a collision with another atom will present a random, abrupt change in the phase of the wave ([9] p84). The mean time between such events is called the coherence time, τ_c, and is expressed by Eq. (6), where λ is the center wavelength, $\Delta\lambda$ is the spread of the line due to broadening and c is the speed of light. For an example 650 nm laser, a spectrum of 1 nm width yields a coherence time of 1.41 ps. For a LED, the spectrum may be wider with a FWHM of 20 nm, giving a coherence time of 0.07 ps.

$$\tau_c \cong \frac{\lambda^2}{c\Delta\lambda} \tag{6}$$

Expanding this to many atoms, we can write the total electric field amplitude, $E(t)$, as Eq. (7), where $a(t)$ represents the amplitude variation and $\phi(t)$ represents the phase variation with time, t, of the phasor addition of all constituent radiating atoms ([9] p84).

$$E(t) = E_0\, exp\,(-i\omega_0 t)a(t)\exp\left(i\phi(t)\right) \tag{7}$$

Detection can be viewed using semi-classical theory [9]. This is the combination of a classical (i.e. wave) treatment of incident radiation with a quantum (i.e. particle) treatment for the atomic detector. If we call m the number of photon counts produced in an experimental integration time, T_e, and repeat the experiment multiple times, we can define a probability of finding m photodetections, $P_m(T_e)$. The mean number of photodetections, $\langle m\rangle$, can be calculated from $\langle m\rangle = \xi\bar{I}T_e$, where ξ models the detector's finite efficiency and $\bar{I}$ is a cycle averaged incident intensity. Loudon ([9] p121) derives $P_m(T_e)$, Eq. (8), which has the same form as the Poisson distribution, with the rate parameter, λ_r, being equal to $\langle m\rangle$.

$$P_m(T_e) = \frac{\langle m\rangle^m}{m!}e^{-\langle m\rangle} \tag{8}$$

This applies to light from stimulated emission, i.e. coherent light, but also for chaotic light when the integration time of the measurement is much longer than the coherence time, such that fluctuations in intensity are averaged. There is therefore a continuum between the Geometric and Poisson distributions depending on the integration time and coherence time ([9] p122). Normally we consider the variance of the Poisson distribution to be equal to the mean, i.e. $\sigma^2 = var(x) = \langle m\rangle$, however fluctuations in light produces a departure from this assumption. The variance of the distribution is given by Eq. (9), where $\bar{I}(t, T_e)$ is the mean intensity that falls on the photodetector during the period from t to $t + T_e$ ([9] p122].

$$\sigma^2 = (\Delta m)^2 = \langle m\rangle + \xi^2 T_e^2\left[\left\langle \bar{I}(t, T_e)^2\right\rangle - \bar{I}^2\right] \tag{9}$$

The first term is photon shot noise and is the primary assumption used for the statistics of coherent light. The second term represents an excess noise linked to *"wave fluctuations"* of both coherent and incoherent sources. The variance simplifies to Eq. (10) ([9] p122). If the measurement time is longer than the coherence time, then the second term is negligible and we can assume that $\sigma^2 = \langle m\rangle$, i.e. pure Poisson statistics. If the measurement time is equal to or less than the coherence time, then the second term increases leading to super-Poisson statistics. Sub-Poisson statistics, i.e. squeezed light states are also valid, however this requires a quantum treatment of the electromagnetic field ([9] p201).

$$\sigma^2 = \langle m \rangle + \frac{\langle m \rangle^2 \tau_c}{T_e} \tag{10}$$

However, for chaotic light, such as a thermal source or a LED, the variance of Eq. (10) simplifies into a different, more Geometric form, given by Eq. (11) ([9] p122 and p199). Here we see that such sources follow super-Poisson statistics. It is now understandable why lasers with fine line widths are used for (i) physical experiments requiring photon counting and (ii) high speed (100 Gb/s) optical communication links where noise about mean signal amplitudes represents a significant contribution to errors.

$$\sigma^2 = \langle m \rangle + \langle m \rangle^2 \tag{11}$$

3. High-sensitivity optical to electrical conversion

There are three processes that lead to photon interaction, and thus detection [14]. These are the (i) photoelectric effect [11, 15], (ii) Compton scattering and (iii) pair production, with the final two phenomena only occurring at high photon energies.

- A photon can be scattered by an atomic electron, a process called Compton scattering. It imparts some energy to the electron, meaning that the photon energy is decreased, and thus the wavelength becomes longer. Compton scattering therefore does not destroy the photon. As noted in [14], this effect is small for energies *"below tens of KeV,"* i.e. 10KeV is 0.124 nm and 1KeV is 1.24 nm.
- In pair-production, the photon energy is enough to result in the creation of an electron-positron pair, i.e. the anti-matter counterpart of the electron. The energy must be higher than $E_p = 2m_e c^2$, where m_e is the electron rest-mass. As this is 1.02 MeV, i.e. 0.0012 nm, this process occurs only for X-ray and Gamma-ray interactions [15].
- Both processes can thus be discounted from many photon counting applications operating in the UV, visible and near-IR [14, 15].

This elimination leaves the photoelectric effect – which is subdivided into the external and internal photoelectric effects [11, 14–16]. We also have further subdivision into the photoconductive effect and photovoltaic effect. Due to the prevalence of these key words in peer-reviewed literature [17, 18], legal patents, company white papers and graduate and undergraduate textbooks, they represent the time and geography independent nomenclature for the field. The internal photoelectric effect is certainly not *"irrelevant"* for solid-state photonic technologies [11, 15, 18].

3.1. The internal and external photoelectric effect

In the internal and external photoelectric processes, a photon is absorbed, thus destroying the photon. If the photon energy, $\boldsymbol{E_p}$, is sufficient to overcome the surface work function, $\boldsymbol{q\phi_b}$, of the material (Eq. (12)), a bound-free transition takes place whereby an electron is promoted from an outer electron orbital and is expelled from the surface into an external vacuum (of permittivity, ε_0) [13, 14, 17]. The remaining energy is accounted for by the kinetic energy of the electron as a free particle [13], whereby it can be accelerated and cause secondary electron emission. This *"external photoelectric effect"* is utilised within photomultiplier tubes for primary photoelectron generation. The work function, $\boldsymbol{\phi}_m$, of Silicon is ~4.6 eV (requiring photons of wavelength shorter than 260 nm, i.e. UV). This prompted PMTs to use metals with lower work functions such as Caesium, Rubidium and Antimony with work functions of 1.95, 2.26, 4.55 eV for wavelengths of 635, 548 and 272 nm respectively. The Schottky effect, whereby an electric field, $\boldsymbol{\varepsilon}$, can lower the potential barrier between the material and vacuum, $\boldsymbol{\Delta\phi}$, is also key in PMTs [19] allowing a reduced 'effective' work function and thus a longer wavelength detection threshold.

$$E_p = h\upsilon \geq q\phi_b = q(\phi_m - \Delta\phi) = q\left(\phi_m - \sqrt{\frac{q\varepsilon}{2\pi\epsilon_0}}\right) \tag{12}$$

In contrast, semiconductors have a band-gap, E_g, between the valence, E_v (i.e. outer-orbital bound electrons) and conduction electrons, E_c (i.e. a cloud of delocalized electrons) [17, 18]. A photon with energy greater than this band-gap $(E_p \geq E_g)$ can promote an electron from the valence to the conduction band [16, 17]. As the absence of an electron in a valence state is described as a hole, the *"internal photoelectric effect"* produces an electron–hole pair [16–20]. This is a bound-bound or intrinsic transition [13, 15]. The electron is still ejected from the atom; however, it is not ejected from the surface [16]. If two electrodes are placed on the material with a slight potential gradient, or if that potential exists due to a p-n doped junction within the material [20], the electron–hole pair are separated and drift apart due to their relative charges. With many photo-generated carriers within the material due to numerous incident photons, the bulk conductivity of the material increases [16, 17], allowing a photocurrent to flow through an external circuit. This is the photoconductive mode. Photons of high energy are highly likely to cause band to band transitions, however as the wavelength increases towards a photon energy close to the band-gap, the likelihood of transition decreases, given by the absorption coefficient, α [20]. This leads to a long-wavelength cut off, λ_c, given by Eq. (13) [19]. For Silicon, this is 1.1 μm, where the absorption coefficient is $1\text{x}10^1$ cm^{-1}, whereas at 400 nm it is $1\text{x}10^5$ cm^{-1}.

$$\lambda_C = \frac{1.24}{E_g} \tag{13}$$

As materials have various band-gap energies, diverse materials can be used to detect different wavelengths (e.g. Silicon 350 nm to 900 nm and Germanium 750 nm to 1.6 μm). Typically, one electron–hole pair is produced per absorbed photon, limiting the quantum efficiency (typically ≤1), and the spectral responsivity. The optically-induced current, I_p, can be calculated using Eq. (14) [19], assuming a detector thickness much larger than the light penetration depth, $(1/\alpha)$. P_{OPT} is the incident optical power, μ_n is the electron mobility, η is the quantum efficiency, L is the distance between the contacts, and τ is the carrier lifetime.

$$I_P = q\left(\eta \frac{P_{OPT}}{hv}\right)\left(\frac{\mu_n \tau \varepsilon}{L}\right) \tag{14}$$

As the average depth of absorption changes with wavelength, the depth of a photodiode p-n junction is chosen to maximise the received photocurrent. The width of the junction is critical in this; however, it also has implications for bandwidth, which is restricted by three phenomena, (i) the capacitance of the junction, (ii) the time delay of carriers generated outside of the junction, diffusing into that junction, and (iii) the drift or transit time, τ_r, of the carriers within the junction (Eq. (15)) [19].

$$t_r = \frac{L}{\mu_n \varepsilon} \tag{15}$$

4. Avalanche multiplication gain: high-sensitivity detection

A challenge central to photon counting, is how to apply enough gain such that a single electron–hole pair can produce an appreciable signal for detection. As all electrical readout schemes include thermal noise [19], the signal, $S_{ph}(t)$, we obtain from the detection of a single-photon can easily be hidden by thermal noise, $n_{th}(t)$. The signal-to-noise ratio (SNR), would be given by: $SNR = |S_{ph}(t)|/|n_{th}(t)|$. If we were to use a traditional amplifier to provide a gain, $g = 1000$, the amplifier would also amplify the input noise and would likely contribute further noise terms, $n_{amp}(t, g)$. This would give an amplified output with a SNR given by Eq. (16), i.e. the SNR would not benefit from amplification.

$$SNR = \frac{g|S_{ph}(t)|}{g|n_{th}(t)| + |n_{amp}(t, g)|} \tag{16}$$

To provide single photon sensitivity, an ideal amplifier would need to (a) minimally amplify noise, (b) provide a gain to separate the signal from noise sources, (c) have a high bandwidth allowing fast temporal resolution of an event, (d) contribute little extra noise and (e) for many applications would need to be both small and low power. This is where avalanche gain or carrier multiplication becomes critical for modern photon counting applications.

4.1. The avalanche upshot

For a p-n junction, there exists a small built-in potential, V_{bi}, which separates photo-generated electron–hole pairs. The avalanche multiplication process occurs at an increased reverse bias voltage, V_r, in comparison to standard photodiodes ([19] p317). At this bias, the total energy difference between the p- and n-type regions becomes: $q(V_{bi} + V_r)$. The resultant increase in the electric field, ε, accelerates a free carrier, labelled 1 in **Figure 1**, to a kinetic energy, E_k, sufficient to overcome the ionisation energy (band-gap), E_g, of the material ([19] p79) (Eq. (17), where m_1 is the effective carrier mass and v_s is the saturation velocity). The actual values are larger than this minimum at 3.6 eV for electrons and 5.0 eV for holes. Upon a collision between a carrier such as a photoelectron and the Silicon crystal lattice, the accelerated carrier ionises

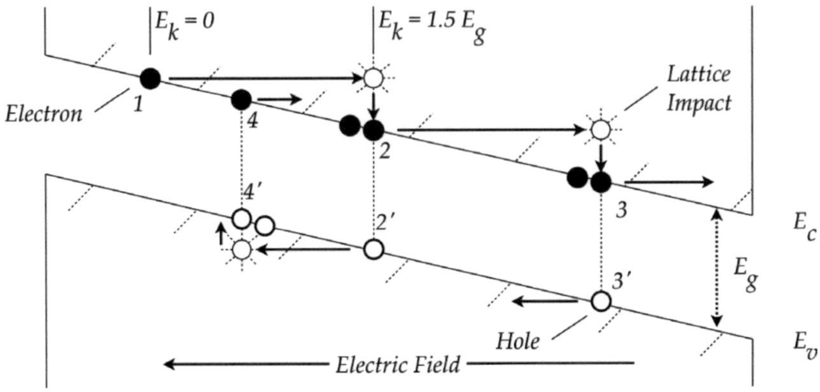

Figure 1. Energy band diagram showing avalanche multiplication. The electric field accelerates an electron (1) to a kinetic energy of E_k. Upon collision with the lattice, this energy is sufficient to ionise a bound electron, promoting it to the conduction band (2 and 2′). Adapted from [19] p79.

another carrier. An electron–hole pair, labelled 2 and 2′, is generated with those carriers then accelerated by the electric field, causing further ionisation [3, 4]. This process continues exponentially creating an avalanche of carriers within the p-n junction.

$$E_k \geq \frac{1}{2} m_1 v_s^2 = (1.5) E_g \tag{17}$$

The generation rate, G_{av}, of electron–hole pairs through impact ionisation can be calculated using Eq. (18) [19]. Here J_n and J_p are the electron and hole current densities, while α_n and α_p are the electron and hole ionisation rates, i.e. the number of carriers generated by an ionising carrier per unit distance. The ionisation rates vary for different materials and with the electric field strength. We can view the generation rate as being time varying. At time $t = -\delta t$, both J_n and J_p are zero, i.e. before a photon liberates a carrier. At $t = 0$, there is a photon absorption giving an electron–hole pair. At $t = +\delta t$, ionisation has increased J_n and J_p, giving more carriers at $t = +2\delta t$, and therefore a consequent increase in generation rate.

$$G_{av} = \frac{1}{q} \left(\alpha_n |J_n| + \alpha_p \left| J_p \right| \right) \tag{18}$$

4.2. Avalanche photodiodes (APDs)

A class of detector called an avalanche photodiode (APD) [1–3], biases the p-n junction such that carrier multiplication achieves a constraint generation rate, and thus a constant gain. Careful biasing and device design is needed to ensure the process does not lead to run-away avalanche and catastrophic junction breakdown. The diode therefore produces a photocurrent dependant on the photon flux, with the avalanche gain, M, and the width of the p-n junction, W_j, being highly dependent on the applied bias. The width also dictates the proportion of the electron–hole pairs that are captured with varying depth into the surface.

There are two main noise sources. The thermal noise, $\langle i_t^2 \rangle$, is given by Eq. (19), where k_B is the Boltzmann constant, T is the temperature, B is the bandwidth of the system and R_{eq} is the parallel combination of the load, series, junction and amplifier input resistances. As avalanche multiplication is inherently random, there is an associated multiplication of the shot noise, $\langle i_s^2 \rangle$, Eq. (20), where $F(M)$ is a noise factor associated with the multiplication, M and I_{PH} and I_D are the photo- and dark- currents respectively. The multiplication excess noise factor is highly dependent on the ratio of the electron and hole ionisation rates (Eq. (21)) ([19] p317), where ideally, we should have more electron than hole multiplication. For Silicon, the ratio of α_p/α_n may be low at 0.04. For a gain of $M = 10$, this gives an excess noise factor of 2.22.

$$\langle i_t^2 \rangle = 4k_B T\left(\frac{1}{R_{eq}}\right)B \tag{19}$$

$$\langle i_s^2 \rangle = 2q(I_{PH} + I_D)M^2F(M)B \tag{20}$$

$$F(M) = M\left(\frac{\alpha_p}{\alpha_n}\right) + \left(2 - \frac{1}{M}\right)\left(1 - \frac{\alpha_p}{\alpha_n}\right) \tag{21}$$

Luckily, the avalanche multiplication also multiplies the current caused by the internal photo-electric effect, I_{PH}. This is given by Eq. (22), where P_{OPT} is the incident optical power in watts and η is the quantum efficiency of the photodiode [19]. Avalanche gain can therefore be a highly effective technique for high-sensitivity detector technologies.

$$I_{PH} = \frac{q\eta P_{OPT}M}{hv} \tag{22}$$

4.3. Single-photon avalanche diodes (SPADs) and other technologies

Several dedicated photodiode structures can be biased further into reverse bias, giving higher gains and therefore greater sensitivities. In some cases, run-away multiplication can be used to create specialised diodes called single-photon avalanche diodes (SPADs) or Geiger-mode avalanche photodiodes (GM-APDs) [4, 5]. The Geiger region lies beyond the avalanche photodiode region but before the breakdown of a guard ring that surrounds the device. It is called this because run-away avalanche, with gain factors in the region of $M = 1x10^6$, leads to a current pulse behaviour similar to Geiger-Muller tubes. In modern SPADs, a single photon, yielding a single electron–hole pair can produce a sizable avalanche photocurrent, and in well-designed circuits can produce a voltage pulse suitable for standard complementary metal-oxide-semiconductor (CMOS) logic [4] with both high temporal accuracy (low jitter) and a short duration (5–20 ns).

Several other photon counting technologies utilise the avalanche breakdown multiplication of carriers. SPADs have been fabricated into large arrays in modern CMOS processes, allowing the advantages of high-speed dedicated on-chip logic and complex signal processing. Silicon photo-multipliers (Si-PMs) can also be made by the parallel combination of avalanche currents across a shared load resistance. Si-PMs employ less complex circuitry, reducing the prospects for single-photon imaging, but often allow higher optical fill-factors for physical experimentation. Both APDs and Geiger-mode devices can be manufactured using III-V materials such as InGaAs/InP for single-photon detection over many wavelengths. Other highly sensitive detectors include electron-multiplying charge-coupled-devices (EM-CCDs) which use avalanche multiplication and micro-channel plate (MCP) detectors which use impact ionisation and the release of secondary electrons.

While the historical literature review of this chapter will centre upon semiconductor sensors, photon detection has a long history of using the traditional photomultiplier tube (PMT). Being vacuum tubes, they are large and require high voltages. However, they are renowned for high temporal resolution and can present a large area detector with low noise and high gain. When the noise is normalised against optically active area, PMTs are often a preferred detector in comparison to solid-state solutions. The choice between detectors is of course a product of the applications, with PMTs being unsuitable for high-speed simultaneous rather than raster-scanned single-photon imaging. The history of these devices is covered well in [21], however three principal references from the 1930s collectively cover the operation principles and early development of the PMT [22–24]. Upon a photon absorption, electrons are emitted from a photo-cathode via the external photoelectric effect. A potential between the cathode and an initial dynode accelerates the electrons. Upon hitting the dynode, secondary electron emission acts to increase the number of electrons that are accelerated to an iteratively-biased set of subsequent dynodes. Thus, at the final anode an initial photo-electron can produce an appreciable anode current.

5. Single-photon avalanche diodes: principles and early history

The bulk of this chapter and its references provide a full, robust literature review dedicated to the early history of semiconductor photon counting. As p-n junctions and the avalanche process are utilised in avalanche photodiodes, single-photon avalanche diodes and silicon-photomultipliers, the sections below will track the development of such devices. Starting from initial physical studies on carrier multiplication and avalanche, development of p-n junction transistors and investigation of noise sources within these early transistors, we ultimately end at the use of artificial structures. These man-made structures allowed the study of multiplication in a more deterministic manner, eventually becoming used for the detection of ionising radiation and light. For the most part, primary sources will be referenced, with others provided in parentheses, to provide the de-facto literature review for this field, particularly for early historical developments. It must be noted that the explosion in literature from the 1970s to present, naturally restricts the scope of history that can be covered.

5.1. Early history: 1900 to 1939

During the early 20th century, the predominant electrical technology was the vacuum tube. With the rapid expansion of radio technologies there was a demand for electronic amplifiers that were capable of high-gain, high-frequency received signal amplification or the output of high power signals to increase transmission distances. Much work also centred on power rectifiers and tube-diodes, predominantly for power supplies, where demand pushed these towards higher rectification voltages and the handling of increased electrical power throughput. Three developments impacted the field, the first being the discovery of unilateral conduction of crystals (Braun in 1874) contributing to the development of crystal detectors and point-contact diodes. The second was the 1930s advancement of solid-state power rectifiers and diodes formed from Copper Oxide or Selenium. The third, was the use of trace gases within evacuated tubes, modifying the electrical properties of the device.

It is here with trace gas evacuated tubes that the story of electron multiplication and avalanche can be traced back to. In 1901 John Townsend, then at Oxford University, showed that initial ionisation of the gas between the cathode and anode of a vacuum tube via X-rays, lead to an increase in current as the electrode potential difference increased. The hypothesis being that an initial ionisation event lead to the exponential collision ionisation of gas atoms [25]. The experimental setup used by Townsend is shown in **Figure 2**, with a schematic of avalanche multiplication shown on the right. While he derived a theory for ionisation based upon the free-path between atoms and eventually the ionisation energy of the molecule, he also made use of earlier experiments (Stoletow, 1890), whereby ionisation was triggered by ultraviolet (UV) excitation of electrons from Zinc (i.e. electrons provided via the external photoelectric effect). Townsend also hypothesised, fitting against experiment, that at low potentials gas ionisation would occur only in favourable occasions, i.e. low probability, but that at high potentials the probability of ionisation increased. Townsend demonstrated differences in conductivity based upon the polarity of the potential and the shape of the electrodes, finding diode behaviour he called 'unipolar conduction'.

In 1903 Townsend extended this analysis to include positive and negative ions and the breakdown or 'sparking' potential of gases. Experimentally, UV light was used to initiate ionisation and liberation of electrons from an electrode plate. Townsend showed modification of the breakdown voltage by collision ionisation showing exceptional agreement with his theory [26]. The theory developed here was later used for both DC and AC gain analysis of p-n

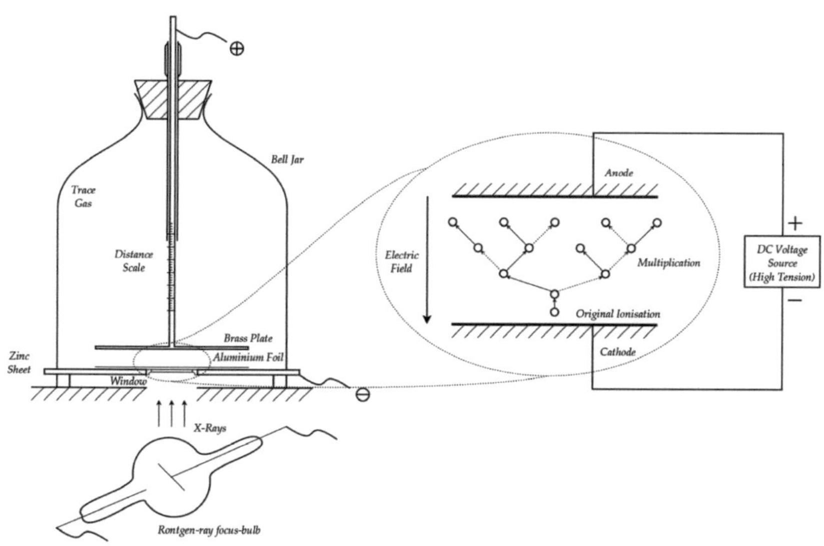

Figure 2. Left: Townsend's trace gas experimental apparatus producing impact ionisation and avalanche between an adjustable gap within an evacuated bell jar. Right: A schematic of avalanche breakdown in the presence of an accelerating electrical field.

junction avalanche behaviour and breakdown voltages. The rapid ionisation, prior to pure gas sparking potential, was to become known as a *"Townsend discharge"*.

This lead directly to the 1913 development of the Thyratron by Langmuir and Meikle. This was an early tube-based run-away avalanche gas tube used for high-voltage power regulation and fast-switching relays [27, 28]. Both a transition from a sharp concentrated breakdown arc into a blue/purple diffuse glow, and a self-maintaining arc suitable for low-frequency rectification were shown. Langmuir developed a theory of electrical conduction in a hard vacuum, without previously assumed positive trace ions, showing a space charge effect that could be reduced in the presence of positive ions from trace gases (e.g. Nitrogen). Langmuir also observed an effect later called 'bifurcation' where the I-V curve splits into two traces and can switch between two conductive states. He writes that, *"... the current rose steadily, until a potential of about 130 volts was reached. With potentials higher than this, the current would rise to a high value, 0.013 amp. per sq. cm or more, immediately on lighting the filament, and the discharge was accompanied by a strong purple glow. Suddenly, the current fell to 0.005 amp. per sq. cm or less, and at the same time the purple glow the purple glow vanished"*.

Rapid gas ionisation, through Townsend discharges, also prompted the 1908 investigations at Manchester University, into the Geiger-Müller tube for the detection of ionising radiation such as alpha particles [29], eventually becoming a matured tube concept in 1926, although suitable references are in German. Interestingly, Müller was not involved in the initial concept. A number of phenomena were observed that would later become key issues in photon detection. These were, (i) variation of the number of pulses matching previously established probability laws (i.e. Poisson variation), (ii) variation in the pulse height resulting from particle path differences and thus changes in the degree of ionisation multiplication (i.e. avalanche multiplication noise), and (iii) low pulse rate 'natural disturbances' in the presence of no alpha-particles (i.e. the dark count rate (DCR) and background radiation). The interesting point here is that both linear-gain and run-away multiplication leading to breakdown were observed experimentally in gases and explained by theory by the late 1920s. This provided a platform for later solid-state semiconductor investigations.

As shown by Nix [16], multiple researchers had observed external and internal photoelectric effects in several materials during the 1870s to the early-1930s. In combination with contemporary ideas of using such materials as detectors similar to gas discharge tubes but in a robust, solid-state form, Nix noted that Adams and Day found photocurrents when Platinum-Selenium contacts were illuminated (1876), following on from W. Smith's findings that Selenium was photo-resistive in 1873 [10]. Other materials were tested, such as Diamond, Silver Sulphide and Lead Sulphide, however Copper Oxide was observed to be photovoltaic in 1927 by Grondahl and Geiger. Copper oxide photoconductivity was then highly studied by multiple authors including Walter Schottky in the early 1930s [16]. The history of other photovoltaic and photoconductive studies during this period is given in the 1967 NASA report by Crossley et al. [30].

During the 1930s, the use of crystals and point-contact diodes increased, although their use was hindered by the mechanical, electrical and thermal variability of the contact formed by an *"active rectifying region"* and the *"cat's whisker."* Clarence Zener, then at Bristol University, theorised a form of dielectric voltage breakdown in semiconductor solids in which electrons can be excited by an electric field and may tunnel to a higher energy band, thus increasing conductivity [31]. This was based on experimental work by Von. Hippel in 1931. By deriving the rate at which this transition occurs, it was clear the rate was dependent on the energy gap of the material and the applied electric field. Zener and his theory explained both the magnitude of the field as per breakdown observations, and the rapid increase in breakdown. He stated that *"Further, the breakdown will occur suddenly as F* (potential) is increased, γ (transition rate) being increased by a factor of 100 (in our example) when F* changes from $1.0x10^6$ to $1.1x10^6$"* [31]. It became clear that this 'Zener effect' predominated when a diode's reverse breakdown was of low order, but that diodes with higher breakdown voltages were often attributable to an avalanche effect similar to Townsend gas discharges. This was key to later findings in the 1950s with respect to noise sources and breakdown effects in early transistors.

5.2. Early solid-state history: 1940 to 1949

During the period of 1940 to 1949, there were two parallel research themes that became critical to both modern technology, and the historical development of photon counting technologies. These were: (i) the progression in rectification diodes and the p-n junction that lead to the invention of the transistor and (ii) the discovery of photo-effects in Silicon and Germanium, the study of these effects and their utilisation for optical applications. In this section, we will begin with diodes and rectifiers as many of the innovations in optical detectors utilised the pure grown ingots, the theory and the progress in solid-state diodes.

5.2.1. Rectification and the p-n junction

The second world war was a direct driver of solid-state point-contact rectifiers by the Allied forces. As noted by Scaff and Ohl in 1947 [32], then at Bell Telephone Laboratories, there was a renaissance in point-contact diodes both from a mechanical robustness and a frequency perspective. As military applications expanded to include radio-frequency (RF) to intermediate-frequency (IF) super-heterodyne detectors (e.g. 3–24 GHz) for Radar and RF to direct-current (DC) rectification, vacuum tubes became limited by the transit time of electrons and the anode-cathode capacitance. Hundreds of materials were tested in the late 1930s and early 1940s, including zinc-oxide, molybdenum-disulphide and iron-sulphide. However, Silicon was found to have the best overall RF characteristics. In perhaps the first use of Silicon wafers cut from pure ingots that used Bell Labs' early impurity doping processes (e.g. Boron), it was found that electrical performance was directly related to the processing of the Silicon (grinding, polishing and etching), the doping profile and the mechanical construction of the point-contact diode housing. Military, commercial and academic standardisation enabled significant progress to be made in operating frequencies, power handling, SNR and operational lifetimes in harsh military applications.

While point-contact diodes were being improved, researchers such as Russel Ohl noticed impurities within the crystals could modify its electrical characteristics. In the early 1940s, positive (p-type) and negative (n-type) dopants were explored in samples of Germanium and Silicon [33]. The work undertaken at Bell Telephone Laboratories led directly to advances with p-n junctions and culminated with the patenting of both the point-contact transistor in 1947 (Bardeen and Brattain) and the p-n junction transistor in 1948 (Shockley). The theory, and indeed the history of this development is covered extensively elsewhere in the literature, however much of the theoretical research in the 1940s is discussed in a review article by William Shockley in 1949 [34]. Here concepts such as the structure of the material's band-gap, trapping of charge carriers, p-n and n-p-n transistor theory and indeed *"patch effects"* due to cracks and discontinuities due to dust and impurities were discussed. The developments since the mid-1940s – which has led to modern integrated circuits, CMOS processes and indeed integrated sensors (e.g. modern photodiodes) – are well covered in the book *"Crystal Fire"* by Riordan and Hoddeson [35].

5.2.2. The p-n junction and Si/Ge photoeffects

While Nix in 1932 [16] had shown that photoconductivity effects had been observed in many semiconductor and insulating materials (also see [30]), photoeffects in Silicon are often noted as being first observed by Russel Ohl at Bell Labs in the February of 1940. In Ohl's 1941 patent [36], a block of Silicon cut from a small, solidified melt was shown to increase in conductivity when a strong light source was incident upon its surface (**Figure 3** left). P-N junctions were present in this sample, prompting Ohl to suggest that: *"Ingots which are suitable for the production of photo EMF cells, possess a characteristic structure which is visible when the surface is suitably prepared in vertical section"*. As he noted striations in the cut Silicon sample, he named these striations *"barriers"*, suggesting that these are critical to operation of the photo-cell.

At the same time as Ohl's work, many authors started to observe photoelectric effects, both photo-conductive and photo-voltaic, within p-n junctions and pure Silicon or Germanium [34]. The earliest of these observations and diode structures are noted by Torrey and Whitmer ([33] p392) as being unpublished datasets from research on crystal rectifiers at the Massachusetts Institute of Technology (MIT) Radiation Laboratory [37] or military records (Miller and Greenblatt, 1945, US National Defence Research Committee) [38]. However, 1946 saw assessment of carrier velocity using modulated light [39], (effectively the carrier transit time), new bridge photodiodes with similar sensitivity as Selenium diodes but with far better stability and temporal response [40] and observations of reverse saturation currents that varied in proportion to optical intensity [41]. Likewise, in 1947 Bray and Lark-Horovitz [42] continued the paradigm that light quanta matching the material could lift electrons from full energy bands to the conductive band, while Benzer [43] showed linear with intensity, photoelectric effects in a diode with both a p-n junction and a metal–semiconductor Schottky barrier. In fact, Benzer demonstrated that the

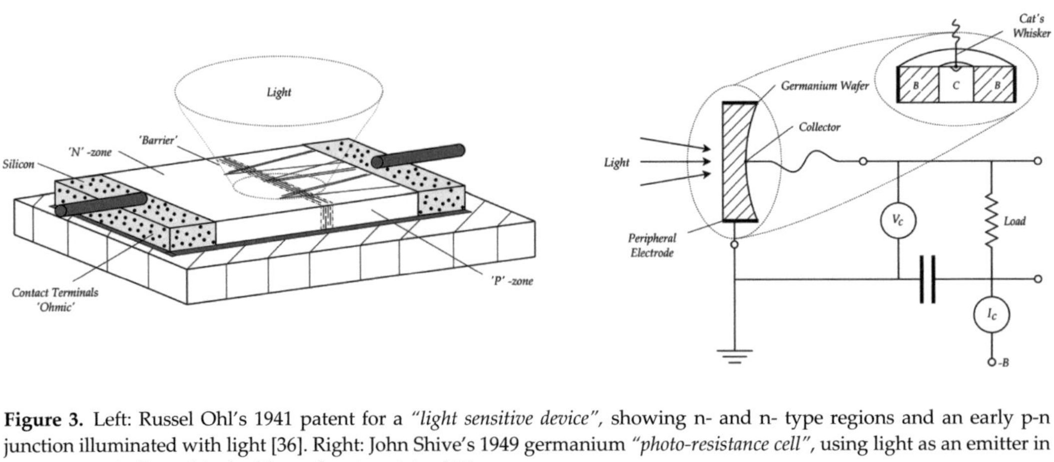

Figure 3. Left: Russel Ohl's 1941 patent for a *"light sensitive device"*, showing n- and n- type regions and an early p-n junction illuminated with light [36]. Right: John Shive's 1949 germanium *"photo-resistance cell"*, using light as an emitter in a geometry similar to his double-surface transistor [44].

overall diode I-V characteristic could be explained by the series combination of the separate junction I-V characteristics [43]. While difficult to verify, the way Shive [44] utilised quotation marks when noting that Benzer observed a *"photo-diode"* effect [43] and the quotation marks used by Benzer himself may indicate the coining of the term as a replacement to *"photo-cell"* or, as used by [42], simply Silicon or Germanium rectifiers that happen to show a photo-effect.

In 1949, authors such as Fan and Becker were exploring the theoretical basis of both the photoconductive and photovoltaic phenomena. They noted that by considering the concentrations of conductive electrons and holes, (i) liberated by thermal excitation, (ii) liberated by the internal photoelectric effect and (iii) the probability that carriers recombine, a suitable model could be derived [45]. This model, shown in Eq. (23), uses H as the rate of transition of electrons from valence to conduction band due to thermal generation and L as the rate of transition under optical excitation. This was shown to fit to experimental values for the open-circuit voltage, V_{OPEN} [46].

$$V_{OPEN} = (kT/e)log(H + L/H) \tag{23}$$

By August 1949, John Shive at Bell Telephone Laboratories proposed a variant of the photo-resistive cell, calling it a *"photo-transistor"* [44] (**Figure 3** right). This portmanteau was likely through Benzer's use of 'photo-diode' to describe optically sensitive p-n junctions [43] and the contraction of 'transresistance' into 'transistor' at Bell Labs at the time [34]. Using perhaps the first dedicated photo-sensitive structure to explicitly use back-side illumination, Shive showed electrical gain of the optical signal at a reverse biased base–collector junction. This gain term was similar in magnitude to that observed between the emitter and collector in bipolar-junction transistors (BJTs), but with the emitter in this case being charges produced and injected photo-electrically [44].

6. Early transistor and microplasma history (1950–1959)

In 1950, two parallel strands of research were underway. Firstly, the growth of high-purity Silicon and Germanium ingots [47] using the lifetime of carriers as a guide to high *"crystal lattice perfection"*. This allowed the reduction of recombination centres [20], which were hindering both diode performance and theoretical studies. To achieve this, Bell Labs improved upon the existing Czochralski method of crystal growth [47], principally due to commercial expansion of solid-state rectification diodes [32]. This was continued by McAfee and Pearson [48] for transistor optimization. The second strand of the research centred on the continued optical and electrical investigation of p-n junctions formed in Silicon and Germanium. For example, Goucher [49] measured, using a pulsed light technique, the photon quantum yield of electron–hole pairs. A departure from unary quantum efficiency at short wavelengths lead Goucher to conclude that there was a thin surface region of recombination centres [20]. Elsewhere in the US, in recognition of future infrared sensors, absorption experiments were being carried out by Fan and Becker at Purdue University [50]. Likewise, singular p-n junctions, rather than structures that included Schottky and p-n junctions in series [43], were being tested at Purdue. These experiments indicated that the junction capacitance, the input resistance of the measurement amplifier and the internal resistance of the unilluminated portions of the diode, were key to maximising photovoltaic voltages [51].

In 1951, Pietenpol used p-n junctions for both rectification (showing a reverse I-V breakdown characteristic) and optical detection [52]. Along with a unity quantum efficiency, he attributed agreement of experiment and theory as indicative of the *"diffusion of current carriers to and from the junction"*. For rectification, a high reverse breakdown voltage (1200 V) and bandwidth (200 kHz), demonstrated competitiveness with existing tube-based rectifiers and Bell Lab's existing point-contact diodes, hence their commercialization and continued research and development. In July 1951, Shockley, Sparks and Teal [53] presented work on n-p-n and p-n-p junction transistors, including phototransistors. This combined experimental results, furthering junction theories presented in [20, 34]. There are several interesting points from this and Shive's earlier 'photo-transistor' [44]. Firstly, trapping of carriers may fall off as carrier injection increases through a saturation effect, directly impacting the carrier lifetime. Secondly, a p-n hook region discussed by Shive could be utilised to obtain n-p-n photo-transistors that were extremely responsive to light, with quantum efficiencies of 100–200 electron–hole pairs per absorbed quanta [53]. The bi-polar junction transistor became the basis for transistor-based analogue and digital circuitry prior to the routine use of CMOS and field-effect transistor (FET) technologies. Despite the success of bi-polar transistors, phototransistors have remained secondary to photodiodes and avalanche photodiodes due to longer response times and appropriate biasing requirements.

Of course, due to early crystal processing techniques, no junction used in experiments was entirely perfect. As such, research on the variety of breakdown effects began in earnest. McAfee and others at Bell Labs [54] started by investigating the Zener breakdown of the junctions, extending Zener's earlier theory to include larger energy gaps. In perhaps the first mention of later avalanche studies, an alternative breakdown mode whereby secondary electrons are produced when the electric field reaches a *"critical"* value was discussed. This was discounted through experimentation, however *"patch effects"*, i.e. a prelude to defects yielding avalanche gain, were directly mentioned for future investigations [54]. Complementing this, the magnitude of breakdown was shown to be sharp, with McAfee noting that, *"a change of voltage of one-half percent is sufficient to cause the current to change by two orders of magnitude"* [48]. The measured slope of the I-V curves however did not fit Zener-Shockley theory suggesting a further multiplication phenomena. Further, the voltage noise was found to significantly exceed thermal noise [48]. The prevalence of recombination to explain optical and electrical behaviours, lead to Shockley and Read's paper on the statistics of electron and hole recombination [55], Hall's paper on the same topic in Germanium [56], (including carrier lifetime against temperature measurement to evaluate the *"activation energy"* of in-band trapping centres) and a method of experimentally probing the p-n junction [57] to obtain capacitance, junction width and voltage dependencies to inform theory. As breakdown, Zener or due to patch effects, restricted the effectiveness of solid-state diodes, Pearson and Sawyer [58] continued investigation using the Silicon crystals grown at Bell Labs. Several important issues became elucidated including that a built-in potential, V_{bi}, must be incorporated into breakdown theory, and that the I-V curve gradient in the Zener region was larger than theory, which while not yet understood, required investigation. The most important issues however were that 'noise' was observed at the Zener knee and that a *"softness"* of the reverse characteristic was also observed [58] (**Figure 4C**). The softness of the knee, defined as an unusual increase in current before true Zener breakdown, was improved by annealing. *"Crystal*

http://dx.doi.org/10.5772/intechopen.72148

lattice defects", i.e. patch effects, were cited as a possible cause for this behaviour. The noise at the Zener knee (i.e. bifurcation of the I-V characteristic), showed clipped voltage pulses as high as 3 V. The noise was also temperature dependent, with pulses which were uniformly random. The noise behaviour varied greatly between units and was cited as being caused by mechanical issues within the junctions. This, along with the patch effects noted by McAfee [54], is the origin of later 'Microplasma' nomenclature for localised breakdown defects within such junctions.

The study of p-n junctions for optical detection continued with Shive [59] forming n-p-n photo-transistors whereby *"the photoelectric absorption-activation process"* generates holes that diffuse into the p-type region, and if trapped by the potential barriers, act to lower the barrier prompting the increased passage of emitter to collector current. Effective quantum efficiencies of 1000 were achieved with the efficiency, Y, being given by Eq. (24) [53], where σ_n and σ_p are the n- and p- section conductivities, L_{pe} is the diffusion distance of holes in the n-type region and ω is the width of the p-type sandwich layer.

$$Y = 1 + \left(\frac{\sigma_n L_{pe}}{\sigma_p \omega}\right) \tag{24}$$

However, once multiplication i.e. e-h pair production greater than the Zener emission of carriers had been found, many authors begun investigations into the effect using light, alpha-particles and thermal liberation for generation of initial carriers. The reason being that multiplication of photo-currents only, could remove the need for the photo-transistor's continuous, optically modulated, emitter-collector current. The paper by McKay and McAfee in 1953 [60] is key as multiplication in slightly wider p-n junctions than previously studied [54], is attributed to an avalanche ionisation effect similar to Townsend avalanche [19, 20]. Indeed in 1967, Emmons [61] noted that this was the first time that Townsend's avalanche theory was applied to the direct-current (DC) analysis of p-n junction multiplication behaviour. McKay and McAfee used avalanche multiplication to apply a gain to a photo-generated current, demonstrating increased quantum efficiency as the voltage approached breakdown, i.e. the avalanche photodiode, although the first such device is attributed to Nishizawa in 1952 (patent *JP1955-8969A* [62]). Linking back to Pearson and Sawyer [58], McKay and McAfee attributed the softness at the Zener knee to multiplication, within the junction, of the thermally generated dark-current, while pulsed experiments using alpha-particles

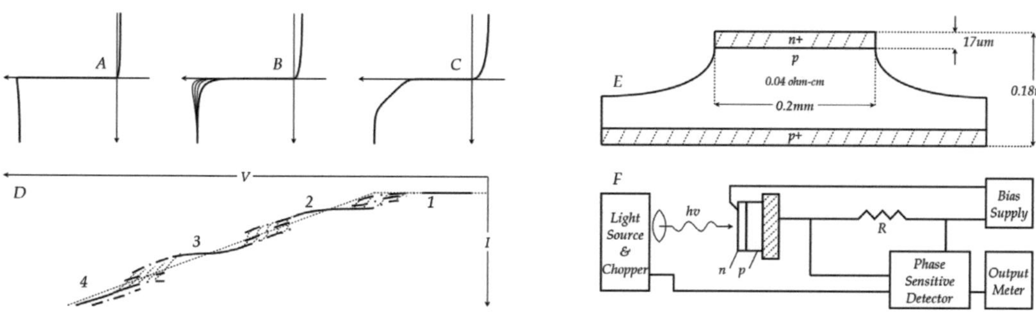

Figure 4. Current–voltage characteristics with; (A) good breakdown, (B) noise at the Zener knee and (C) softness prior to breakdown, adapted from [58], (D) activation of multiple microplasmas with a saw-tooth reverse bias sweep, adapted from [73], (E) cross-section of a uniform n+/p junction diode, and (F) experimental setup for optically-induced carrier multiplication studies, both adapted from [72].

showed that avalanche occurred on time scales less than $2x10^{-8}$ s. Crucially, even in 1953 McKay and McAfee noted that *"For wider junctions, the multiplication factor, M may become infinite for fields below those required for field emission"* [60]. This paper therefore not only indicates the origins of the avalanche photodiode, but also alludes to junctions for Geiger-mode avalanche devices such as SPADs. The pivotal work undertaken in McKay and McAfee's 1953 paper prompted more theoretical treatment by McKay [63] and Wolff [64] in 1954. Likewise, as p-n junctions as transistors were beginning to be used as fast switches, Kingston [65] theoretically investigated the switching time, showing dependence on the structure and the minority carrier lifetime. The minority carrier is the opposite to the dominant carrier within a doped region. For an n-type region, the dominant carrier is the electron, hence holes are classified as minority carriers.

In 1955, Miller [66] showed that avalanche breakdown also occurred in Germanium. Through investigation of the carrier ionisation rates, he presented agreement with Wolff's theory. Interestingly, it was noted that while breakdown voltages should be static, the multiplication factor could be different depending on if an electron or hole initiated the avalanche. Crucially, Newman [67] discovered reproducible, defect-correlated light spots of approx. $10\mu m$ diameter within an avalanching junction. Soft breakdown was hypothesised to be due to breakdown of small patches at a spectrum of voltage levels below the Zener breakdown voltage, with the light being due to radiative relaxation of high-energy carriers produced during avalanche. Chynoweth and McKay confirmed this at Bell Labs finding further *"localised light-emitting spots"*, which were correlated to scratches, defects and the spatial location of the main p-n junction. They called these *"Microplasmas"* [62, 63], which was to become the de-facto nomenclature during the late 1950s and 1960s.

Herein, researchers began investigation of Microplasma defects as noise sources within the wider p-n junction [63, 64]. The pulse behaviour was studied by Rose [69] finding long quiescent periods at which the device could be held at a voltage above the breakdown voltage, *"but awaits the entry of a chance carrier into the region"*. He also found that (a) microplasmas were much smaller than the $10\mu m$ observed in [67], (b) that current pulses could equate to a local heating effect, (c) that current pulse duration varied exponentially, (d) that random fluctuation to zero carriers could explain the *"turn-off"* behaviour, and (e) that an equivalent circuit was an unreliable tool due to variation in the microplasma dimensions. The initiation of avalanche was also of interest, with Chynoweth and McKay investigating the kinetic energy required for impact ionisation in Silicon [71]. There was significant debate and variation in results for this kinetic energy in the literature. While 2.3 eV was speculated by Wolff [64], Chynoweth and McKay showed that the energy was 2.25 eV for electrons and the lower hole ionisation rates could be explained by an estimated threshold energy of 2.8 eV for holes [71]. Values of 1.5 eV for electrons and 3.5 eV for holes were also proposed (Miller, 1957). In contrast, modern texts point to a minimum theoretical value given by Eq. (17), i.e. 1.65 eV, and measured values of 3.6 eV and 5.0 eV for electrons and holes respectively ([19] p79).

Between 1958 and 1959, studies split into two domains. The first was the study of single microplasmas, in comparison to uniform junctions that exhibited many such defects. Senitzkey and Moll [72] achieved this using small area diodes ($\sim 200\mu m$), with a sharp characteristic rather than the softness observed in some diode I-V characteristics and by initiating a defect by introducing an Aluminium impurity at a known position. The link between dislocations and breakdown was confirmed definitively by Chynoweth and Pearson [70], although at the time it was not possible to confirm if avalanche at the location was due to increased electric field due to dopant non-uniformity, carrier tunnelling due to traps in the energy bands, band-gap narrowing due to lattice distortion or indeed large crystal misalignments. The microplasmas, bi-stable turn-on turn-off bifurcation noise in the pre-breakdown region and noise pulses from multiple microplasmas found in [68, 71], were then verified by I-V and light emission studies in 1959 [73] (**Figure 4D**).

The second domain was transient studies, both of microplasma turn-on, turn-off and the AC behaviour of diodes using avalanche. In 1958 Read [74] devised a high-frequency (5 GHz) microwave oscillator, utilising avalanche multiplication during the positive portion of a sinusoidal input signal. In doing so, he investigated the transit time, build-up and signal frequency response utilising Townsend's 1901/1903 theory of impact ionisation for AC analysis. As noted in 1967 [61], Read hypothesised that avalanche build-up time constants will limit the AC bandwidth of such diodes. Avalanche transistors, and their transient behaviour ([75, 76], and references therein) contributed to noise performance investigations of the avalanche process, while Champlin [77] continued the microplasma bi-stable noise studies of Rose and McKay. He demonstrated that both current and voltage pulses could be modelled in an analytical manner. Under low series impedance conditions (5 Ω), many quantities of his model (probability rate for non-conducting to conducting transitions etc.) could be assumed to be time-independent allowing a Markov model to be used. In high series impedance cases (10 kΩ), time and voltage independence could not be assumed, producing a non-Markovian process, which Champlin noted as departing from previous models by Rose, although for many situations close agreement was found.

The 1950s were a period of significant progress in device design and fundamental research into breakdown behaviour. Inherent in this experimentation was the use of what would now be regarded as passive quench passive reset (PQPR) circuits [4], or rather, circuits that used a series resistance either as a current sense resistance (low-impedance), or in order to develop an appreciable voltage pulse suitable for counting (high-impedance). Initially, series resistances of 1 Ω were used [65, 67] however the general use of a load resistance to make measurements was quite standard [30, 38, 47]. Depending on the experiment, this resistance increased [59, 66] sometimes to values as high as 10 kΩ [77] while 50 Ω was used to match the 50 Ω input of test equipment [72]. By the end of the 1950s, artificial, single-microplasma diodes had been used for study of avalanche multiplication and microplasmas as unwanted noise sources in p-n junction transistors. Models had also been proposed for microplasma bi-stability and the avalanche mechanism in semiconductors (e.g. the McKay, Wolff, Rose and Champlin models), and the link between lattice dislocations, doping imperfections and microplasmas had been definitively proven. Light and ionising-radiation applications had been explored, initially as methods of injecting carriers, but studies of photo-transistor and photo-carrier avalanche multiplication showed that such diodes could be important for the optical detection challenges at the time.

7. Artificial microplasmas and early applications: 1960 to 1969

Deep physical investigation of microplasma physics and of course the theory to explain their behaviour continued to be researched as intensely as the previous decade. However, the 1960s can effectively be characterised as the starting period for both applications of the avalanche mechanism and the increased investigation of diode structures. As the number of researchers and open topics increased dramatically in this period, the discussion below will be split into two chronologically-ordered sections. The first will discuss the progression of microplasma modelling and experimental observations, while the second will discuss the evolution of the physical device and the applications to which it was applied.

7.1. Microplasma experimental observations and theories

Based upon earlier experiments [52, 57] and models [69], McIntyre proposed an extended microplasma model [78]. This was tailored to linear and step junctions upon which most observations had been made, in comparison to the p-i-n junction [69]. Deriving the turn-off probability, McIntyre conjectured that turn-off is due to the chance fluctuation to zero of carrier-pairs, thus preventing ionisation. Investigating turn-on mechanisms, he proceeded with a photomultiplier analogy yielding Binomial and Poisson theories, and an election candidate ballot box analogy. Despite some correlation, each departed from experiment.

While optical emission investigations continued, it was suggested in [79] that there were four classifications depending on if microplasma pulses, light and multiplication were observed. However, at least one classification was in doubt. This was the combination of microplasma pulses without multiplication, however this may have been due to measurement methodology issues [80], as carrier multiplication was suggested elsewhere [62, 63]. Two categories were suggested by Goetzberger and Stephens [81], those with (i) bright light emission but low breakdown voltage and (ii) dim light output and high breakdown voltage. Disagreement with results in [79] were attributed to non-observable light emission due to the depth of some junction defects. Goetzberger and Stephens concluded that microplasmas were preferentially located at lattice dislocations, but this may not be the causal factor. They also concluded that microplasmas are, *"caused by some kind of imperfection that itself has a statistical distribution of its properties"* [81].

Haitz and Goetzberger [80] proposed an improved method of investigating multiplication within microplasmas, refining experiments in [79]. Indeed, multiplication can occur within microplasmas (up to an observed ratio of $1x10^6$), refuting the classifications in [79]. To continue efforts in classification, two kinds of avalanche were noted, the first through microplasma action, and the second through entire-area avalanche breakdown observed in custom fabricated *"guard-ring"* diodes by Batdorf et al. [82]. Proposing a theory for multiplication, Haitz and Goetzberger note that rather than continuous multiplication, photon arrivals cause a microplasma to turn on again, thus multiplication is by virtue of an increased time in which the microplasma is conducting. They thus relate microplasma photocurrent multiplication to an ionisation counter, re-affirming Ruge and Keil's 1963 link between avalanche gain, current pulses and existing Geiger-Muller detectors [83] (see later).

Exploring avalanche breakdown with a microplasma-free junction [84], three interesting phenomena were discovered. Firstly, a theory in which statistical variation of donors and acceptors in the junction (Shockley, 1961) leads to non-uniformity in breakdown voltage and thus the avalanche breakdown of the whole area, was supported by experiment. Secondly, striations were observed through light emission which correlated to distinct annular non-uniformities in grown wafers. Thus, the diodes were not truly *"uniform"*, prompting further work on crystal growth and general p-n junction regularity. And thirdly, the pulse multiplication model in [80] was verified for high (>500) avalanche multiplication factors.

In 1964 and 1965, Haitz published two influential papers on the electrical behaviour and noise contributions of microplasmas and the avalanche mechanism [79, 80]. By proposing an equivalent circuit, Haitz modelled the phenomenological rather than actual nature of the microplasma (**Figure 5A**) [85]. This uses an internal resistance, R_s, in series with a bi-stable microplasma switch, *S*, and a breakdown voltage extrapolated from the common multiplication vs. voltage curve, V_b (**Figure 5B**). Haitz derived the current and voltage forms for on–off transients, giving the now standard view of the breakdown-quench-recharge cycle (**Figure 5C**). To quote [85], *"As long as the microplasma is nonconducting, the diode capacity, C, is charged to the applied voltage,* V_a*. As soon as a carrier triggers an avalanche, the microplasma switches on to a current,* $I_a = (V_a - V_b)/R_s$*, This turn-on is very fast and is estimated by Senitzky and Moll* [72] *to be of the order of* 10^{-11} *sec. Due to the voltage drop across* R_L *the capacity is discharged and voltage and current drop simultaneously to the operating point given by the intersection of the V-I characteristic and* R_L *load line"*. In [86], Haitz discusses four principal noise contributions, (i) thermal carrier generation, which is now known as dark count rate (DCR) noise, (ii) re-emission of carriers trapped during previous avalanches, i.e. after-pulsing, (iii) Zener/Shockley band-to-band tunnelling and (iv) minority carrier diffusion from elsewhere in the substrate, triggering an avalanche (see also Tager, 1964). Continuing his studies on noise, Haitz investigated an optical cross-talk mechanism [81, 82], although the re-absorption of light emitted by radiative recombination during avalanche had been discussed by Newman in 1955 [67] and Champlin in 1959 [77]. This supplemented the coupling experiments conducted by Ruge and Keil in 1963 [89], along with Conradi (1963), where the distances between and non-clustering of triggered microplasmas precluded thermal phenomena, thereby giving credence to Haitz's 1962 hypothesis of optical coupling. Haitz [88] fabricated a wafer of over 100 artificial microplasmas (discussed later), using a diode with a background count rate of approx. 1 pulse/sec (at $V_b + 200mV$). Through experiment, minority carrier and lattice phonon [19] mechanisms were

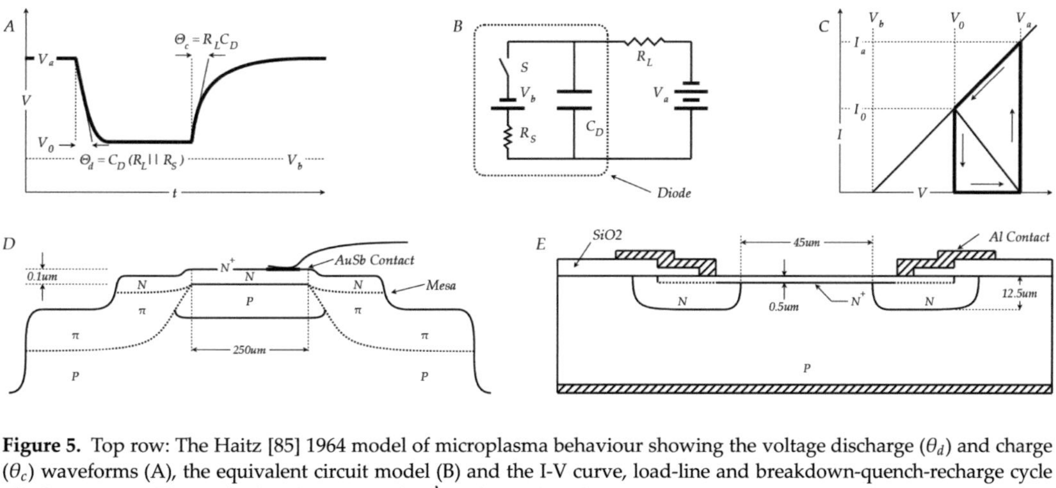

Figure 5. Top row: The Haitz [85] 1964 model of microplasma behaviour showing the voltage discharge (θ_d) and charge (θ_c) waveforms (A), the equivalent circuit model (B) and the I-V curve, load-line and breakdown-quench-recharge cycle (C). Bottom row: The Batdorf et al. 1960 [82] diode with a π-doped guard ring and mesa structure (D) and the Anderson et al. 1965 [105] planar structure (E).

discounted for distances of $\geq 100\mu m$, as the coupling was still present between separate Silicon slices. Interestingly, an analysis by Ashkin and Gershenzon in 1963, of the refractive index of the space charge layer in a p-n junction in comparison to the bulk Silicon suggested a light waveguide which was denoted as a *"pipe"* [87]. For closer distances, minority carriers were proposed as a mechanism whereby the avalanche spread laterally [88].

While the pulsed mechanism was of interest for light and gamma detection [88], continuous time gain used in APDs still required examination. In 1964 Lee, Logan et al. [90] reinvestigated the kinetic energy ionisation rates for electrons and holes as there had been several inconsistences shown between previous work and more recent analysis by Baraff in 1962. In particular, a simplified analysis was combined with a refinement in the cleanliness of test diode growth and the use of a local multiplication uniformity rather than uniformity of emitted light approach. They noted that better agreement to an ionisation energy in the range $E_g \leq E_k \geq 1.5\ E_g$, could be due to microplasma-free junctions and a method that allows purer injected hole currents. Analysis of avalanche as applied to APDs continued in 1966 with several critical papers. McIntyre [91] concentrated on inferring the SNR that could be obtained for applications requiring high photodiode gain. The noise of the process was shown to increase with the cube of the multiplication factor, M, however McIntyre noted that if most of the carriers entering the high-field region have a high ionisation rate, the noise factor decreases. Melchior and Lynch [92] commented that multiplication in the diode is limited by its noise in comparison to receiver noise. Baertsch [93], showed discrepancies with McIntyre's noise theory, particularly (a) reduction of noise if the primary photocurrent in a Silicon APD was due to holes, and (b) increase in noise greater than McIntyre's theory at high multiplication values. This departure was blamed on electronic noise or the differing ratio of electron and hole ionisation rates. One of the complications in all experimental studies within the 1960s, was the handling of diode structure, doping or bulk-material induced changes in characteristics, and the associated departures from theory. Indeed, with a different diode structure Baertsch [94] presented significantly better agreement to McIntyre's theory, highlighting the difficulties in obtaining reproducible results. Sze and Gibbons [95] re-calculated the ideal (microplasma-free) breakdown voltages for both abrupt and linearly graded junctions for several bulk materials including Silicon, Germanium and alternative III-V alloy diodes. This allowed other researchers to estimate the departure of their devices from the ideal, entirely uniform breakdown characteristics. This was used as a baseline for investigation into junction curvature in fabricated planar diodes [96], continuing the work by Gibbons and Kocsis in 1965. The results showed that for abrupt junctions, the radius of curvature, significantly impacted the breakdown voltage through the modification of the electric field intensity. If the curvature was equal to the junction depth, this produced a more marked dependence. The breakdown in this region was always lower than that in the planar region, producing edge breakdown effects and structural dependence on results. In comparison, linearly graded junctions had only a small dependence on curvature.

Towards the end of the 1960s, theoretical work returned to the question of the avalanche mechanism [97], noise [98], transient and frequency behaviour [61] and the avalanche turn-on mechanism [100]. In [97] the existing models of avalanche breakdown were extended. The variation in multiplication was stated to be due to differences in the electron and hole ionisation rates, and not spatial non-uniformity of the electric field within the junction. In 1967, Haitz [98] provided extension and experimental verification of an avalanche noise theory by Hines [99]. Here, (i) assumption that the avalanche region is small in comparison to the total drift region, (ii) an assumed power law for the ionisation rate voltage dependence and most importantly, (iii) neglection of the spreading and thermal resistance of the bulk material, were each accounted for via correction factors. Haitz observed peaks in the noise spectrum with avalanche current, attributing these to variation in doping [84] as per Shockley (1961) and not microplasma action. Consequently, he updated the noise model to include noise generated by a Poisson spatial distribution of dopant impurities within the diode, and reiterated the lateral spreading mechanism [88].

By the middle of 1967, Emmons noted [61] that depending on the ratio of electron and hole ionisation rates, there need not be a reduction in the bandwidth of a diode due to avalanche multiplication, (as proposed by Read [74] in 1958 and Lee and Batdorf in 1964). These earlier works assumed that: (i) the electron, α_n and hole α_p ionisation rates were equal, (ii) the velocities were also equal and (iii) that Maxwell's time-varying electric field, polarisation *"displacement current"* could be neglected. Through analysis, Emmons showed, (using equations for DC [60] and AC [74] cases of Townsend gas discharge [25]), that if the avalanche multiplication was kept below the ratio of the ionisation rates, i.e. that $M_0 < \alpha_n/\alpha_p$, then bandwidth was not dependent on avalanche multiplication and indeed that noise was minimised. His application at the time was to find the conditions that would *"produce the closest solid-state approximation to the vacuum-tube photomultiplier"*.

In 1968, Nishizawa already credited with the invention of the APD in 1952 [62], was working with Kimura on microplasma turn-on mechanisms [100]. Treating turn-on as a stochastic phenomenon, they modelled it using a 2-state Markov process, concluding that the turn-on probability was a strong function of field intensity in the p-n junction and the rate at which carriers are generated near the junction. This turn-on probability was then utilised by Melchior and Goetzberger in 1969 to form a *"gating"* or *"quenching"* technique using a sinusoidal excess bias waveform [101]. Thus, by the end of the 1960s, many facets of the theory of microplasmas in Silicon and Germanium, and the avalanche multiplication process had been confirmed and informed by experimental observation.

7.2. Junction and artificial microplasma structures and applications

To support both theoretical and experimental work (previous sub-section), and to investigate applications for such high-sensitivity photodiodes, significant effort was made in the 1960s to remove microplasmas from p-n junctions. This required innovation in planar technologies (dopant implantation and diffusion, masking and etching etc.), and investigation of diode structures (topologies, guard-rings and substrates etc.). Eventually, applications emerged using true microplasma-free uniform avalanche multiplication, and the technologies surrounding APDs and GM-APDs became more mature.

In July of 1960, Batforf and Chynoweth et al. [82] proposed the use of a planar *"guard ring"* that surrounds the active area (also see [102]). The reasoning being that if there were no dislocations within the junction, and if doping was uniform, then the periphery of the junction would be the next preferred breakdown site [70]. Their diode, shown in **Figure 5D**, used a lightly doped p-type region (π) around a known diameter circular n-type area diffused into a p-substrate. While the nomenclature did not spread, they called this $\sim 250\ \mu m$ diameter deterministic test structure, a *"Macroplasma."* This was a step towards planar technology although it incorporated some surface etching similar to previous mesa (i.e. table) structures (~ 200 μm diameter) [72]. Goetzberger and Stephens [81] note the use of open tube systems for diffusion, a multiple predeposit-wash technique and guard rings for uniform doping and minimal surface defects. It is difficult to ascribe guard rings to particular authors as Senitzky and Moll [72] used an effective virtual guard ring formed by removal of surrounding Silicon (forming their mesa structure), but inducing surface issues. Modern processing often forces device designers down either a planar or mesa structural path.

At this point, many structures were fabricated to study uniform breakdown, however in 1963, Ruge and Keil [83] set about using microplasmas and avalanche gain for gamma radiation detection. They compared the voltage-pulse output and the linearity with incident radiation as equivalent to Geiger-Muller radiation detectors (re-iterated in [80]). Despite their application, at this point (a) gamma detection in Silicon had been proposed albeit with low signal amplitudes that necessitated high-gain, low-noise amplifiers, and (b) alpha-particle radiation had already been used for the study of ionisation rates in the avalanche process [60]. Also in 1963, Goetzberger et al. [84] noted several technical processing advances that were required for minimal- or zero- microplasma densities in fabricated diodes. Finding that their process lead to microplasmas originating at surface effects, electrolytic polishing was utilised in the reverse process of common electroplating. The material deposition technique was also refined using Helium as a carrier gas as more reactive gases lead to the formation of undesirable precipitates and a phenomenon called surface pitting. The multiple pre-deposit and washing technique was again suggested as depositing phosphorus lead to a glass being formed (SiO_2 and phosphorus pentoxide), which hindered diffusion into the substrate and promoted non-uniform doping.

Between 1964 and 1965, significant device and application progress was made on avalanche photodiodes. In [103], Johnson noted significant SNR and signal amplitude enhancement in uniform breakdown APDs. This was then extended by Johnson in 1965 [104] noting that at high multiplication factors (high-M) and light levels, the SNR became dominated by shot noise, while thermal noise dominated the low-light, low-M case. Johnson therefore suggested that for modulated optical signals, the modulation depth must be large to maximise receiver SNR. While Johnson was at Texas Instruments Inc., researchers including Anderson and McMullin (under the supervision of Goetzberger) continued work at Bell Labs on microplasma-free APDs at microwave frequencies [105]. Testing custom n^+-n-p diodes (which incorporated n-type guard rings, (see [84]), at frequencies up to 10GHz they noted that, (a) electrons have an advantage in terms of ionisation coefficient, although the bandwidth becomes limited by the electron diffusion time, and (b) the SNR becomes limited by photon shot noise. Their tested diode structure is shown in **Figure 5E** with an n^+ to p-type junction. Melchior and Anderson [106] then noted that an optimum SNR could be obtained if the *"multiplication is such that the multiplied shot noise is just equal to the sum of the series resistance and receiver noise"*. They warned that M-factors greater than this may improve optical sensitivity, but the SNR remains dependent on receiver noise.

As previously mentioned, Haitz fabricated an array of over 100 diodes [88], using a 10μm n^+-p active junction and a wide radius n^- to p-substrate guard ring. Firstly, this came at a time when arrays of diodes were being investigated for solid-state imaging by Schuster and Strull

(October 1965) [107]. This is often cited as the first photodiode array, however the Haitz array, (which did not include readout circuitry suitable for imaging), was published in the April of that year. Secondly, Haitz used avalanche and microplasma pulses as a direct equivalent of modern SPAD arrays and Si-PMs, whereas Schuster and Strull utilised photo-transistors equivalent to modern CMOS image sensors with in-pixel electrical amplification. Thirdly, the depth of the guard ring diffusion used by Haitz, allowed a deep region under the junction where any minority carriers would be quickly absorbed in the n^- regions, thus reducing the diode background pulse rate.

In January of 1966, Lee and Batdorf [108] presented an overview of research on avalanche multiplication, the time dependence of AC-signals and the recent technological developments for applications such as high-speed APDs and the Read microwave oscillator [74]. They also noted that efforts to remove the guard ring, but to still constrain avalanche using oxidisation and surface treatment techniques, had proved fruitless. To this day, guard rings are still a requirement but can present a fill factor issue, thereby limiting optical sensitivity. Respectively in February and June, Huth at General Electric [109] and Haitz and Smits at Texas Instruments in collaboration with Bell Labs [110], published results, noise analysis and application discussions for Germanium APDs for x-ray, alpha-particle and IR-optical detection. In Europe, Keil and Bernt [111] were investigating microplasmas in Silicon for infrared detection, building upon previous infrared absorption experiments [50]. In the December of 1966, Melchior and Lynch [112] again addressed signal and noise responses in APDs at modulated speeds up to 10GHz, building upon [106]. As light detection using the increased pulse rate of a microplasma includes turn-on turn-off noise, Melchior and Anderson suggested large area, uniform avalanche detectors as of primary interest for high (but not single-photon) sensitivity. As such, they proposed an amalgamation of the mesa structure and the planar guard ring in an attempt to reduce the reverse current and device noise, while promoting uniform multiplication across a $40\mu m$ diameter optically active area. They thus reduced the curvature of the junction, minimising edge breakdown [96], while protecting the junction from edge/surface effects [72].

A highly pertinent question for applications research at the time was how to optimise these avalanche photodiodes. Ruegg at Stanford Electric Laboratories [113] proposed design parameters, a diode structure and manufacturing processes for an optimised device. This was an initial *"reach-through"* APD. While presenting the structure, he stated that an optimised diode can be formed by using a depletion layer that reaches through the device to the illuminated surface, and with a depletion layer width that is approximately equal to the penetration depth. Such devices are still used and have been further refined [3–5]. While investigation and improvement of manufacturing processes was continuing to remove microplasmas, it was clear that to improve both performance and yield, circuit and electrical innovation was also needed. In 1967, a bias voltage mechanism called *"AC-pumping"* was proposed [114]. This was in fact the first use of a *"gating"* signal, whereby microplasmas and avalanche are suppressed electrically. When the gating signal is high, the bias is above breakdown allowing avalanche gain upon the reception of a minority carrier. When the gating signal is low, the diode does not exhibit microplasma pulses. When gating at high-speed ($\geq 100MHz$), the diodes can be significantly improved, including diodes which previously showed correct microplasma-free uniform avalanche behaviour. This effect was studied in greater depth in 1969 [115], where the nomenclature of a *"quenching technique"* was possibly coined for the first time with respect to man-made avalanching junctions. The square- and sinusoidal-wave gating techniques are still utilised, often in quantum key distribution systems where noise can be suppressed and performance can be increased [3].

Following from Sze and Gibbons' work on junction curvature [96], Kao and Wolley [116], and separately, Sigmund [117] proposed structures optimising the junction for spherical or circular edges. In [116] Y. Kao and E. Wolley investigated guard rings with both (i) different spacing between the active junction edge and guard ring, and (ii) multiple concentric guard rings. Finding that this reduced surface breakdown, they also found that it improved junction curvature. In [117], Sigmund used an alloying technique to back-fill mechanical depressions in a n-type Silicon wafer. The annealed, alloyed structure formed was a back-side illuminated cone on p-type Silicon within the n-type wafer. The tip of this cone, orientated towards the illuminated side, formed a spherical junction, with a radius of curvature of $2\mu m$, with a depth from the illuminated side of $2\mu m$ and an active region diameter of approx. $3\mu m$. While this was a novel diode structure, Sigmund also proposed a capacitive readout, passing only higher frequencies to a buffer that separated the diode circuit from the impedance of testing equipment. While not covered here due to the explosion in publications per annum in this field, the 1960s represented the drive of physicists and engineers in the 1970s and 80s to explore device structures for high-performance optical detection applications. Many of the authors above were supported through companies such as Bell Labs, Texas Instruments (TI), General Electric (GE), International Business Machines (IBM) and Shockley Transistor. This highlights the commercial drives for solid-state, high-gain, high-bandwidth photodiodes.

8. Conclusions

This chapter has introduced photon detection and some background into the nature of the 'wave' or 'particle' we wish to detect. Discussing three viewpoints for the question of: *"What is a photon?"* we have linked this with phenomena that allow the detection of photons: the internal/external photoelectric effect, Compton scattering and pair-production.

This chapter has also discussed the historical development of avalanche multiplication, with a focus on solid-state semiconductors detectors. Two classes of photon-counting detectors have resulted, (i) linear gain devices such as APDs and EMCCDs, and (ii) Geiger-mode devices such as SPADs and Si-PMs. The impact ionisation mechanism allows an initial electron–hole pair, generated via internal photoelectric effects, to be multiplied into many current carriers. We have seen that acceleration by an electric field imparts kinetic energy onto an initial carrier, where, upon a collision with the semiconductor lattice, the energy is sufficient to promote an electron from the valence to conduction band. John Townsend hypothesised (1901) that breakdown in a vacuum tube containing trace gases, could be explained by the exponential ionisation of the molecules. Developments in solid-state rectifiers later lead to the photodiode (1940), the p-n junction and the transistor (1947). Naturally, there were a variety of non-ideal behaviours to investigate, with junction breakdown being found to sometimes be localised at crystal defects or regions with different doping levels. These *"Microplasma"* defects, were hypothesised to involve impact ionisation prompting both (i) methods to remove them and (ii) direct utilisation of the mechanism to provide electrical gain (1960s).

History illuminates the path of science and engineering allowing us to see what has been attempted by previous researchers. By citing primary sources, this chapter aims to provide a literature review for the period of 1900 to 1969, i.e. the early history of modern photon counting technologies such as single-photon avalanche diodes and silicon photo-multipliers. As the field has grown exponentially, and has widened with respect to technologies, modern developments (1970 to present) and other gain mechanisms require similar historical studies. However, we have included several recent review articles and texts that point to modern trends and performances [1–5, 17, 19]. To finalise this chapter and allow the reader to follow the literature into the 1970s to 2000s period, (i) a reference from 2003 showing the beginnings of integration of SPADs with planar integrated circuits in CMOS technologies [118], and (ii) a 2007 two-part technical review of operation principles, features and electronics for SPAD arrays [119, 120], are provided.

Author details

Edward M.D. Fisher

Address all correspondence to: e.fisher@ed.ac.uk

The Institute of Digital Communications (IDCOM), The School of Engineering, The University of Edinburgh, Scotland, UK

References

[1] Hadfield R. Single-photon detectors for optical quantum information applications. Nature Photonics. 2009;**3**(12):696-705

[2] Eisaman M, Fan J, Migdall A, Polyakov SV. Single-photon sources and detectors. Review of Scientific Instruments. 2011;**82**(7). DOI: 10.1063/1.3610677

[3] Campbell J. Recent advances in telecommunications avalanche photodiodes. IEEE Journal of Lightwave Technology (JLT). 2007;**25**(1):109-121

[4] Cova S, Ghioni M, Lacaita A, Samori C, Zappa F. Avalanche photodiodes and quenching circuits for single-photon detection. Applied Optics. 1996;**35**(12):1956-1975

[5] Renker D. Geiger-mode avalanche photodiodes; history, properties and problems. Nuclear Instruments and Methods in Physics Research A. 2006;**567**(1):48-56

[6] Hacking I. Representing and Intervening: Introductory Topics in the Philosophy of Natural Science. 1st ed. Cambridge, UK: Cambridge University Press; 1983

[7] Hawking S, Mlodinow L. The Grand Design. 1st ed. London: Bantam Press; 2010

[8] Kuhn T. The Structure of Scientific Revolutions. 3rd ed. Chicago: University of Chicago Press; 1962

[9] Loudon R. The Quantum Theory of Light. 3rd ed. New York: Oxford University Press; 2000

[10] Satchel J, Cassidy D, Renn J, editors. A. Einstein – The collected papers of Albert Einstein (English translation). The Swiss Years 1900–1909. 1990;**2**(1):89

[11] Tada K. Einstein's photon hypothesis and its impact on science and technology. Association of Asia Pacific Physical Sciences (AAPPS) Bulletin. 2005;**15**(2):32-38

[12] Lamb WE. Anti-photon. Applied Physics B: Lasers and Optics. 1995;**60**(2):77-84

[13] Roychoudhuri C, Kracklauer AF, Creath K, editors. The Nature of Light: What Is a Photon? 1st ed. Florida, USA: CRC Press; 2008

[14] Henley E, Garcia A. Subatomic Physics. 3rd ed. Singapore: World Scientific; 2007

[15] Durini D editor. High Performance Silicon Imaging: Fundamentals and Applications of CMOS and CCD Sensors. 1st ed. Cambridge, UK: Woodhead Publishing; 2014

[16] Nix FC. Photoconductivity. Reviews of Modern Physics. 1932;**4**(4):723-766

[17] Antoncik E, Gaur N. Theory of quantum efficiency in silicon and germanium. Journal of Physics C: Solid State Physics. 1978;**11**(4):735-744

[18] Christensen O. Quantum efficiency of the internal photoelectric effect in silicon and germanium. Journal of Applied Physics. 1976;**47**(2):689-695

[19] Sze SM. Semiconductor Devices: Physics and Technology. 2nd ed. New York, USA: John Wiley and Sons; 2001

[20] Shockley W. Electrons and Holes in Semiconductors with Applications to Transistor Electronics. 1st ed. D. Van Nostrand Company: Princeton, New Jersey, USA; 1950

[21] Lubsandorzhiev B. On the history of photomultiplier tube invention. Nuclear Instruments and Methods in Physics Research A. 2006;**567**(1):236-238

[22] Iams H, Salzberg B. The secondary emission phototube. Proceedings of the IRE. 1935; **23**(1):55-64

[23] Zworykin V, Morton G, Malter L. The secondary emission multiplier – A new electronic device. Proceedings of the IRE. 1936;**23**(3):351

[24] Kubetsky L. Multiple amplifier. Proceedings of the IRE. 1937;**25**(4):421-433

[25] Townsend J. The conductivity produced in gases by the motion of negatively charged ions. Philosophical Magazine (London). 1901;**1**(2):189-227

[26] Townsend J. The genesis of ions by the motion of positive ions in a gas, and a theory of the sparking potential. Philosophical Magazine (London). 1903;**6**(35):598-618

[27] Langmuir I. The effect of space charge and residual gases on thermionic currents in high vacuum. Physical Review. 1913;**2**(6):450-486

[28] Meikle GS. The hot cathode argon gas filled rectifier. General Electric Review. 1916;**XIX (19)**(4):297-304

[29] Rutherford E, Geiger H. An electrical method of counting the number of α-particles from radio-active substances. Proceedings of the Royal Society. 1908;**81**(546):141-161

[30] Crossley P, Noel G, Wolf M. Review and Evaluation of Past Solar Cell Development Efforts. In: Report #NASA-CR-86979 Contract #NASW1427. Astro-Electronics Applied Research Laboratory, National Aeronautics and Space Administration (NASA); 1967

[31] Zener C. A theory of the electrical breakdown of solid dielectrics. Proceedings of the Royal Society of London. 1934;**144**(885):532-529

[32] Scaff J, Ohl R. Development of silicon crystal rectifiers for microwave radar receivers. Bell System Technical Journal. 1947;**26**(1):1-30

[33] Torrey H, Whitmer C. Crystal Rectifiers. New York: McGraw-Hill Inc; 1948

[34] Shockley W. The theory of p-n junctions in semiconductors and p-n junction transistors. Bell System Technical Journal. 1949;**28**(3):435-489

[35] Riordan M, Hoddeson L. Crystal Fire: The Invention of the Transistor and the Birth of the Information Age. 1st ed. Norton and Company Inc: W. W; 1998

[36] Ohl RS. Light-Sensitive Electric Device. US Patent 2402662. Issued 25-06-1946;Filed 27-05-1941

[37] Fox M, Pearsall C, Powell V. Manufacturing Procedure for the Radiation Laboratory High Burnout Crystals. Radiation Laboratory Report #501; 1943

[38] Miller P, Greenblatt M. Photoeffects in Pure Silicon. National Defence Research Committee (NDRC). 1945;Division 14 (RADAR) Report #412 (University of Pennsylvania)

[39] Brown F. The velocity of propagation of the transmitted photo-effect in silicon crystals. Physical Review: Minutes of the Meeting at Cambridge, April 25–27 Proceedings ofthe American Physical Society. 1946;**69**(11–12):686

[40] Teal G, Fisher J, Treptow A, New Bridge A. Photo-cell employing a photo-conductive effect in silicon: Some properties of high purity silicon. Physical Review: Minutes of the Meeting at Cambridge, April 25–27, Proceedings of the American Physical Society. 1946; **69**(11–12):686

[41] Benzer S. High-voltage and photo-sensitive characteristics in germanium. Physical Review: Minutes of the Meeting at Cambridge, April 25–27, Proceedings of the American Physical Society. 1946;**69**(11–12):683

[42] Bray R, Lark-Horovitz K. Photo- and thermo- effects in p-type germanium rectifiers. Physical Review: Minutes of the Meeting at Minneapolis, November 29–30, Proceedings of the American Physical Society. 1947;**71**:141

[43] Benzer S. Excess-defect germanium contacts. Physical Review. 1947;**72**(12):1267-1268

[44] Shive J. A new germanium photo-resistance cell. Physical Review. 1949;**76**(4):575

[45] Fan H. Theory of photovoltaic effect of p-n barrier in a semi-conductor. Physical Review: Minutes of the Meeting at Cleveland, March 10-12, Proceedings of the American Physical Society. 1949;**75**(1):1631

[46] Becker M, Fan H. Photovoltaic effect of p-n barriers produced in germanium by α- and deuteron bombardment. Physical Review: Minutes of the Meeting at Cleveland, March 10–12, Proceedings of the American Physical Society. 1949;**75**(1):1631

[47] Teal G, Little J. Growth of germanium single crystals. Physical Review: Minutes of Meeting at Oak Ridge March 16–18, Proceedings of the American Physical Society. 1950;**78**(5):647

[48] McAfee K, Pearson G. The electrical properties of silicon p-n junctions grown from the melt. Physical Review. 1952;**87**:190

[49] Goucher F. The photon yield of electron-hole pairs in germanium. Physical Review. 1950;**78**(6):816

[50] Fan H, Becker M. Infra-red absorption of silicon. Physical Review. 1950;**78**(2):178

[51] Becker M, Fan H. Photovoltaic effect of p-n junctions in germanium. Physical Review. 1950;**78**(3):301-302

[52] Pietenpol W. P-N junction rectifier and photo-cell. Physical Review. 1951;**82**(1):121

[53] Shockley W, Sparks M, Teal G. P-N junction transistors. Physical Review. 1951;**83**(1):151-162

[54] McAfee K, Ryder E, Shockley W, Sparks M. Observations of zener current in germanium p-n junctions. Physical Review. 1951;**83**(3):650-651

[55] Shockley W, Read W Jr. Statistics of the recombinations of holes and electrons. Physical Review. 1952;**87**(5):835-842

[56] Hall R. Electron-hole recombination in germanium. Physical Review. 1952;**87**(2):387

[57] Pearson G, Read W, Shockley W. Probing the space-charge layer in a p-n junction. Physical Review. 1952;**85**(6):1055-1057

[58] Pearson G, Sawyer B. Silicon P-N junction alloy diodes. Proceedings of the IRE. 1952; **40**(11):1348-1351

[59] Shive J. The properties of germanium phototransistors. Journal of the Optical Society of America. 1953;**43**(4):239-244

[60] McKay K, McAfee K. Electron multiplication in silicon and germanium. Physical Review. 1953;**91**(5):1079-1084

[61] Emmons R. Avalanche-photodiode frequency response. Journal of Applied Physics. 1967;**38**(9):3705-3714

[62] Nishizawa J. P-I-N Photo-Diode. Japanese Patent #JP1955-8969A; 1952

[63] McKay K. Avalanche breakdown in silicon. Physical Review. 1954;**94**(4):877-884

[64] Wolff P. Theory of electron multiplication in silicon and germanium. Physical Review. 1954;**95**(6):1415-1420

[65] Kingston R. Switching time in junction diodes and junction transistors. Proceedings of the IRE. 1954;**42**(5):829-834

[66] Miller S. Avalanche breakdown in germanium. Physical Review. 1955;**99**(4):1234-1241

[67] Newman R. Visible light from a silicon p-n junction. Physical Review. 1955;**100**(2):700

[68] Chynoweth A, McKay K. Photo emission from avalanche breakdown in silicon. Physical Review. 1956;**102**(2):369-376

[69] Rose D. Microplasmas in silicon. Physical Review. 1957;**105**(2):413-418

[70] Chynoweth A, Pearson G. Effect of dislocations on breakdown in silicon p-n junctions. Journal of Applied Physics. 1958;**29**(7):1103-1110

[71] Chynoweth A, McKay K. Threshold energy for electron-hole pair-production by electrons in silicon. Physical Review. 1957;**108**(1):29-34

[72] Senitzkey B, Moll J. Breakdown in silicon. Physical Review. 1958;**110**(3):612-620

[73] Chynoweth A, McKay K. Light emission and noise studies of individual microplasmas in silicon p-n junctions. Journal of Applied Physics. 1959;**30**(11):1811-1813

[74] Read W. A proposed high-frequency, negative-resistance diode. Bell System Technical Journal. 1958;**37**(2):401-446

[75] Shockley W, Gibbons J. Theory of transient build-up in avalanche transistors. Transactions of the American Institute of Electrical Engineers (AIEE). 1959;**77**(6):993-998

[76] Hamilton D, Gibbons J, Shockley W. Physical principles of avalanche transistor pulse circuits. Proceedings of the IRE. 1959;**47**(6):1102-1108

[77] Champlin K. Microplasma fluctuations in silicon. Journal of Applied Physics. 1959;**30**(7):1039-1050

[78] McIntyre R. Theory of microplasma instability in silicon. Journal of Applied Physics. 1961;**32**(6):983-995

[79] Kikuchi M. Visible light emission and microplasma phenomena in silicon p-n junction. Journal of the Physical Society of Japan. 1960;**15**(10):1822-1831

[80] Haitz R, Goetzberger A, Scarlett R, Shockley W. Avalanche effects in silicon p-n junctions I: Localized photomultiplication studies on microplasmas. Journal of Applied Physics. 1963;**34**(6):1581-1590

[81] Goetzberger A, Stephens C. Voltage dependence of microplasma density in p-n junctions in silicon. Journal of Applied Physics. 1961;**32**(12):2646-2650

[82] Batdorf R, Chynoweth A, Dacey G, Foy P. Uniform silicon p-n junctions: I: Broad area breakdown. Journal of Applied Physics. 1960;**31**(7):1153-1160

[83] Ruge I, Keil G. Microplasmas in silicon p-n junctions as detectors for gamma radiation. The Review of Scientific Instruments. 1963;**34**(4):390-392

[84] Goetzberger A, McDonald B, Haitz R, Scarlet R. Avalanche effects in silicon p-n junctions: II: Structurally perfect junctions. Journal of Applied Physics. 1963;**34**(6):1591-1600

[85] Haitz R. Model for the electrical behaviour of a microplasma. Journal of Applied Physics. 1964;**35**(5):1370-1376

[86] Haitz R. Mechanisms contributing to the noise pulse rate of avalanche diodes. Journal of Applied Physics. 1965;**36**(10):3123-3131

[87] Haitz R. Microplasma interaction in silicon p-n junctions. Solid State Electronics. 1964; **7**:439-444

[88] Haitz R. Studies on optical-coupling between silicon p-n junctions. Solid State Electronics. 1965;**8**:417-415

[89] Ruge I, Keil G. Mutual interaction between microplasmas in silicon p-n junctions. Journal of Applied Physics. 1963;**34**(11):3306-3308

[90] Lee C, Logan R, Batdorf R, Kleimack J, Wiegmann W. Ionisation rates of holes and electrons in silicon. Physical Review. 1964;**134**(3A):761-773

[91] McIntyre R. Multiplication noise in uniform avalanche diodes. IEEE Transactions on Electron Devices. 1966;**13**(1):164-163

[92] Melchior H, Lynch W. Signal and noise response of high speed germanium avalanche photodiodes. IEEE Transactions on Electron Devices. 1966;**ED-13**(12):829-838

[93] Baertsch R. Low-frequency noise measurements in silicon avalanche diodes. IEEE Transactions on Electron Devices. 1966;**13**(3):383-385

[94] Baertsch R. Noise and ionisation rate measurements in silicon photodiodes. IEEE Transactions on Electron Devices. 1966;**13**(12):987

[95] Sze S, Gibbons G. Avalanche breakdown voltages of abrupt and linearly graded p-n junctions in Ge, Si, GaAs and GaP. Applied Physics Letters. 1966;**8**(5):111-113

[96] Sze S, Gibbons G. Effect of junction curvature on breakdown voltages in semiconductors. Solid State Electronics. 1966;**9**(9):831-845

[97] Ogawa I. Carrier multiplication in semiconductor detectors. Nuclear Instruments & Methods. 1967;**49**(2):325-333

[98] Haitz R. Noise of a self-sustaining avalanche discharge in silicon: Low-frequency noise studies. Journal of Applied Physics. 1967;**38**(7):2935-2946

[99] Hines E. Noise theory for the read type avalanche diode. IEEE Transactions on Electron Devices. 1966;**13**(1):158

[100] Kimura C, Nishizawa J. Turn-on mechanism of a microplasma. Japanese Journal of Applied Physics. 1968;**7**(12):1453-1463

[101] Melchior H, Goetzberger A, Nicollian E, Lynch W. Electrical suppression of avalanche currents in semiconductor junctions. Solid State Electronics. 1969;**12**:449-462

[102] Goetzberger A. Uniform avalanche effect in silicon three-layer diodes. Journal of Applied Physics. 1960;**31**(12):2260-2261

[103] Johnson K. Photodiode signal enhancement effect at avalanche breakdown voltage. Proceedings of the International Solid-State Circuits Conference (ISSCC). 1964:64-65

[104] Johnson K. High-speed photodiode signal enhancement at avalanche breakdown voltage. IEEE Transactions on Electron Devices. 1965;**12**(2):55-63

[105] Anderson L, McMullin P, D'Asaro L, Goetzberger A. Microwave photodiodes exhibiting microplasma-free carrier multiplication. Applied Physics Letters. 1965;**6**(4):62-64

[106] Melchior H, Anderson L. Noise in High Speed Avalanche Photodiodes. In: Proc. International Electron Devices Meeting; 20–22 October; Washington DC; 1965. p. 11.2

[107] Schuster M, Strull G. A monolithic mosaic of photon sensors for solid-state imaging applications. IEEE Transactions on Electron Devices. 1966;**13**(12):907-912

[108] Lee C, Batdorf R, Wiegmann W, Kaminsky G. Technological developments evolving from research on read diodes. IEEE Transactions on Electron Devices. 1966;**ED-13**(1):175-180

[109] Huth G. Recent results obtained with high field, internally amplifying semiconductor radiation detectors. IEEE Transactions on Nuclear Science. 1966;**13**(1):36-42

[110] Haitz R, Smits F. Noise analysis for silicon particle detectors with internal multiplication. IEEE Transactions on Nuclear Science. 1966;**13**(3):198-207

[111] Keil G, Bernt H. Infrared detection by avalanche discharge in silicon p-n junctions. Solid-State Electronics. 1966;**9**(4):321-325

[112] Melchior H, Lynch W. Signal and noise response of high speed germanium avalanche photodiodes. IEEE Transactions on Electron Devices. 1966;**ED-13**(12):829-838

[113] Ruegg H. An optimized avalanche photodiode. IEEE Transactions on Electron Devices. 1967;**14**(5):239-251

[114] Goetzberger A, Melchior H. Electrical suppression of microplasmas. IEEE Transactions on Electron Devices. 1967;**14**(9):633

[115] Melchior H, Goetzberger A, Nicollian E, Lynch W. Electrical suppression of avalanche currents in semiconductor junctions. Solid-State Electronics. 1969;**12**:449-462

[116] Kao Y, Wolley E. High-voltage planar p-n junctions. Proceedings of the IEEE. 1967;**55**(8):1409-1414

[117] Sigmund H. Photoelectrical properties of spherical avalanche diodes in silicon. Infrared Physics. 1968;**8**(1):259-264

[118] Sciacca E, Giudice A, Sanfilippo D, Zappa F, Lombardo S, Consentino R, Franco C, Ghioni M, Fallica G, Bonanno G, Cova S, Rimini E. Silicon planar technology for single-photon optical detectors. IEEE Transactions on Electron Devices. 2003;**50**(4):918-925

[119] Zappa F, Tisa S, Tosi A, Cova S. Principles and features of single-photon avalanche diode arrays. Sensors and Actuators A. 2007;**140**(1):103-112

[120] Tisa S, Zappa F, Tosi A, Cova S. Electronics for single-photon avalanche diode arrays. Sensors and Actuators A. 2007;**140**(1):113-122

Chapter 2

High Sensitivity Photodetector for Photon-Counting Applications

Fabio Acerbi and Matteo Perenzoni

Additional information is available at the end of the chapter

http://dx.doi.org/10.5772/intechopen.71940

Abstract

In the last years, there has been a large development of low-light applications, and many of them are based on photon counting using single-photon detectors (SPDs). These are very sensitive detectors typically with an internal gain. The first candidate SPD was the photomultiplier tube (PMT), reaching a very high gain (~10^6), but there have been a large development of many other solutions, like solid-state solutions. Among them, single-photon avalanche diodes (SPADs) have been used in spectroscopy, florescence imaging, etc., particularly for their good detection efficiency and time resolution (tens of picoseconds). SPADs have been developed in silicon and III–V materials, for the NIR wavelength range. SPADs can be used as single high-performance pixels, or in arrays. SPAD arrays have imaging capabilities, with high sensitivity. Another kind of array is the silicon photomultiplier (SiPM), where all the pixels are connected to a common anode and a common cathode. SiPMs are used in nuclear medicine, physics experiments, quantum-physics experiments, light detection and ranging (LIDAR), etc., due to their high detection efficiency combined with large sensitive areas, and high dynamic range. SiPMs with many small cells present several advantages and nowadays the SPAD pitch can be reduced down to 5 μm.

Keywords: photon counting, single-photon, dynamic range, photodetector, high sensitivity, imaging

1. Introduction

In the last years, there has been a large development and an increasing interest in photodetectors for low-light applications. Single-pixel detectors, like photomultiplier tube (PMT) [1, 2], and single-photon avalanche diode (SPAD) [3], can be extended to the more complex single-photon imagers [4]. These technologies enabled several new applications, physical experiment, and techniques. They also significantly increased the image quality and the sensitivity in biomedical and medical imaging techniques [5, 6].

Photon counting is a statistically accurate technique to measure faint light signals, based on the measurement of a random sequence of pulses generated by detection of single photons. In analog "linear" detection technique, the incoming light is composed by many photons and the output of the photodetector is typically continuous and proportional to the intensity of the light signal. In photon counting, the intensity of the light is so low that the output signal from the photodetector is composed by separated pulses (see **Figure 1**). Photon counting can give a better signal-to-noise ratio [1]. The baseline fluctuation and the electronic noise in the front-end are no more relevant, since the detection is based on a thresholding on the photo-generated pulses, resulting in a more "digital" approach. Photodetectors are single-photon detectors (SPDs); they have particular characteristics like a high internal gain. They produce a prompt and strong signal in response to a small photo-generated charge like the one from a single photon. This gives them the potentiality for a very good time resolution, which is exploited, for example, in time-correlated single photon counting (TCSPC) technique.

1.1. Time-correlated single photon counting

TCSPC is a technique to reconstruct the temporal shape of faint light signal. It is based on the detection of single photons of periodical light signal and on the measurements of their

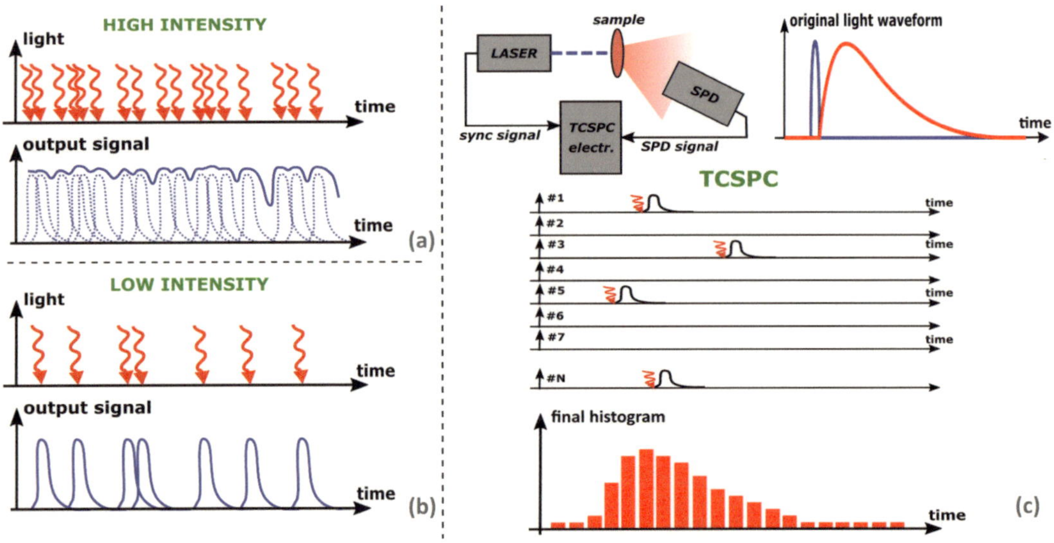

Figure 1. Representation of photon flux and detector output in case of high light intensity (a) and of low light intensity (b) and representation of the time-correlated single photon counting (TCSPC) technique (c) for the reconstruction of low light level periodic light signals.

detection times. For each photon, the detection time is measured, building a histogram. After many detections, the histogram will represent the waveforms of the detected optical signal. The TCSPC principle is represented in **Figure 1**, showing example of detection cycles and the resulting histogram. The assumption is that the light signal intensity is low enough so that the probability of having one photon at the detector in each cycle is much less than one; thus, the probability to have more than one photon is negligible. This is to avoid distortion because the TCSPC system can detect only one event per cycle. Each measurement starts with the arrival of a pulse from the reference signal (START) and stops with the arrival of the signal related to the photon detection (STOP).

2. Single-photon detectors

Photon counting requires a very-high-sensitivity detector (down to single photon level), thus single-photon detector (SPD). Moreover, the signal produced from the detector has to have an amplitude high enough (i.e., higher than the electronic noise) to be "sensed" by the front-end electronics. In a semiconductor-based detector, for example, the absorption of a photon produces a single electron-hole pair, which is typically too low compared to the sensitivity of a front-end circuit. An internal charge multiplication mechanism is needed inside the detector. This can be obtained, for example, by secondary electron emission or by avalanche multiplication. Single-photon detectors can be divided into the following groups:

- Vacuum based: photomultiplier tubes (PMTs), micro-channel plates (MCPs), etc.
- Solid-state: electron-multiplying CCD (EMCCDs), single photon avalanche diodes (SPADs), SPAD array, silicon photomultipliers (SiPMs), quanta-imagers (QIS), etc.
- Cryogenic-temperature based: superconducting nanowire single-photon detectors (SNSPDs), transition-edge sensors (TESs), etc.

Vacuum-based detectors are mature technologies with big active areas and low noise, but they can be bulky, they require high biases, and are sensitive to magnetic fields [7]. Cryogenic-based detectors have good performance: low noise and high detection efficiency; thus, they are often used for quantum-physics experiments [8, 9]. They require typically a multi-stage cryostat, which can be bulky. Solid-state solution are compact and requires low biases (<100 V). They have a good detection efficiency and time resolution, but usually a higher noise. In the following, some of the main SPDs are better described.

The photomultiplier tubes are one of the first photodetector used for photon counting [1]. It can reach gains up to 10^6 or 10^7. The PMT has more than 50 years of history and has been used in a variety of applications due to its great versatility. A PMT is a vacuum tube that consists of an input window, a photocathode, focusing electrodes, an electron multiplier section (dynodes), and an anode. Incoming photons can be absorbed in the photocathode material and an electron-hole pair is generated. The electron can escape the material due to photoelectric effect, being then focused on the primary dynode and accelerated by the electric field. The impact creates several other secondary electrons, which are then all accelerated toward the successive dynode, and so on, until all the generated electrons are collected by the anode. Different version have been developed during the years: PMT-MCP (microchannel plate) exploits microchannels to obtain the electron multiplication instead of the dynodes. Electron, extracted from the photocathode "bounces" in this microchannel and produces secondary electrons. MCP-PMTs are among the fastest photon counting detectors [2]. Moreover, MCP can be used to build position-sensitive detectors, when coupled with a multianode structure or it can be used to create image intensifiers, typically used in front of a CCD imager (charge-coupled device) for night-vision or to build very sensitive imagers.

Among cryogenic-based SPD, superconducting nanowire single-photon detectors (SnSPDs) [10, 11] and transition edge sensors (TESs) [12] are the most used. SNSPDs are nanostructured devices based on long stripes of an ultrathin superconducting film, operated well below the critical temperature (TC), and biased with a subcritical current. The absorption of a photon produces a hot-spot region, in which the superconductivity is suppressed. The hot spot grows in size until electrons diffuse out of the spot. The current locally exceeds the critical limit, thus generating secondary hot spots. The superconductivity is destroyed and a resistive barrier is formed; thus, a voltage pulse can be detected. After a certain delay, superconductivity is restored. Recently, the performance of SNSPD improved significantly, also due to better cryostat solutions. Cryogenic-based single-photon detectors demonstrate high quantum efficiency (QE) at visible and near-infrared (NIR) wavelengths [10], low dark count rate, picosecond pulse-to-pulse timing jitter, and a gigahertz counting rate. SnSPDs are also recently being used with "imaging" capabilities [13].

Solid-state detectors are typically the preferable choice in applications like consumer electronics, in portable instrumentation or to build imagers. Most of solid-state SPDs are based on avalanche multiplication process, like in single-photon avalanche diodes (SPADs). The single pixel of an array of SPAD or, in general one single SPAD, can have a sensitive area between tens and hundreds of micrometers. This is significantly low compared to PMTs, but comparable with cryogenic-based detectors. To extend the sensitive area, an array of pixels can be realized, creating a bigger detector. All these pixels can be connected together, like in silicon photomultipliers (SiPMs) or each pixel can be read-out separately, creating an imager.

3. Single-photon avalanche diode

SPADs are photodetectors essentially based on a p-n junction, designed to be biased above the breakdown voltage [3]. In such conditions, the electric field is so high (typically $>10^5$ V/cm) that a single carrier injected or generated into the depletion layer can trigger a self-sustaining avalanche multiplication process. As represented in **Figure 2**, when the

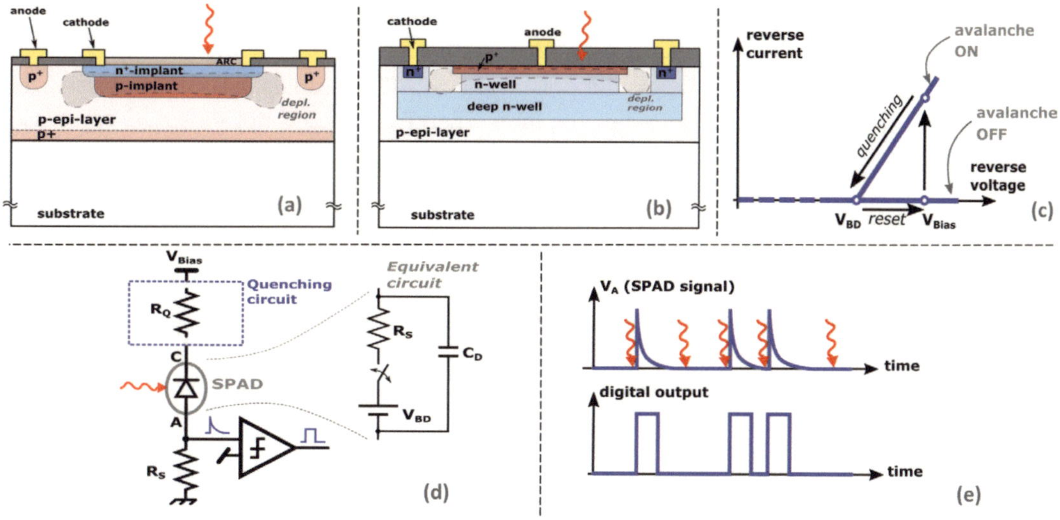

Figure 2. Schematic representation of a SPAD cross-section, in custom technology (a) and in CMOS technology, exploiting well and deep-well implants (b). The typical reverse I–V curve of a SPAD, with the metastable behavior is represented (c) along with the typical SPAD biasing, quenching and front-end circuit (d) and the resulting output digitalized signal (e).

avalanche is triggered, the current rises swiftly (nanoseconds or sub-nanosecond rise time) to a macroscopic level (milliampere range). If the primary carrier is photo-generated, the leading edge of the avalanche pulse marks (with good time resolution) the arrival time of the detected photon. The current continues to flow until the avalanche is quenched by lowering the bias voltage down to or below the breakdown voltage ("quenching") [3]: this lower electric field is no more able to accelerate the carriers at a sufficient energy. After the avalanche quenching, the bias voltage must be restored in order to be able to detect another photon (reset phase). All these operations require a suitable electronics: this circuit is usually referred to as quenching circuit.

The simpler quenching circuit is just a series resistor, with a relatively high resistance value. With this resistor, when the current in the SPAD increases, due to avalanche buildup, the voltage drop at the quenching resistor rises, and thus the voltage at the SPAD consequently decreases, reaching values close to V_{BD}, eventually quenching the avalanche. Then the SPAD is reset through the same resistor, restoring the bias to V_{BIAS} value, with a time constant:

$$\tau_{reset} = R_Q \cdot C_{SPAD} = R_q \cdot (C_D + C_{totCathode}) \quad (1)$$

On the other SPAD terminal, there is the avalanche-sensing part. The front-end circuit has to sense the avalanche and to provide an output pulse per each detection. This can be done with a sensing resistor, a transistor or a trans-impedance amplifier, followed by a voltage discriminator, to obtain a digital pulse.

Alternatively to passive quenching, active quenching or mixed active/passive quenching solutions can be used [3]. With active solutions, a transistor is used to force the bias of one of the SPAD node to either quench or reset the bias at the SPAD terminals [14]. In such way, the recharge is faster and the dead-time (i.e., the time when the SPAD is not sensitive) can be set and it is well-defined.

3.1. SPAD parameters

The active area of silicon SPAD is generally circular, to have uniform electric fields, with a diameter between 10 μm and 500 μm [15]. With SPADs and with other photon-counting detectors, due to the "digital" operation mode (different from "linear" mode), there are specific parameters identifying the performance:

- *Photon detection efficiency* (PDE), i.e., the ability to detect photons. This is the ratio between the number of detected photons and the photons arriving at the detector. PDE is calculated as the product of: (i) the quantum efficiency (QE) and (ii) the avalanche triggering probability (P_T). The latter is the probability that photo-generated carriers can reach the high-field region and trigger a self-sustaining detectable avalanche.

- The noise is typically divided into "primary" noise and correlated noise. The primary noise represents all the avalanche pulses due to thermally generated carriers (or generated by tunneling or field-assisted thermal generation). The *dark count rate* (DCR) is typically in the order of 10–1000 counts per second.

- The correlated noise for a SPAD is represented by the *afterpulsing*. During the avalanche, a large amount of carriers flows through the depleted region and some of them can be trapped in deep-levels (traps), being subsequently released with a delay, causing retriggering of another spurious avalanche, not related to photon absorption but to a previous avalanche, thus "correlated noise." The time distribution of carrier release follows normally an exponential (or multiexponential) distribution. Its time constant depends on temperature, being slower at low temperatures. The afterpulsing probability depends on the SPAD itself, on the quenching circuit, and on the dead-time.

- The *dead-time* is the time interval after an avalanche, where the SPAD is not sensitive to another photon. This interval is necessary to recharge the SPAD and to let the traps to release the carriers without triggering a spurious avalanche. This is typically in the order of tens of nanosecond. Differently from active quenching, with passive quenching, the recharge is exponential, and thus it is not easy to identify a precise dead-time. The recharge time-constant can be used as a parameter.

- The time resolution of the SPAD, i.e., the ability of precise time-tag the photon arrival time, is another important parameter. The "*timing jitter*" or "*single-photon time resolution*" (SPTR) quantifies the time spread between the photon arrival and the pulse detection by the front-end electronics. This spread is due to the different absorption position and the statistical avalanche buildup time. It is in the order of a few tens of picoseconds [15]. To measure the timing jitter, it can be used a pulsed laser, attenuated to single-photon level. Using TCSPC technique, the resulting time-resolution histogram shape is generally Gaussian, with an exponential tail, as shown in **Figure 3**. The tail is due to photons absorbed in the neutral region: once photo-generated the carriers diffuse randomly and can reach the depleted region but with a certain delay. The tail in the timing jitter histogram is particularly detrimental in some applications [16].

3.2. SPAD for the near-infrared range

Some applications require detecting single photons with a wavelength in the near-infrared (NIR) range, above 1000 nm, for example when using laser emitting in the telecom wavelengths, at 1310 nm and 1550 nm. There are SPADs made by semiconductor materials different from silicon. In particular, there has been recently a development of SPADs made with III/V materials like InGaAs/InP [17, 18] or InGaAs/InAlAs [19] or InGaAsP/InP [20]. InGaAs/InP SPADs are used to detect photons at 1550 nm. They are based on a separate absorption, grading, charge and multiplication (SAGCM) heterostructure as shown in **Figure 4**. The absorption layer is made of InGaAs and the multiplication layer is of InP. Between them, there

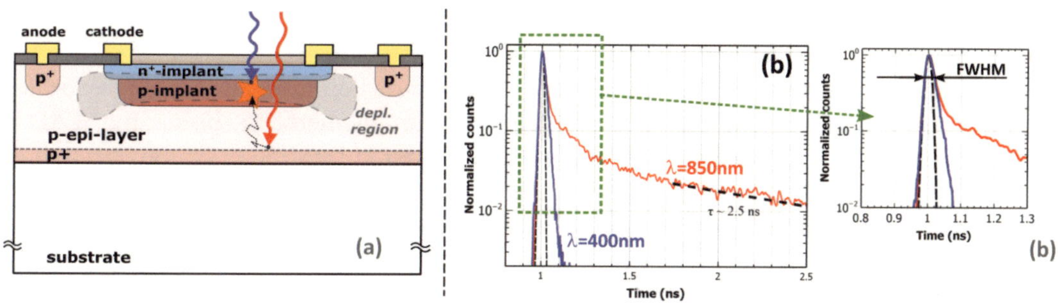

Figure 3. Representation of two photon absorption cases in the depleted region and in the neutral region beneath it (a) and examples of the relative timing histograms (b). At λ = 400 nm, all the photons are absorbed in the depleted region, whereas at 850 nm, they are absorbed mostly in the neutral region, creating the tail.

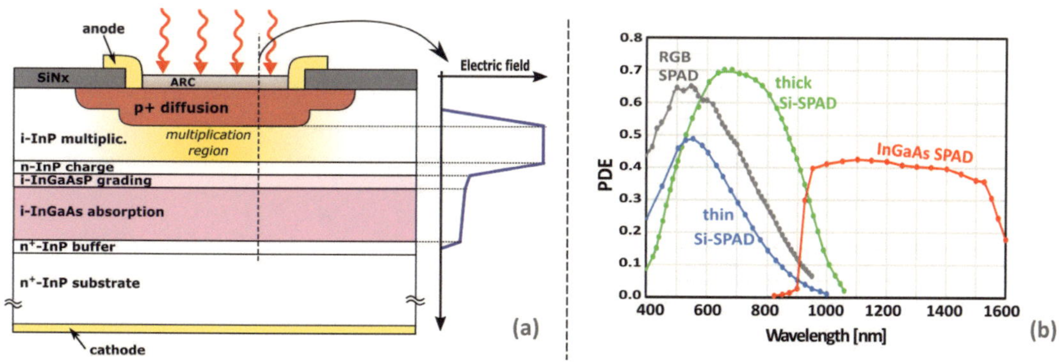

Figure 4. Schematic cross-section of an InGaAs/InP SPAD, with different layers (a). Comparison of photon detection efficiency (PDE) of some silicon SPADs and InGaAs SPADs (b): "thin" silicon SPAD [23], "thick" silicon SPAD [23], the FBK RGB technology (SiPM PDE divided by the FF) [25] and the typical state-of-the-art InGaAs/InP SPADs [24].

is one or more layers of InGaAsP, called "grading layer." This is needed to smooth the hetero-barrier in the valence band between the InGaAs and the InP layers. On the top of it, there is a "charge layer" that shapes the electric field profile.

InGaAs has an energy gap of about 0.75 eV [17]. The PDE of InGaAs/InP SPAD is typically around 40% between 1000 nm and 1550 nm (see **Figure 4(b)**), being limited by the thickness of the absorption layer. This cannot be too high in order to limit the noise. The primary noise is typically higher in InGaAs/InP SPADs compared to silicon ones. The technology of the III/V material is commonly less mature. The main noise source is thermal generation in the InGaAs layer down to about 225 K, whereas at lower temperatures, it is trap-assisted-tunneling in the multiplication layer. The afterpulsing probability is also higher in InGaAs/InP SPADs, compared to silicon. The typical dead-time can be in the order of few microseconds. However, despite these limitations, InGaAs/InP SPADs have been used in several applications. Advantages such as good PDE, compactness, and low power consumption make this a competitive solution for NIR single-photon counting. They are typically used inside compact modules and cooled at 230 K (or lower) to decrease the primary noise [24].

The high afterpulsing probability can be a limiting factor in several applications. Some solutions have been proposed that aim to reduce the amount of charge flowing per each avalanche. The first is based on very fast quenching circuit, reducing the bias at the SPAD quickly once an avalanche is detected [21]. Another solution is based on fast gating the detector, for example, using sinusoidal signals with a frequency of few gigahertz: the bias at the SPAD is modulated and, once an avalanche is triggered, it can last at maximum for the duration of half a gating period [18]. This technique is called *sinusoidal gating*. Another approach is based on the integration of a quenching resistor directly on the SPAD, very close to the active area. This approach limits the overall capacitance at the node between the SPAD and the quenching resistor, thus the avalanche charge. Some implementations of this kind of detector are called negative-feedback avalanche diode (NFAD) [18].

4. SPAD array and low-light imagers

SPADs can be arranged in arrays, 1D or 2D, manufactured in custom process or CMOS process. SPAD array in CMOS process has the advantage of having all the quenching, control and read-out electronics in each pixel. Unfortunately, usually with the CMOS process, the SPAD performance is not as good as with a custom process, due to the non-optimized implants and electric fields. In particular, they have a higher DCR.

SPAD arrays are one of the solutions for low-light-level imaging [4, 5]. With such technology, it is possible to have a sensitivity down to the single-photon level, working in photon-counting mode. With the proper in-pixel circuitry, each pixel counts the number of photons detected in the integrating period. This information is stored and then downloaded, to build an image. Scientific imaging applications often require such low level of sensitivity, typically with the addition of time-resolving capabilities. As few examples: fluorescence lifetime imaging microscopy (FLIM), Raman spectroscopy, time-resolved near-infrared spectroscopy, and consumer applications like three-dimensional (3-D) imaging based on time-of-flight [26]. In particular, fluorescence microscopy is an established technique for the analysis of biological processes and relies on the measurement of the fluorescence intensity upon an excitation [22] at different wavelengths. Fluorescent light emission can occur in a variety of temporal scales, from nanoseconds upward. Its lifetime measurements add valuable information and require a specific kind of detectors, like a SPAD array with time-tagging circuitry per pixel. In the same way, 3D imaging, based on time-of-flight (TOF), is an application that is rapidly emerging in many fields [26], to have a three-dimensional image of the scene or to measure the distance from the objects. 3D imaging can be obtained by means of direct TOF or indirect TOF. The first one is based on the direct detection of the time-of-arrival of the reflected light pulse.

The first approach to have SPAD pixels with time-tagging capabilities is the addition of a time-to-digital converter (TDC) in each pixel [27]. The SPAD array will have in each pixel the photon-number information as well as the time-of arrival information (typically of the first photon), as shown, for example, in **Figure 5(a)**. Devices realized with this approach, showed good performance in many biological and 3D-ranging applications. However, the TDC generally occupies a significant part of the pixel area, reducing the fill factor (FF). There are also special implementations without the TDC, but performing distance measurements with indirect time-of-flight (iToF) method, by using three different counters per pixel [4].

In general, to improve the FF in a CMOS implementation, it is necessary to reduce the read-out circuitry occupation and complexity. The SPAD structure has to be placed very close to the read-out electronics, but being electrically isolated with proper implants, as shown in **Figure 5(b)**. To minimize the pixel circuitry, it is possible to implement just the time-to-amplitude conversion

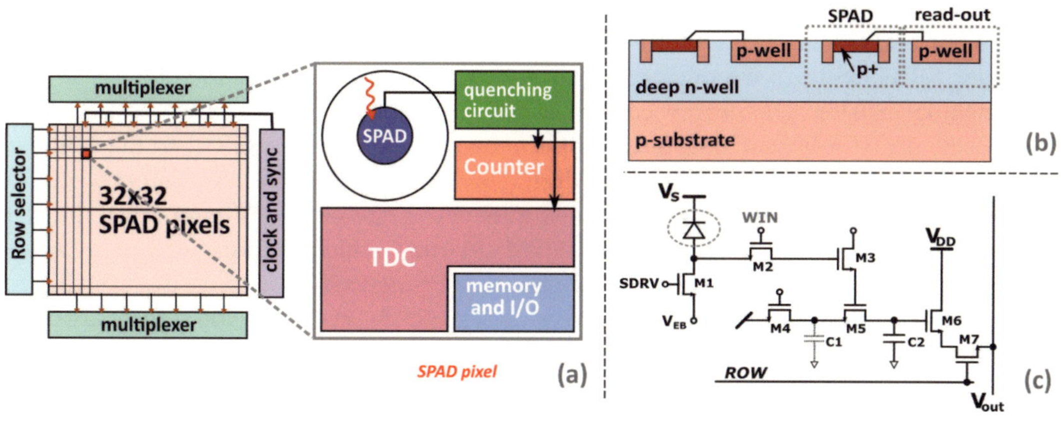

Figure 5. Example of 32 × 32 SPAD array, where each pixel contains a SPAD, the quenching circuitry, counters, and the time-to-digital converter (TDC) [27] (a). Example of possible SPAD CMOS implementation with read-out electronics isolated inside the deep p-well (b) [28]. Example of analog time-gated SPAD pixel, with a reduced electronics-complexity to obtain a higher FF (c) [5].

(TAC) in pixel [29, 30], or using an all-analog approach [5, 28], by means just of few transistors and capacitances. **Figure 5(c)** shows an example of analog time-gated SPAD pixel, based on analog time-gating and analog counter [5].

An alternative implementation of low-level imager is the so-called "quanta image sensor" (QIS) [31]. This is based on the concept of extending the sensitivity of a "classical" image sensor, which is not based on avalanche multiplication process. The pixel is composed of a pinned photodiode (PPD), collecting the photo-generated charge, which then is transferred to a small floating diffusion (FD). FD has a small dimension and its capacitance is very small, thus the conversion gain between charge and voltage amplitude is very high, enough to measure a single photo-electron. In a single-bit QIS, the output after each acquisition is a binary bit plane, where each bit represents the presence or absence of at least one photoelectron. A series of bit planes has to be generated, with a high-speed readout, eventually being able to create an image.

5. Silicon photomultiplier

Silicon photomultipliers (SiPMs) are arrays of many single-photon avalanche diodes (SPADs), all connected in parallel [25, 32, 43]. Each SPAD is called microcell (or cell) and has a square area with a side between 10 μm and 100 μm, whereas the overall SiPM active area can be typically between 1×1 mm^2 and 10×10 mm^2. This is one of the main advantages of SiPM over other SPDs: they can have big active area, but preserving the good performance of the single SPAD, with the additional advantage of being photon-number resolved (i.e., being able to count the number of photons arriving at the detector simultaneously, with a high dynamic range). These characteristics are becoming more and more important in a large number of applications [33–37] (**Figure 6**).

The silicon photomultipliers can be divided into: (i) analog SiPM [25] (see (**Figure 6a**), (**Figure 6b**), and (**Figure 6c**)) and (ii) digital SiPM [37, 38] (see (**Figure 6d**) and (**Figure 6e**)). In analog SiPM (aSiPM), the microcell, composing the array, is just made of a SPAD and a quenching resistor. The output current is the sum of the currents from the triggered cells in the array; thus, the output (amplitude or charge) is proportional to the number of detected photons (see (**Figure 6b**)). In digital SiPM (dSiPM), each microcell is typically composed by a SPAD and a more-complex quenching circuit (see (**Figure 6d**)). The cell provides a digital output to the internal dSiPM circuitry, which eventually digitally sum-up the signals from the microcells and can contain a time-to-digital converter (TDC) (see (**Figure 6e**)) to time-stamp the photons arrival times [37, 38]. In dSiPM, the signals are digital starting from the single microcell, thus less affected by the front-end noise. However, due to the more complex quenching and due to the presence of the control circuitry, the fill factor is lower as well as the photon detection efficiency.

5.1. SiPM applications

SiPMs have obtained a growing attention as alternative to the traditional photomultiplier tubes in the detection of low photon fluxes due to a number of advantages typical of solid-state detectors, such as compactness, ruggedness, ease of use, low operational voltage, and insensitivity to magnetic fields [32]. One of the most common applications is the detection of fast scintillation light in particle detectors that are used in nuclear medicine [32, 38] and in high-energy physics experiments [33, 34]. In these applications, SiPMs are coupled with scintillator crystals, which convert high-energy particle, X or gamma ray into visible or NUV light, being detected by the SiPM. An example of 4 × 4 mm^2 SiPM with some scintillator crystals (with different heights) is shown in **Figure 7a**. SiPMs allowed significant advancements in positron-emission tomography (PET) [35] and other medical applications, especially due to their excellent time resolution [37]. This made possible to develop PET scanners with improved performance as

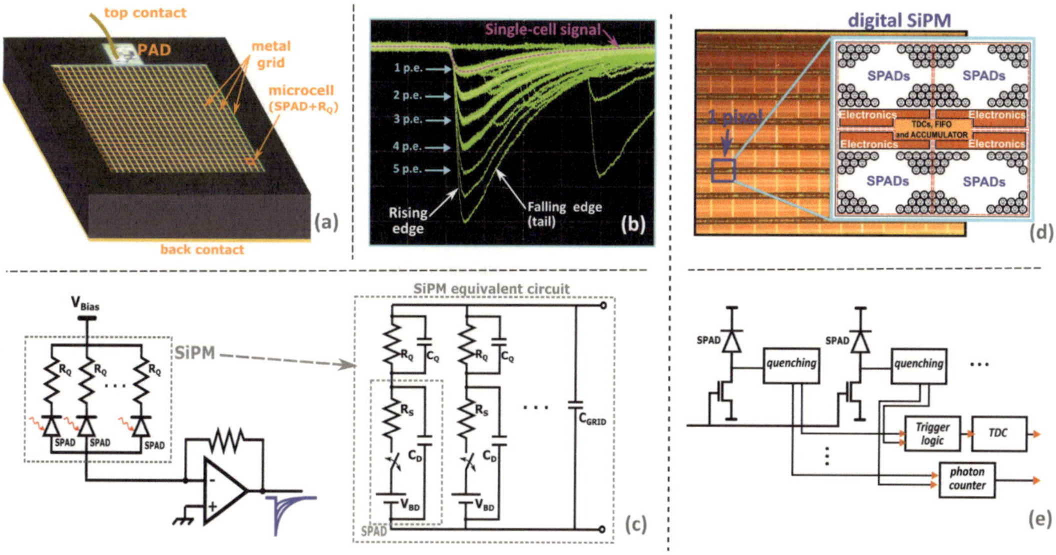

Figure 6. Example of silicon photomultiplier chip (SiPM), with back contact and common top PAD (a). Example of SiPM signal, acquired with oscilloscope in persistence mode (b). Typical circuit for the readout of a SiPM, with trans-impedance amplifier (c) with the SiPM equivalent circuit, composed by quenching resistor (R_Q), quenching capacitance (C_Q), i.e., parasitic capacitance of the quenching resistor through the SPAD, and the metal grid equivalent capacitance (C_{GRID}). Example of digital SiPM with TDC per each subpixel [38] (d) and schematic of dSiPM with one global TDC (e).

Figure 7. Picture of a test SiPM with several scintillator crystals (to be mounted on the top of it) (a). Example of setup for the measurement of coincidence resolving time (CRT), with two SiPMs with crystals detecting two coincident gamma rays (b). Example of a SiPM TILE with 6 × 6 element of 4 × 4 mm^2 SiPMs (c).

regards the time-of-flight (TOF) technique, which significantly improves image quality. In these applications, SiPMs are typically grouped in TILEs, to cover areas of few centimeters squared. An example is shown in **Figure 7c**.

SiPMs are also becoming an interesting choice in other applications based on single-photon or few-photon detection. For example: (i) light detection and ranging (LIDAR) [39], where many cells with good detection efficiency are highly desirable, (ii) optical spectroscopy [40], where high detection efficiency and big sensitive areas are very useful, (iii) fluorescence detection, (iv) flow cytometry, (v) Cherenkov detection for physics experiments [36], etc.

5.2. From SPAD to SiPM

Moving from a single SPAD to a silicon photomultiplier, there are some additional parameters and other noise sources that have to be considered.

- The *fill factor* of the cell has to be included in the photon detection efficiency (PDE) calculation. The cell FF is the ratio between the sensitive area and the total area of the cell. Nowadays, typical FF for analog SiPMs are between 40 and 80%.
- *Optical crosstalk* between the cells is an additional source of correlated noise [43]. During the avalanche in one cell, not only the carriers can be trapped but also the secondary photons are produced [3]. Being emitted isotopically, some can reach the neighboring cells, triggering another spurious avalanche. They can be absorbed in the depleted region or in the neutral region. In the former case, there is a *direct crosstalk* (DiCT) (see **Figure 8**), giving an instantaneous triggering of the neighboring cell. In the latter case, the crosstalk event happen delayed in time (typically few nanoseconds or tens of nanoseconds), creating a *delayed crosstalk* (DeCT). The direct crosstalk produces a current pulse that has twice the amplitude of a single-cell event.
- Another source of correlated noise is the *diffused afterpulsing*, as shown in **Figure 8**. The secondary photon can be reabsorbed in the same cell, and the photo-generated carrier can diffuse and reach the depleted region with a certain delay, producing an afterpulsing.
- An important parameter for SiPMs is the *gain* of the cell, i.e., the number of carriers produced in response to a single photo-generated carrier. The presence of an integrated resistor right above each cell reduces the amount of carriers flowing per each avalanche and reduces the amount of carriers flowing per each avalanche and makes this quantity well defined. The gain is typically between few 10^5 and 10^6, similarly to a photomultiplier tubes.
- Finally, the *single-photon time resolution* (SPTR) [42] is important for SiPMs. Differently from SPAD, SPTR of analog SiPM is mainly limited by the effect of electronic noise from the front-end circuit and by the transit-time spread. The former is the effect of the baseline fluctuation due to the noise: given a limited signal slope, the baseline variation induces a threshold crossing time variation, thus worsening the time resolution. The latter is the effect of different lengths of the path connecting the cells to the common PAD.

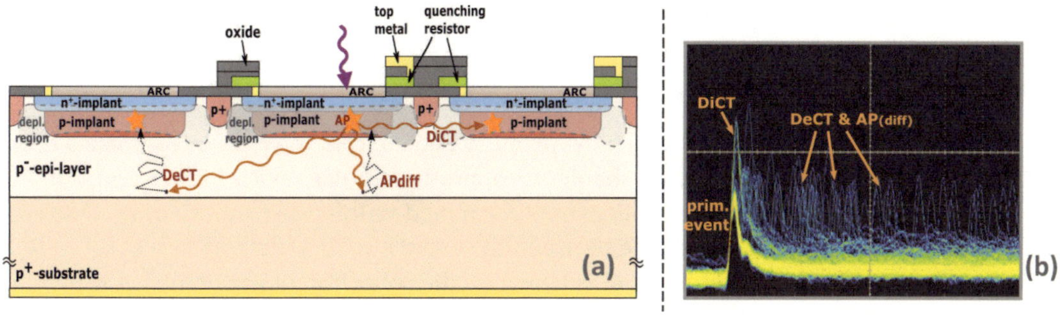

Figure 8. Representation of SiPM cross-section and of correlated noise source (a) [43] and typical oscilloscope persistence trace showing the direct and delayed crosstalk and afterpulsing events (b).

The integrated quenching resistor also reduces the amplitude of the *single-cell signal*. Considering the SiPM equivalent circuit, when the avalanche switch is closed, the bias at the internal node (between SPAD and quenching resistor) is lowered, discharging C_D and C_Q. Then, it is recharged through R_Q. However, the detectable signal is only the current that flows through the anode and cathode pins of the SiPM. The SiPM signal is composed by a fast peak (capacitive coupling through C_Q) and a slower component, due to the recharge current through R_Q. Both are "filtered" by the presence of C_{GRID}. The bigger the SiPM the larger is C_{GRID}, thus the higher is the filtering effect on the signal [42].

PDE is one of the most important parameter for a SiPM. It has been significantly improved over the last years. One possible improvement is having the PDE spectrally peaked in the region of interest of the specific application, for example, in the blue or in the green wavelength region. An Examples are the RGB [25] and NUV [43] technologies from FBK, made with p-on-n junction or n-on-p junction type. Another improvement is the increment of the cell FF, which can be obtained reducing the *border region*, i.e., the not-sensitive region between one cell and the neighboring one.

5.3. Front-end and read-out

The signal coming out from the SiPM is the superposition of many pulses, either in light or in dark. Depending on the application, it is possible to measure directly the current level produced by the SiPM, or count the avalanche pulses (photon-counting mode). If the count rate is low, the pulses are clearly distinguishable, but increasing the count rate, they start to overlap, making the counting more difficult. To avoid this situation, some techniques have been developed:

- *High-pass filtering or pole-zero cancellation*. At the front-end level, it is possible to filter the signal to remove the slow tails of the signals.
- *DLED (delayed leading-edge discrimination) method* [44]. This method consists of subtracting from the signal its delayed replica, creating some sort of high-pass filtering.

- At the device level, one producer added a third output, called *fast-output* [45]. In the SiPM, there is a capacitive pick-up in each microcell, connected between the internal nodes and a common output. This produces a faster signal which is used for timing purpose.

Another problem is how to distinguish the primary events from the correlated noise. Direct crosstalk events are easily distinguishable: they produce pulses with higher amplitudes, but afterpulsing and delayed crosstalk events are mixed within the primary ones. One efficient way is to evaluate the inter-time between the events, with a method described in Ref. [46], and used in Ref. [47]. This method is based on the collection of a train of many events, filtering and peak-detection. For each event, the inter-time and the amplitude (normalized to single-cell amplitude) are extracted. Plotting the amplitude vs. inter-time and the histogram of the inter-times, it creates a plot like in **Figure 9(a)**. By fitting the events with high inter-times, with an exponential function, it is possible to identify and distinguish the primary events.

5.4. SiPM performance

The performance of SiPMs has been significantly improved over the last years [48–52]. Producers pushed the technology limits to obtain a higher PDE, now at levels of 50–60% (peak), with FF around 70–80%. In particular, some examples are: PDE of 34% at 400 nm, with 15 μm cell pitch [48], PDE of 49% at 420 nm, with cell size of 35 μm [51], PDE of 43% at 400 nm, with cell pitch of 25 μm [50], and PDE of 33% at 520 nm, with a cell pitch of 50 μm [52]. The primary noise has been reduced, being now at the level of 50–100 kcps/mm^2 [48–52] at T = 20°C (see **Figure 9b**). The correlated noise has also been reduced due to: (i) an improvement in the silicon materials [43] and (ii) a better cell-to-cell isolation with trenches [46, 49, 53]. In particular, some examples are: CT probability of 10%, at 8 V of excess bias, with 15 μm cell pitch [48], which increases to 35%, with 7 V of excess bias when the cell pitch is 30 μm, or CT probability of just 3% in a 50 μm cell, with 3 V of excess bias, due to better cell isolation [49]. The reduction of the correlated noise and the uniformity of gain between cells give the SiPMs

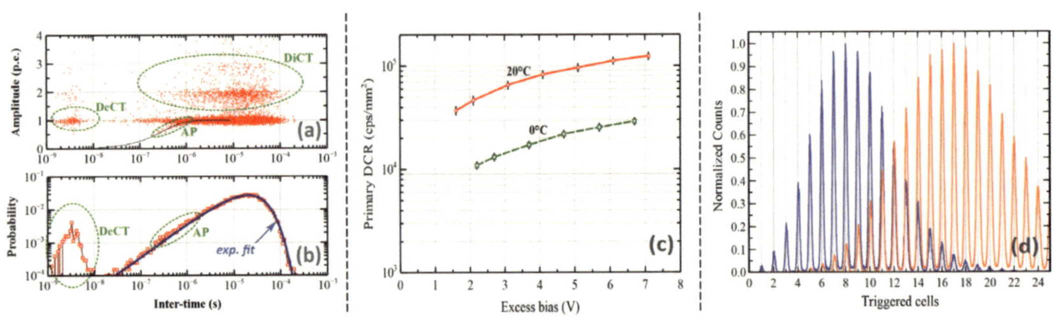

Figure 9. Example of amplitude vs. inter-time plot (a) and of inter-time histogram (b) [46]. Typical primary dark count rate of a SiPM, at different temperatures (c) [43] and charge spectrum (d).

a good photon-number resolution. **Figure 9c** shows a typical charge spectrum obtained with a 40-μm cell-pitch SiPM [43].

5.5. SiPM with small cells

The recent efforts to increase the PDE lead to an enhanced FF [48, 49, 54], meaning a smaller border region between cells. This gives a higher optical crosstalk. One efficient way to limit the correlated noise is to reduce the cell gain, giving a smaller amount of carriers flowing, thus smaller probability of trapping and emitting secondary photons. A lower gain can be obtained by means of small cells [54]. However, reducing the cell pitch normally means smaller FF, thus lower PDE. To have both smaller cells and a good PDE, the border structure of the cell has to be redesigned [48].

SiPM with small cells also have other advantages: a higher cell density and a shorter recovery time due to the smaller diode capacitance. Both these features increase the dynamic range of the SiPM, which for some applications is very important. Few examples are: (i) for calorimetry in high-energy physics experiments with high luminosity, (ii) in LIDAR and (iii) for prompt gamma imaging in proton therapy [55]. Short recovery time means reduced pile-up of the avalanche pulses, thus higher maximum count rate. Another interesting feature of SiPMs with small cell is their improved radiation hardness. The smaller is the cell size the lower is the performance degradation caused by the effects of radiations. Indeed, with a lower gain there is a smaller correlated noise, thus the noisy cells that have an increased primary dark count rate (DCR) due to radiation damage produce a lower number of correlated pulses, reducing the total overall DCR of the SiPM. With a lower total DCR and reduced gain, the SiPM has a smaller power consumption, even when damaged by radiation. Furthermore, due to the higher number of cells with a lower correlated noise, even after radiation damages there are a larger number of cells ready to be triggered by a light signal, thus a smaller PDE reduction.

5.5.1. High density silicon photomultipliers

One interesting example of SiPM with small cells is the so-called "high density" (HD) SiPM technology, developed in FBK [48, 54]. In HD technology, there are deep trenches (few micrometers deep), with high aspect ratio between the cells to electrically isolate them (as shown in **Figure 10**). The border region at the edge of each active area in the cells have been reduced to less than 2 μm, making possible to have a high FF of about 80% for a 30-μm cell-pitch SiPM and more than 50% for a 12-μm cell-pitch SiPM. For the FBK previous non-HD technology, the FF was about 60% for a 40-μm pitch SiPM [41]. The gain decreases as the cell pitch reduces: it is about 2.4×10^6 for the 30-μm pitch SiPM and 3×10^5 for the 12-μm pitch SiPM, at 5 V of excess bias. Due to the small active area and the trench isolation, the direct crosstalk probability is around 9% for 15 μm cell-pitch SiPM, which have a FF of 62%, whereas it was about 35% for non-HD SiPM with 40 μm cell-pitch (FF of 60%). The PDE, in HD technology, with n-on-p junction is peaked at 550 nm, reaching values of ~40% at 4 V of excess bias, and ~50% at 10 V of excess bias, for the 30 μm pitch SiPM-HD, and 25% for the 12 μm pitch SiPM-HD (see **Figure 10**).

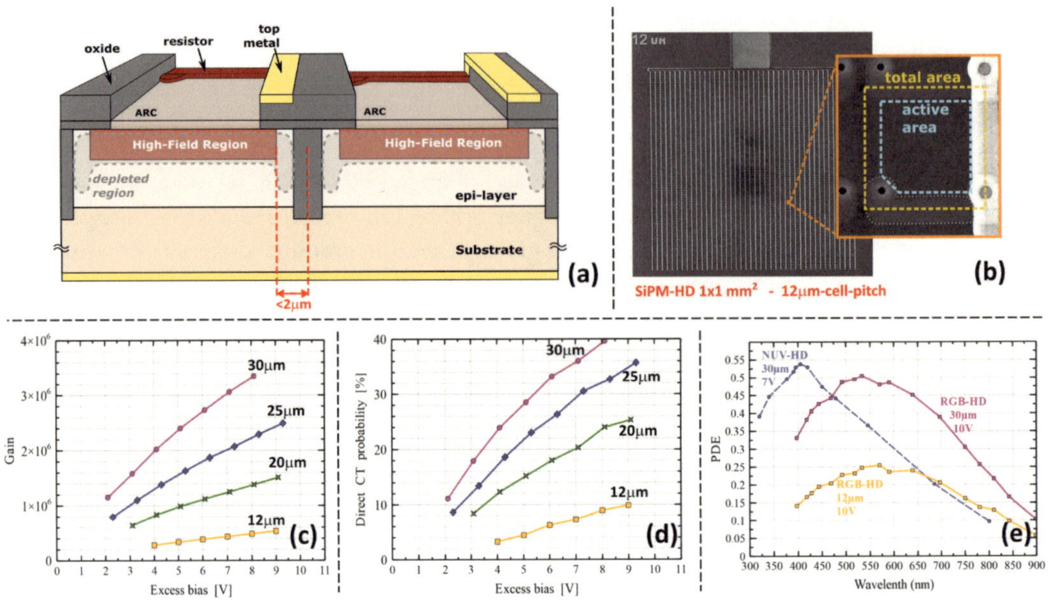

Figure 10. Schematic representation of RGB-HD SiPM structure (a). SEM image of a 12-μm cell pitch SiPM with details of the single microcell (b). Measured performance of high density SiPMs: gain (c), direct crosstalk probability (d), and PDE vs. wavelength (e) of RGB-HD and one NUV-HD (e).

5.5.2. Ultra-high density silicon photomultipliers

The HD technology has been further improved developing the "ultra-high density" (UHD) technology [56]. UHD SiPMs have very small cells and high cell density. All the feature size of the manufacturing process have been reduced (e.g., contact dimension, resistor width, etc.). The cells have a circular active area, to avoid corners with smaller electric field, and they are arranged in a honeycomb configuration (see **Figure 11**). The border region is now less than 1 μm. UHD SiPMs have been produced with cell pitch between 5 μm and 12.5 μm. With a pitch of 5 μm, the FF is about 40%, and it is higher than 70% for the 12.5-μm cell-pitch SiPM. The cell density is between 7400 cell/mm^2 and 46,190 cells/mm^2. Moreover, very small cells mean low gain, low correlated noise, and very fast recovery time.

The realization of very small cell sizes poses different challenges not only in the design and in the microfabrication process, but also due to intrinsic problems. The "border region" at the edge of the high-field region is no longer negligible but starts to play a very important role. **Figure 12** shows the TCAD simulation of electric field inside a 10 μm SiPM cell. The effective region, where the electric field is high, is smaller than the nominal one (defined by layout). Moreover, the depleted region close to the trench extends laterally toward the center of the cell. This makes the carriers photo-generated in that region to drift laterally toward the peripheral region, instead of drifting vertically, thus they are not detected. These effects are collectively called "border effect." This issue worsens the

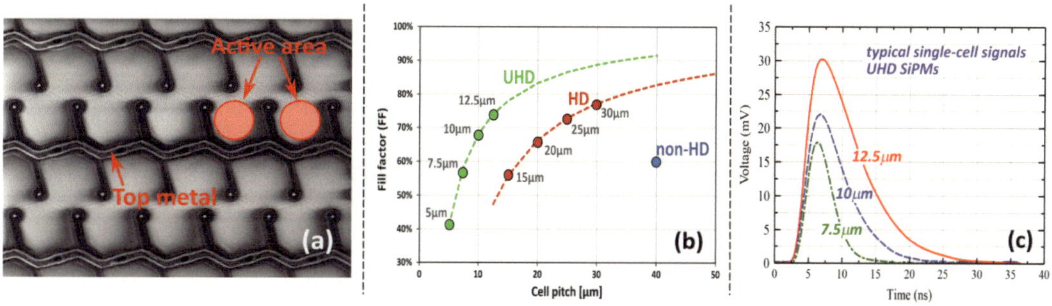

Figure 11. SEM image of 10-μm cell ultra-high density SIPM (a), showing active areas, metal and polysilicon resistors. Nominal FF of UHD, HD and non-HD technologies from FBK (b). Typical single-cell signals of UHD SiPMs (c).

performance of small cells, and it is more and more important as the cell pitch decreases. To reduce this issue, a new version of the cell has been designed, with modified doping profile. The new version has a wider high-field region and a reduced lateral depletion underneath the junction.

UHD SiPM with new structure have a higher PDE: an UHD SiPM with 7.5 μm pitch reaches a PDE of about 30% in the peak, with 6 V of excess bias at a wavelength of 470 nm, whereas a 10-μm pitch SiPM have a PDE peak of ~40% in the same conditions. SiPM with 5 μm cell pitch reaches a PDE higher than 15%. Due to the small cell capacitance, the signals from UHD SiPMs are very fast, in the order of few nanoseconds FWHM, as shown in **Figure 11**. The noise of UHD SiPM is generally in the order of 100 kcps/mm^2, but in the new structure, it is higher. This is probably due to an increased electric field; but, this technology is relatively new and there is room for improvement. The gain of the cells, thus the crosstalk probability, is low (even without absorbing material in the trenches). In a 7.5 μm cell, the gain is about 2×10^5, at 6 V of excess bias, and the direct crosstalk probability is smaller than 5%.

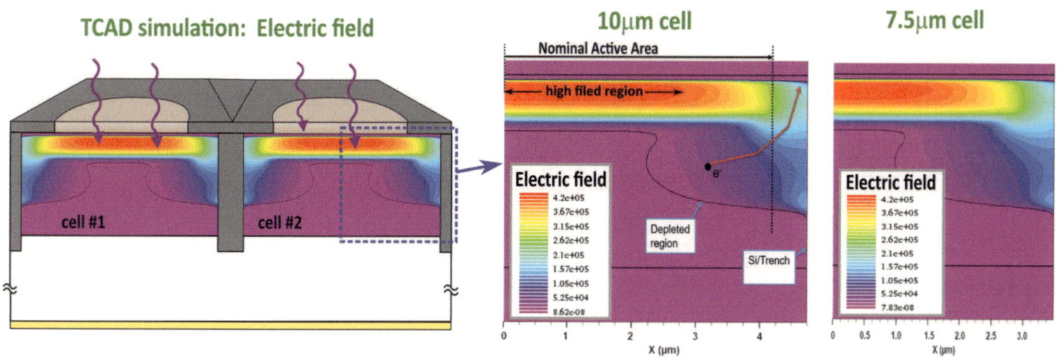

Figure 12. TCAD simulations of electric fields inside a 10-μm pitch and a 7.5-μm pitch microcells of the ultra-high density SiPMs, with depleted region highlighted.

6. Conclusions

We have reviewed some of the most interesting photodetectors technologies for photon counting. Solid-state solution, like single-photon avalanche diodes (SPADs), is able to reach high detection efficiencies and good time resolution, in the order of few tens of picoseconds. SPAD can be made in silicon or III/V materials, for the detection in the NIR wavelength range. Building a silicon SPAD in CMOS technology, it is possible to integrate some electronics into each pixel, to count the photons and to time-tag them. An array of such kind of SPAD pixels can be exploited to create low-light imagers. SPAD imagers are nowadays used in several biomedical applications (e.g., FLIM, Raman spectroscopy, etc.) and in 3D ranging. This technology is continuously evolving with a pixel density getting higher due to the CMOS technology development. Another interesting detector based on SPAD arrays is the silicon photomultiplier (SiPM). Here, all the SPADs are connected in parallel, in analog or digital way. It has single-photon sensitivity but, differently from single-SPADs, it is able to reach big active areas (few millimeters squared) and it is able to count the number of photons arriving simultaneously with good photon-number resolution. SiPM performance has been significantly improved over the last years, reaching a high FF and a high detection efficiency. This promising technology is now starting to be used not only in the typical applications (e.g., nuclear medicine and physics experiments) but also in biomedical and 3D ranging applications. SiPM technologies are evolving in the direction of smaller cells (SPADs), which is advantageous for SiPM performance, but it requires improvements in the manufacturing processes. For example, there are new technologies for SiPMs with cell-pitches smaller than 10 μm, down to 5 μm. This provides a higher cell density and a larger dynamic range. Some mixed solutions are also emerging, with imagers made by an array of many mini-SiPMs. This allows to have imaging capabilities but with a high dynamic range per pixel.

Author details

Fabio Acerbi* and Matteo Perenzoni

*Address all correspondence to: acerbi@fbk.eu

Fondazione Bruno Kessler (FBK), Trento, Italy

References

[1] Hamamatsu editor. Photomultiplier Tubes—Basics and Applications. 3rd ed. https://www.hamamatsu.com/resources

[2] Hamamatsu Photonics KK. R3809U-50 datasheet [Internet]. Available from: http://www.hamamatsu.com/eu/en/product [Accessed: Sep 2017]

[3] Cova S, Ghioni M, Lacaita AL, Samori C, Zappa F. Avalanche photodiodes and quenching circuits for single photon-detection. Applied Optics. 1996;**32**(2)

[4] Bronzi D, Villa F, Tisa S, Tosi A, Zappa F, Durini D, Weyers S, Brockherde W. 100 000 frames/s 64 × 32 single-photon detector array for 2-D imaging and 3-D ranging. IEEE Journal of Selected Topics in Quantum Electronics. 2014;**20**(6)

[5] Perenzoni M, Massari N, Perenzoni D, Gasparini L, Stoppa D. A 160 × 120 pixel analog-counting single-photon imager with time-gating and self-referenced column-parallel A/D conversion for fluorescence lifetime imaging. Journal of Solid State Circuit. 2016;**51**(1)

[6] Michalet X, Colyer RA, Scalia G, Ingargiola A, Lin R, Millaud JE, Weiss S, Siegmund OHW, Tremsin AS, Vallerga JV, Cheng A, Levi M, Aharoni D, Arisaka K, Villa F, Guerrieri F, Panzeri F, Rech I, Gulinatti A, Zappa F, Ghioni M, Cova S. Development of new photon-counting detectors for single-molecule fluorescence microscopy. Philosophical Transactions of the Royal Society B. 2012;**368**

[7] Jeon EJ, Kim JY, Kim YD, Ma KJ, Nam JT. Magnetic field effects on the photocathode uniformity of Hamamatsu R7081 photomultiplier tubes. Nuclear Instrument and Methods in Physics Research A. 2013;**697**:46-51

[8] Sansoni L et al. A two-channel, spectrally degenerate polarization entangled source on chip. NPJ Quantum Information. 2017;**3**(5). DOI: 10.1038/s41534-016-0005-z

[9] Dauler EA, Stevens MJ, Baek B, Molnar RJ, Hamilton SA, Mirin RP, Nam SW, Berggren KK. Measuring intensity correlations with a two-element superconducting nanowire single-photon detector. Physical Review A. 2008;**78**(5)

[10] Marsili F, Verma VB, Stern JA, Harrington S, Lita AE, Gerrits T, Vayshenker I, Baek B, Shaw MD, Mirin RP, Nam SW. Detecting single infrared photons with 93% system efficiency. Nature Photonics Letters. 2013;**7**:210-214. DOI: 10.1038/NPHOTON.2013.13

[11] Redaelli L, Bulgarini G, Dobrovolskiy S, Dorenbos SN, Zwiller V, Monroy E, Gérard JM. Design of broadband high-efficiency superconducting-nanowire single photon detectors. Superconductor Science and Technology. 2016;**29**(6). DOI: 10.1088/0953-2048/29/6/065016

[12] Lita AE, Miller AJ, Nam SW. Counting near-infrared single-photons with 95% efficiency. Optics Express. 2008;**16**(5)

[13] Zhao Q-Y et al. Single-photon imager based on a superconducting nanowire delay line. Nature Photonics. 2017;**11**:247-251. DOI: 10.1038/NPHOTON.2017.35

[14] Gallivanoni A, Rech I, Ghioni M. Progress in quenching circuits for single photon avalanche diodes. IEEE Transaction on Nuclear Science. 2010;**57**(6):3815-3826

[15] Ghioni M, Gulinatti A, Rech I, Zappa F, Cova S. Progress in silicon single-photon avalanche diodes. IEEE Journal of Selected Topics in Quantum Electronics. 2007;**13**(4): 852-862

[16] Contini D et al. Effects of time-gated detection in diffuse optical imaging at short source-detector separation. Journal of Physics D: Applied Physics. 2015;**48**. DOI: 10.1088/0022-3727/48/4/045401

[17] Acerbi F, Anti M, Tosi A, Zappa F. Design criteria for InGaAs/InP single-photon avalanche diode. IEEE Photonics Journal. 2013;**5**:6800209

[18] Zhang J, Itzler MA, Zbinden H, Pan J-w. Advances in InGaAs/InP single-photon detector systems dor quantum communications. Light: Science & Applications. 2015;**4**(e286). DOI: 10.1038/lsa.2015.59

[19] Meng X et al. InGaAs/InAlAs single photon avalanche diode for 1550 nm photons. Royal Society Open Science. 2016;**3**(3):150584. DOI: 10.1098/rsos.150584

[20] Jiang X, Itzler MA, Ben-Michael R, Slomkowski K. InGaAsP–InP avalanche photodiodes for single photon detection. IEEE Journal of Selected Topics in Quantum Electronics. 2007;**13**(4):895-905. DOI: 10.1109/JSTQE.2007.903001

[21] Acerbi F, Frera AD, Tosi A, Zappa F. Fast active quenching circuit for reducing avalanche charge and afterpulsing in InGaAs/InP single-photon avalanche diode. IEEE Journal of Quantum Electronics. 2013;**49**(7):563-569. DOI: 10.1109/JQE.2013.2260726

[22] Wang XF, Periasamy A, Herman B, Coleman DM. Fluorescence lifetime imaging microscopy (FLIM): Instrumentation and application. Critical Reviews in Analytical Chemistry. 1992;**23**(5):369-395

[23] Michalet X et al. Development of new photon-counting detectors for single-molecule fluorescence microscopy. Philosophical Transactions of the Royal Society B: Biological Sciences. 2013;**368**(1611). DOI: 10.1098/rstb.2012.0035

[24] Micro photon devices. InGaAs SPAD freerunning datasheet [Internet]. Available from: www.micro-photon-devices.com/Docs/Datasheet/InGaAs_Datasheet_freerunning.pdf [Accessed: Sep. 2017]

[25] Serra N, Ferri A, Gola A, Pro T, Tarolli A, Zorzi N, Piemonte C. Characterization of new FBK SiPM technology for visible light detection. Journal of Instrumentation. 2013;**8**(P03019). DOI: 10.1088/1748-0221/8/03/P03019

[26] Remondino F, Stoppa D. TOF Range-Imaging Cameras. Heidelberg, Germany: Springer; 2013

[27] Villa F et al. CMOS imager with 1024 SPADs and TDCs for single-photon timing and 3-D time-of-flight. IEEE Journal of Selected Topics in Quantum Electronics. 2014;**20**(6). DOI: 10.1109/JSTQE.2014.2342197

[28] Pancheri L, Massari N, Stoppa D. SPAD image sensor with analog counting pixel for time-resolved fluorescence detection. IEEE Transactions on Electron Devices. 2013; **60**(10):3442-3449

[29] Stoppa D, et al. A 32 × 32-pixel array with in-pixel photon counting and arrival time measurement in the analog domain. In: Proc. of ESSCIRC 2009; Athens, Greece. 2009

[30] Parmesan L, et al. A 256 × 256 SPAD array with in-pixel time to amplitude conversion for fluorescence lifetime imaging microscopy. In: Proc. of the international Image Workshop 2015; Vaals, The Nederlands. 2015

[31] Fossum ER. The quanta image sensor: every photon counts. Sensors. 2016;**16**:1260. DOI: 10.3390/s16081260

[32] Renker D. Geiger-mode avalanche photodiodes history, properties and problems. Nuclear Instruments and Methods A. 2006;**567**:48-56

[33] Sefkow F. The CALICE tile hadron calorimeter prototype with SiPM read-out: Design, construction and first test beam results. In: IEEE, editor. Nuclear Science Symposium Conference Record, 2007. NSS '07; 26 Oct–3 Nov 2007; 2007. DOI: 0.1109/NSSMIC. 2007.4436327

[34] Berra A, et al. Shashlink calorimeters with embedded SiPMs for longitudinal segmentations. IEEE Transaction on Nuclear Science. 2017;**64**(4)

[35] Townsted DW. Multimodality imaging of structure and function. Physics in Medicine and Biology. 2008;**53**:R1-R39

[36] Gamal A, Paul B, Michael C, Roland H, Johann M, Herbert O, Ken S. Application of Geiger-mode photosensors in Cherenkov detectors. Nuclear Instruments and Methods in Physics Research Section A. 2011;**639**(1):107-110

[37] Gundacker S, Auffray E, Jarron P, Meyer T, Lecoq P. On the comparison of analog and digital SiPM readout in terms of expected timing performance. Nuclear Instruments and Methods in Physics Research A. 2015;**787**:6-11

[38] Braga LHC et al. A fully digital 8 16 SiPM Array for PET applications with per-pixel TDCs and real-time energy output. IEEE Journal of Solid-State Circuits. 2014;**49**(1): 301-314

[39] Son KT, Lee CC. Multiple-target laser range finding receiver using a silicon photomultiplier array. IEEE Transactions on Instrumentation and Measurement. 2010;**59**: 3005-3011

[40] Dalla Mora A et al. Fast silicon photomultiplier improves signal harvesting and reduces complexity in time domain diffuse optics. Optics Express. 2015;**23**(11)

[41] Seifert S et al. Simulation of silicon photomultiplier signals. IEEE Transaction on Nuclear Scince. 2009;**56**(6):3726-3733

[42] Acerbi F et al. Characterization of single-photon time resolution: From single SPAD to silicon photomultiplier. IEEE Transaction on Nuclear Science. 2014;**61**(5):2678-2686. DOI: 10.1109/TNS.2014.2347131

[43] Acerbi F et al. NUV silicon photomultipliers with high detection efficiency and reduced delayed correlated-noise. IEEE Transaction on Nuclear Science. 2015;**62**(3): 1318-1325

[44] Gola A, Piemonte C, Tarolli A. The DLED algorithm for timing measurements on large area SiPMs coupled to scintillators. IEEE Transactions on Nuclear Science. 2012; **59**(2):358-365

[45] Yeom JY et al. Fast timing silicon photomultipliers for scintillation detectors. IEEE Photonics Technology Letters. 2013;**25**(15):1309-1312

[46] Piemonte C, et al. Development of an automatic procedure for the characterization of silicon photomultipliers. In: Nuclear Science Symposium and Medical Imaging Conference (NSS/MIC), 2012 IEEE; 27 Oct-3 Nov 2012

[47] Otte AN, Garcia D, Nguyen T, Purushotham D. Characterization of three high efficiency and blue sensitive silicon photomultipliers. Nuclear Instruments and Methods in Physics Research A. 2017;**846**:106-125

[48] Piemonte C et al. Performance of NUV-HD silicon photomultiplier technology. IEEE Transaction on Electron Devices. 2016;**63**(3):1111-1116

[49] Hamamatsu. MPPC (Multi-Pixel Photon Counters)—S13360 Series [Internet]. Available from: www.hamamatsu.com/resources [Accessed: Sep. 2017]

[50] Ketek. SiPM WB Series [Internet]. Available from: https://www.ketek.net/sipm/sipm-products/wb-series/ [Accessed: Sep. 2017]

[51] SensL. J-Series Family [Internet]. Available from: http://sensl.com/products/j-series/ [Accessed: Sep. 2017]

[52] Excelitas. C30742-11 Series [Internet]. Available from: http://www.excelitas.com/Downloads/DTS_C30742-11-050_Series_SiPM.pdf [Accessed: Sep. 2017]

[53] Sul W-S, Lee C-H, Cho G-S. Influence of guard-ring structure on the dark count rates of silicon photomultipliers. IEEE Electron Device Letters. 2013;**34**(3):336-338. DOI: 10.1109/LED.2012.2236296

[54] Piemonte C et al. Characterization of the first FBK high-density cell silicon photomultiplier technology. IEEE Transactions on Electron Devices. 2013;**60**(8):2567-2573

[55] Regazzoni V et al. Characterization of high density SiPM non-linearity and energy resolution for prompt gamma imaging applications. Journal of Instrumentation. 2017;**12** (P07001)

[56] Acerbi F et al. High efficiency, ultra high-density silicon photomultipliers. IEEE Journal of Selected Topics in Quantum Electronics; March-April 2018;**24**(2)

Section 2

Fundamentals

Chapter 3

Quantum Non-Demolition Measurement of Photons

Keyu Xia

Additional information is available at the end of the chapter

http://dx.doi.org/10.5772/intechopen.72871

Abstract

According to Heisenberg's uncertainty principle, measurement of a quantum observable introduces noise to this observable and thus limits the available precision of measurement. Quantum non-demolition measurements are designed to circumvent this limitation and have been demonstrated in detecting the photon flux of classical light beam. Quantum non-demolition measurement of a single photon is the ultimate goal because it is of great interest in fundamental physics and also a powerful tool for applications in quantum information processing. This chapter presents a brief introduction of the history and a review of the progress in quantum non-demolition measurement of light. In particular, a detailed description is presented for two works toward cavity-free schemes of quantum non-demolition measurement of single photons. Afterward, an outlook of the future in this direction is given.

Keywords: QND measurement, single photon, four-wave mixing, Rabi oscillation

1. What is quantum non-demolition measurement?

Measurement of observables is at the very heart of quantum measurement. In the classical macroscopic world, measurement of a classical object can be conducted without introducing perturbation to the detected object. Repeating measurement of a classical object can improve the precision to arbitrarily accurate. Counterintuitively, the measurement of an observable of a quantum object cannot be arbitrarily precise in the microscopic world according to the well-known Heisenberg's uncertainty principle [1], which roots in the wave nature of quantum mechanics. For non-commuting operators, A and B, described as physical quantities in the quantum formalism, a very precise measurement of A, resulting in a very small uncertainty ΔA, will be associated with a large value of uncertainty, ΔB, in B. Measuring a quantum object will inevitably cause perturbation in the measured object. This perturbation due to measurement is called as the "measurement back action." This quantum back action, in turn, enlarges uncertainty of the observables. As a result, it limits the available precision in a series of repeated measurements. Then a natural question is what is the limitation of sensitivity in measurement set by quantum mechanics.

In response to this question, Braginsky and Vorontsov introduced in the 1970s the concept of "quantum non-demolition measurement" (QND) to evade the unwanted quantum back action in measurement [2]. Through studying the detectable minimum force on a quantum oscillator, they concluded that "Nondestructive recording of the n-quantum state of an oscillator is possible in principle." Their measurement strategy opened a door for circumventing the issue of back action in quantum measurement. Thorne, Drever, Caves, Zimmermann, Sandberg, Unruh, and others developed the concept of QND measurement further [3–5]. The key point in the QND measurement is to keep the back-action noise confined to the unwanted observable quadrature, without being coupled back onto the quantity to be measured.

Although a great number of efforts have been made in various systems, quantum optics is particularly well suited for implementing QND measurement. The reason is threefold: (1) there are optical sources with very good quality; (2) photon detectors can be extremely sensitive, even being able to detect a single photon; and (3) a quantum system can be initialized with very high accuracy. The photon number and phase are two complementary observables of quantum light. They are associated with non-commuting operators. It means that QND measurement of photon number of a quantum field will inevitably add quantum noise to the phase quadrature. If only, in principle, the photon number of field remains unchanged during measurement, the measurement is QND. Of course, the real implementation of experiment may be imperfect, and this imperfection can cause noise to the variable of interest.

Throughout this chapter, we focus on the measurement of light according to the principle of quantum optics. In particular, we introduce the measurement of photon number of a light beam. In the conventional "direct" measurement, the light is absorbed. Therefore, the measurement completely changes the observable of photon number and causes a very large back action onto the light beam. In a QND measurement of photon number, it is required that the amount of photon number is *measured without changing*. Of course, the measurement still adds perturbation to the light. However, the perturbation is only confined to the phase of the photon but is not added to the photon flux of interest in measurement. In a restricted mathematical language, the condition for QND measurement is that $\langle A_s \rangle_i = \langle A_s \rangle_{i+1}$ and $\langle \Delta A_s \rangle_i = \langle \Delta A_s \rangle_{i+1}$ for two successive detections of observable A_s.

2. Classical measurement by absorbing photons

In the classical world, measurement of light always absorbs photons and then gets energy from them. In this way, the photon carried by a light beam disappears and is destroyed completely. This type of photon detector includes eyes, photoelectric converter, semiconductor photon detector, superconducting photon detector, and so on.

Eyes are photon detectors we use most often (**Figure 1**). It converts the energy of light into electric current and stimulates the nerve. Photons of light enter the eye through the cornea, that is the clear front "window" of the eye. Then light is bent by the cornea, passes freely through the pupil, the opening in the center of the iris, the eye's natural crystalline lens, and then is focused into a sharp point on the retina. The retina is responsible for capturing all of the light rays, processing them into light impulses through millions of tiny eye nerve endings, and then

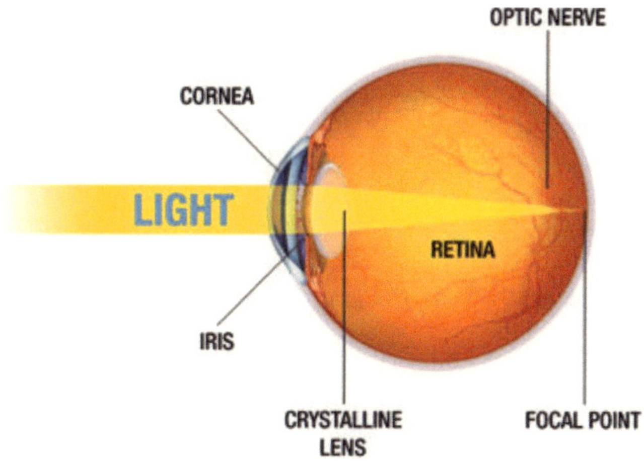

Figure 1. Sketch for seeing photons with eyes (from www.nkcf.org).

converting these light impulses to signals which can be recognized by the optic nerve. In doing so, eyes convert light into bioelectric signals.

Semiconductor photon detector is a sensitive man-made photodetector, which is made by using semiconductor materials. Two principal classes of semiconductor photodetectors are in common use: thermal detectors and photoelectric detectors. Thermal detectors convert photon energy into heat. Most thermal detectors are rather inefficient and relatively slow. Therefore, photoelectric detectors are widely used for optics. The operation of photoelectric detectors is based on the photoeffect. Similar to eyes, the detector absorbs photons from light, generating electronic current pulse which can be measured. The semiconductor photon detector is the most used photodetector in industry. The most common semiconductor-based devices are single-photon avalanche diode (SPAD) detectors and can reach sensitivity at the single photon level. The SPAD detector is reversely biased above the avalanche breakdown voltage in the Geiger mode. When a photon is captured by this SPAD detector, the absorbed photon generates an electron-hole pair which causes a self-sustaining avalanche, rapidly generating a measurable current pulse (**Figure 2**).

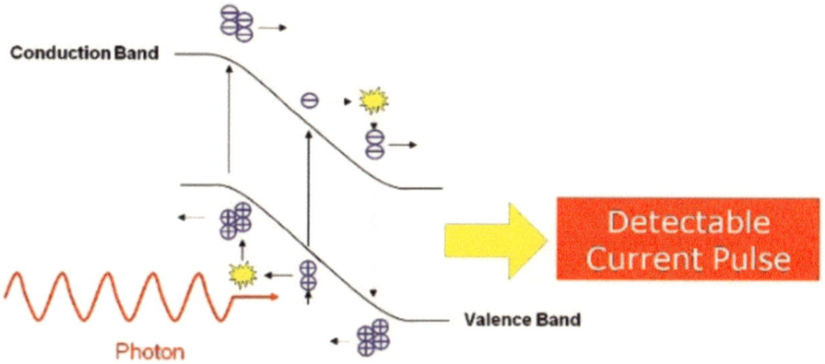

Figure 2. Schematic diagram for semiconductor photon detectors (from www.single-photon.com).

Superconducting nanowires have been used to detect single photons. It exploits a different principle in comparison with eyes and semiconductor photon detectors. It is designed in this way [6, 7]: a patterned superconducting nanowire is cooled below the transition temperature of the superconducting material. The superconducting nanowire is biased by an external current slightly smaller than the critical current at the operating temperature. When a single photon hits the nanowire, it creates a transient normal spot in the resistive state. As a result of loss of superconductivity, a nonzero voltage is induced between two terminals of the nanowire. Measuring this induced voltage can tell the arrival of the single photon. To date, superconducting single photon detectors have achieved a detection efficiency of more than 90% [8, 9].

The abovementioned are three representatives of photon detectors. All of them destroy photons in signals.

3. Measuring light intensity without absorption

QND measurement of light needs to keep the quantum average of the observable and its uncertainty unchanged after detection. In general quantum measurement, the observable of a signal system, A_s, is measured by detecting the change of observable, A_m, of a "meter" system. The concept can be explained by describing the measurement as a joint Hamiltonian [10]

$$H = H_s + H_M + H_I, \tag{1}$$

where H_s is the unperturbed Hamiltonian of the signal system to be measured, H_M is that of the meter system, and H_I describes the way in which the meter measures the signal. The motion for A_s and A_M under measurement is

$$-i\hbar\frac{dA_s}{dt} = [H_s, A_s] + [H_I, A_s], \tag{2}$$

$$-i\hbar\frac{dA_M}{dt} = [H_s, A_M] + [H_I, A_M]. \tag{3}$$

QND measurement requires (i) $[H_s, A_s] = 0$, which is normally satisfied; (ii) $[H_I, A_s] = 0$; and (iii) $[H_I, A_M] \neq 0$. The second condition guarantees that the back action is isolated from A_s. The third one implies that a measurement can induce change in the meter system.

It is quite straightforward to get the cross-Kerr effect in mind for QND measurement of photon flux, $n_s = A_s^\dagger A_s$, of light beam [10, 11]. The Hamiltonian describing the cross-Kerr interaction is as follows:

$$H_I = \chi A_s^\dagger A_s A_M^\dagger A_M, \tag{4}$$

where χ is the strength of nonlinear interaction. Obviously, $[H_I, A_s] = 0$ is met.

The condition $[H_I, A_M] \neq 0$ holds if the phase of probe field is measured. The intuitive picture of QND measurement of photon flux, n_s, with the cross-Kerr effect can be well explained in **Figure 3**. The signal and probe laser fields co-propagate in a Kerr nonlinear medium with length L. Due to the cross-Kerr optical nonlinearity, the refractive index of medium is dependent on the intensity, $I_s \propto n_s$, of the signal field. Its change is proportional to n_s and subsequently causes a phase shift, $\Delta\phi_M = \phi'_M - \phi_M$, to the probe field. Obviously, this phase shift $\Delta\phi_M$ is proportional to the photon number of signal field. Measuring $\Delta\phi_M$ can determine the intensity of the signal field without absorbing its photon.

The concept of QND measurement based on the cross-Kerr effect has been demonstrated in experiments for classical light including many photons [12]. However, QND measurement at the single photon level is still a challenging problem. The difficulty is twofold. Technically, the nonlinearity of normal materials is too weak to induce a large phase shift per photon. Although the cross-Kerr nonlinearity can be improved by orders by using atom system, typically, a single photon can only cause an mrad scale phase shift [13]. It is worth noting two recent experiments in cross-phase modulation [14, 15], which demonstrated the pi phase shift at the single photon level via the cross-Kerr nonlinearity of atoms. At first sight, the methods may be able to apply to QND measurement of single photons. Actually, they are yet to meet the criteria of QND measurement.

In the first work [14], by storing a single photon in a cloud of Rydberg atoms, Tiarks et al. achieved a π phase shift imprinted onto a probe field including only 0.9 photon. However, the efficiency of storing and retrieving signal photon is very low, that is only 0.2. The signal photon suffers a big loss and has a small possibility to survive after inducing the phase shift. In this, this scheme cannot be used for QND measurement of single photons.

Alternatively, Liu et al. used a double-Λ system to induce a giant cross-Kerr nonlinearity to achieve the π phase shift per photon [15]. In their configuration, the signal and the probe fields, each including eight photons, share a common ground state, while they couple to their individual dark states created by other two control fields. As a result, a giant cross-Kerr nonlinearity between them is created. A π cross-phase shift is induced at the single photon level. However, the reported scheme is still classical but has yet to reach the quantum regime

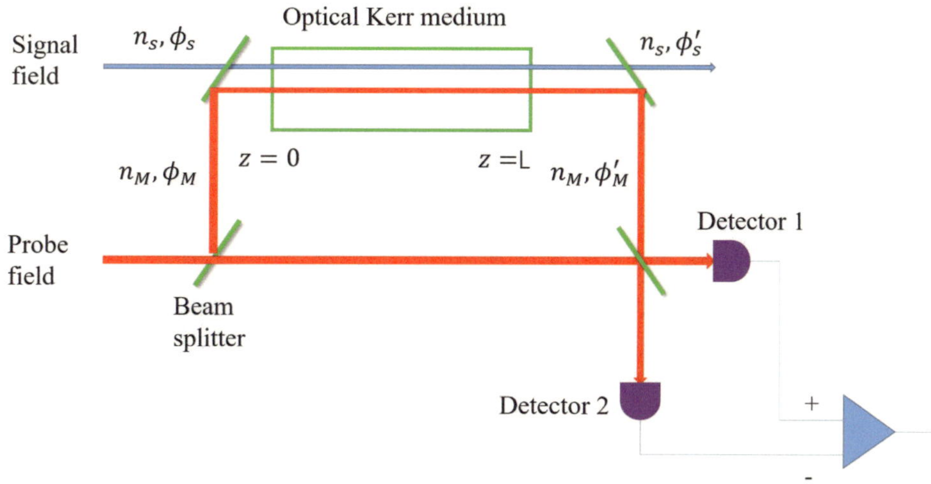

Figure 3. Configuration for the QND measurement of the signal photon number via cross-Kerr nonlinearity [10].

for detecting a single photon in a QND way. There are two prerequisites in this scheme. First, to ensure the atoms are transparent, the probe field needs to be known. Other than the optimal phase, the absorption is considerable. But this phase is unknown for a signal photon to be detected. It means a large loss for the signal photon to be detected. Second, the phase shift is obtained in the steady state where the probe field is classically treated as a constant field. This is not the case for a single photon as it is a quantum field. Therefore, it is hard to do genuine QND measurement at the single photon level.

At the fundamental level, the cross-Kerr-based QND measurement is found invalid when a continuous spatiotemporal multimode model [16] or a finite response time [17–19] is considered. In this sense, although many important progresses have been achieved, QND detection of a moving single photon still needs proposals.

4. Non-demolition measurement of photons with cavities

With the progress of cavity electrodynamics, in particular the ultrastrong coupling between a microwave cavity and an artificial atom, QND measurement of single mw photons have been realized via qubit-photon CNOT gate [20], ac Stark effect [21–23], and the intrinsic phase shift in Rabi oscillation [24]. Photon blockade has been demonstrated as a new effect to implement QND measurement of a single optical photon trapped in a high-quality optical cavity [25].

The first breakthrough of QND measurement of single photons was accomplished by Haroche et al. exploiting the intrinsic π phase shift after a full Rabi oscillation of an atom [24]. The principle can be understood using the schematic diagram as shown in **Figure 4**. The atom is first prepared in Rydberg state with the ground state $|g\rangle$, the excited state $|e\rangle$, and an auxiliary state $|i\rangle$ by B. R1 and R2 conduct the Ramsey interferometer measurement. R1 drives the Rydberg atom into a superposition state of $C_g|g\rangle + C_i|i\rangle$. The mw cavity C induces a phase shift dependent on the photon number in it. It is off resonance with $|g\rangle \leftrightarrow |i\rangle$, but on resonance

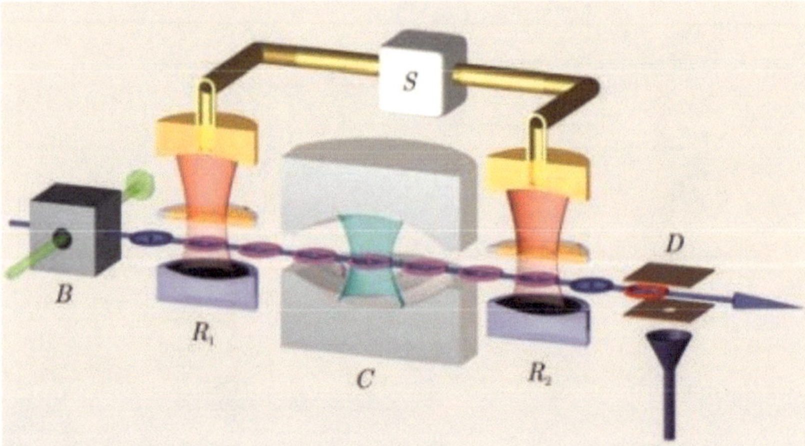

Figure 4. Schematic diagram for QND measurement of a single microwave photon via the intrinsic phase shift of a full Rabi oscillation [24].

with $|g\rangle \leftrightarrow |e\rangle$. It is designed to cause a full Rabi oscillation if the cavity includes one photon and results in a π phase shift to $|g\rangle$ yielding $-C_g|g\rangle + C_i|i\rangle$. While in the empty-cavity case, the atomic state is unchanged. In short, the atomic coherent changes its phase by π if there is one photon in C. R2 mixes the atomic state again, probing after C the superposition phase shift. The final atomic population can be detected with a state-selective detector. The probability of finding the atom in $|g\rangle$ is a cosine function of the phase shift and thus gives information about the phase shift. In this way, Haroche et al. implemented the QND measurement of a single mw photon.

5. Cavity-free schemes for non-demolition measurement of single photons

The concept of QND measurement and its realization in measuring classical light intensity have been introduced earlier. QND measurement of single photons is the ultimate goal. Single "static" photon in cavity has been detected nondestructively. Measuring "moving" single photons without destroying it is still far to be achieved. Two important progresses toward this direction are presented in the following.

5.1. QND measurement via Rabi-type photon-photon interaction

As mentioned earlier, although the optical cross-Kerr effect has been proposed for implementing intensity QND measurement of light, detection of light at the single photon level in a QND way is still a challenging task. In the cross-Kerr-based proposals [10], the signal photon changes the refractive index n_I of medium. The change of n_I causes a phase shift of the co-propagating probe photon. The interaction between the signal and probe photons is "Ising" type. Its application for single-photon QND measurement is questionable at the fundamental level [16–19]. A "Rabi" type photon-photon interaction created from four-wave mixing (FWM) was proposed for a photon-photon controlled quantum phase gate [24]. The proposal treated the moving fields as a single mode and suggested equal group velocity for both the signal and probe pulses. The work did not circumvent the issues raised in [16–19]. Instead, Xia and his coworker studied this type of photon-photon interaction for QND measurement of a single photon taking into account the quantum nonlocality [26]. In the proposal, the four-wave mixing occurs in an optical nonlinear medium. One of the light modes in four-wave mixing is a strong coherent laser. This coherent laser is used to coherent pump the nonlinear process and perform an effective three-wave mixing process involving the signal mode, a_s, the probe mode, a_p, and an auxiliary mode, a_a. The Hamiltonian describing the interaction among these three modes takes the form

$$H_I = \frac{g(E_c)}{2} a_a a_p^\dagger a_s^\dagger + \frac{g(E_c)}{2} a_a^\dagger a_p a_s, \tag{5}$$

where $g(E_c)$ indicates the nonlinear coupling strength that can be tuned by the intensity of the pump field E_c.

To induce a Rabi-type interaction, the auxiliary mode is initially in a vacuum state. The signal field has at most one photon. The probe field is assumed to be weak that, to a good approximation, it can be considered as the superposition state of $|\alpha_p\rangle \approx |0_p\rangle + \alpha_p|1_p\rangle$ with $\alpha_p \vee \ll 1$.

Here, the probe field is truncated up to $|1_p\rangle$. Focusing on the space spanned by the associated state of the probe and auxiliary modes, as shown in **Figure 5(a)**, these two modes form a ladder-type quantum system. The ground state is $|0_p, 0_a\rangle$, and the first and second excited states are $|1_p, 0_a\rangle$ and $|0_p, 1_a\rangle$, respectively. The incoming signal photon will drive the transition between $|1_p, 0_a\rangle$ and $|0_p, 1_a\rangle$. This photon-driven transition between photonic states is a photonic counterpart of atomic Rabi oscillation. For a weak probe field $|\alpha_p\rangle$, the initial state is $|0_p, 0_a\rangle + \alpha_p|1_p, 0_a\rangle$. Similar to the Rabi oscillation in atoms, the state $|1_p, 0_a\rangle$ will suffer a π phase shift after a full Rabi oscillation. As a result, the probe field passing through the medium becomes $|0_p\rangle - \alpha_p|1_p\rangle \approx |-\alpha_p\rangle$. Effectively, the probe field is shifted by phase of π. The concept is depicted in **Figure 5(b)**. Such full Rabi oscillation can be conducted by controlling the pump field intensity or the length of nonlinear medium.

To determine the phase shift of the probe field, a strong local bias is overlapped on the transmitted probe field via a highly reflective beam splitter. By properly choosing the bias field, the transmitted probe field presented to the detector is displaced by $|-\alpha_p\rangle$, yielding $|-2\alpha_p\rangle$ in the presence of a single signal photon or $|0\rangle$ in the absence of signal field. Simply observing the photon "click" on the single-photon detector can determine whether a single signal photon passes through the medium without destroying it. This accomplishes the QND measurement of a single signal photon. Of course, this measurement will cause disturbance in the phase of signal field. However, the photon flux is concerned, and the noise added to the phase quadrature is not unwanted.

To evaluate the performance of the QND measurement, only one investigates the response of system to the initial case of a single signal photon input, $|1_s\rangle$, and a weak probe field, $|\alpha_p\rangle$. Numerical simulation of corresponding quantum Langevin equation shows the transmitted signal and probe fields, and the displaced field presented to the detector for the input $|1_s\rangle$ and $|\alpha_p\rangle$, as shown in **Figure 6**. It is found that the transmitted signal field keeps its initial state with a very high fidelity, while the transmitted probe field on the detector, shifted by a phase of π due to the presence of signal photon, can be well distinguished from the transmission without phase shift in the absence of signal photon.

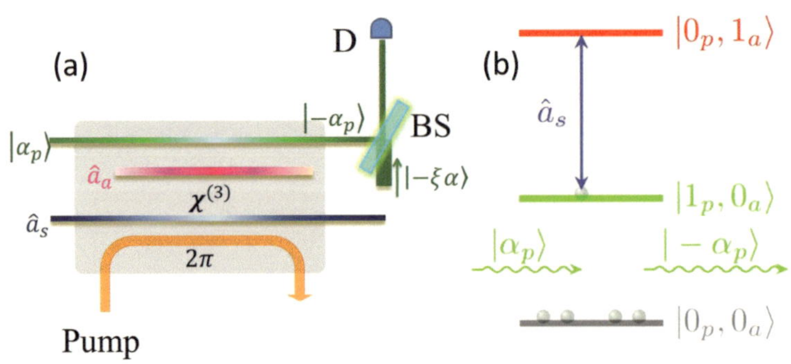

Figure 5. Schematic for detection of a single moving photon. (a) Configuration for QND detection of a single moving photon via four-wave mixing in a nonlinear medium. (b) Level diagram describing the interaction between the signal, auxiliary, and probe photons [26].

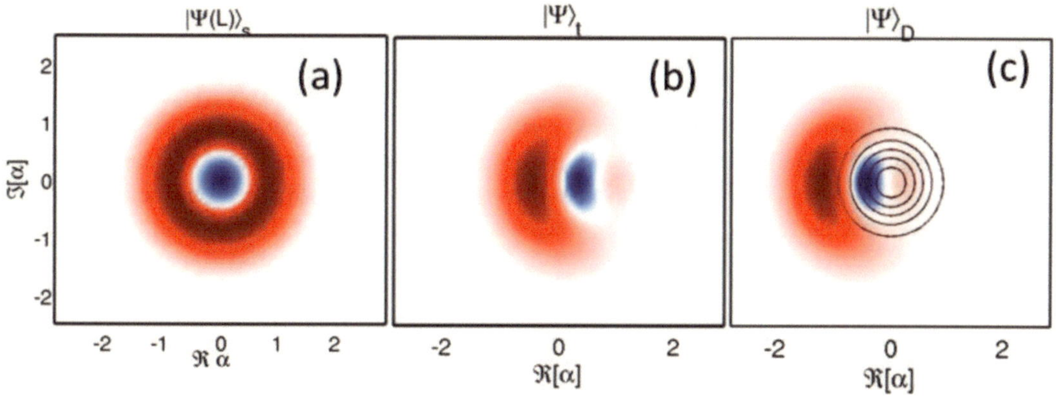

Figure 6. Wigner functions of the transmitted and detected states for a probe field with $|\alpha_p|^2 = 0.6$. In (a) [(b)] transmitted signal (probe) state after interacting (a Full Rabi oscillation) for the length of the media; (c) detected state of probe field presented to detector. The concentric circles show the Wigner function contours of the detection field in the absence of signal input [26].

In the presence of a single signal photon, the field presented to the detector is $|-2\alpha_p\rangle$. In this case, even an ideal photon detector can have a "dark count," that is, no detection, because the state $|-2\alpha_p\rangle$ includes a small occupation in vacuum state $|0\rangle$. This dark count causes error in detection of signal photon. The resulted error probability is given by $P_{err}(\alpha_p) \approx e^{-4|\alpha_p|^2}$. It decreases exponentially as the intensity of probe field increases. However, the fidelity of transmitted signal field decreases as well. Therefore, a weak probe field is preferable for achieving a high fidelity, while a relative strong probe field is required to reduce the detection error. An optimal trade-off is $|\alpha_p|^2 = 0.6$, yielding $P_{err} = 0.09$ and a fidelity of 0.9 (**Figure 7**). To reduce the error probability and improve the fidelity, a cascade configuration is needed. In such configuration, the transmitted signal field of the former QND measurement is fed into the latter. The transmitted probe field is detected in each measurement. For an N-cascade configuration, the error probability decreases exponentially as a function of N, but the fidelity

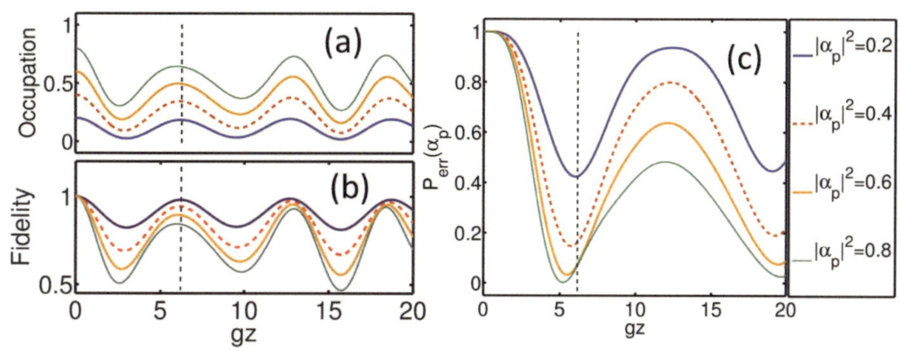

Figure 7. Evolution of the occupation (a), the fidelity (b) and the detection error probability (c) for different probe field, $|\alpha_p|^2$. The black dashed lines at $gz = 2\pi$ are the guides to eye [26].

decreases linearly. A four-cascade detection unit can already achieve $\frac{F^4}{P_{err}^4} > 23.75$ for a very weak probe field of $|\alpha_p|^2 = 0.2$.

The measured photon and the probe photon are "moving" pulse-shaped wavefunctions. The quantum Langevin equation describes the motion of system in the single mode regime, in which both the signal and the probe photons are treated as a single mode. In the real experiment, they are moving pulse including continuous spatiotemporal modes and can be confined in a one-dimensional (1D) waveguide. Therefore, a model accounting for the interaction of continuous spatiotemporal modes is required. The method developed by Fan et al. can model the interaction of the signal and probe photons in 1D real space [27]. In the Fan's method, the photons are the wavefunctions of quantum fields propagating in 1D real space. The probability density of photon appearing at certain time (position) is the squared absolute value of wavefunctions. For the purpose of single-photon QND measurement, only one needs the fidelity and phase shift of a photon-pair input state $|1_p, 1_s\rangle$ after propagating a certain distance. Starting from the vacuum auxiliary field, it can be excited during the propagation of the probe and signal fields. One can define an associate wavefunction $\varnothing_{ps}(t; z_p, z_s)$ for the state $|1_p, 1_s\rangle$, and the wavefunction $\varnothing_a(t; z_a)$ for the state $|1_a\rangle$. These wavefunctions imply that the photons $|1_p\rangle$ and $|1_s\rangle$ ($|1_a\rangle$) appear(s) at z_p and z_s (z_a) at time t with probability density of $|\varnothing_{ps}(t; z_p, z_s)|^2$ ($|\varnothing_a(t; z_a)|^2$). The nonlinear medium can be assumed to possess a spatial nonlocal response distribution with an interaction length of σ that $f_g(z_a, z_p, z_s) = \frac{1}{\sqrt{\pi\sigma^3}} e^{-\left[(z_a - z_p)^2/2\sigma^2\right]} e^{-\left[(z_a - z_s)^2/2\sigma^2\right]}$. Following Fan's treatment, the evolution of the photonic wavefunctions is governed by the partial differential equations [27, 28]

$$\frac{\partial \varnothing_{ps}}{\partial t} = -v_p \frac{\partial \varnothing_{ps}}{\partial z_p} - v_s \frac{\partial \varnothing_{ps}}{\partial z_s} - \frac{ig_0}{2} \int_0^L f_g(z_a, z_p, z_s) \varnothing_a dz_a, \tag{6}$$

$$\frac{\partial \varnothing_a}{\partial t} = -v_a \frac{\partial \varnothing_a}{\partial z_p} - \frac{ig_0}{2} \int_0^L f_g(z_a, z_p, z_s) \varnothing_{ps} dz_p dz_s, \tag{7}$$

where g_0 is the coupling amplitude, v_a(v_p, v_s) is the group velocity of the auxiliary (probe, signal) field in the 1D waveguide. g_0 is not important because the coupling strength in experiment can be tuned via the pump laser intensity. The photon pulses are assumed to be long enough that the group velocity of each mode is constant in time, and the perfect phase and energy matching are satisfied.

Solving Eqs. (2) and (3)) can simulate the evolution of the fields in medium. Without loss of generality, a Gaussian input is applied. For a single-photon pulse which is a quantum field, the photon can appear everywhere within the pulse with a probability density determined by the wave packet. This is the nonlocal nature of a single photon pulse. When the probe and signal fields propagate at the same group velocity in the medium as previous schemes, they have no necessity to interact with each other. Actually, with a large probability, they propagate independently as they never meet each other. The signal photon couples the probe photon only if they appear at the same position. As a result, only the central part of $\varnothing_{ps}$ reverses its sign, implying a

http://dx.doi.org/10.5772/intechopen.72871

pi phase shift, see **Figure 8(a)**. To circumvent this issue raised by the nonlocality of single photon pulse, the probe field pulse is delayed with respect to the signal field pulse but propagates at a higher velocity. To do so, the signal mode can be slowed down via the electromagnetically induced transparency (EIT) technique. In such an arrangement, the probe field pulse scans over the signal field pulse. No matter where the probe and signal photons appear within the pulses, they will interact with each other once. It can be seen from **Figure 8(b)** that a π phase shift can be clearly induced after the probe pulse passes through the entire signal pulse. The fidelity is very high about unity. Another advantage of this arrangement over the former is that the phase shift will not change once the probe field passes the signal field, see **Figure 8(b)**.

By comparing two models, it can be seen that when the probe field has at most one photon, a unit fidelity for the transmitted signal mode is achieved. If the probe contains higher Fock states, then interaction with these high Fock states of probe mode prevents to achieve perfect non-demolition of the signal mode.

Rubidium vapor embedded in a hollow-core photonic crystal fiber [12] or a hollow antiresonant reflecting optical waveguide [29] can be a good experimental implementation for this QND measurement scheme. This setup, to a good approximation, can be modeled as a 1D nonlinear medium. The four-wave mixing can be effectively conducted using a diamond-level configuration as shown in **Figure 9**. The signal field can be slowed via EIT with the fifth level, $4d_{3/2}$.

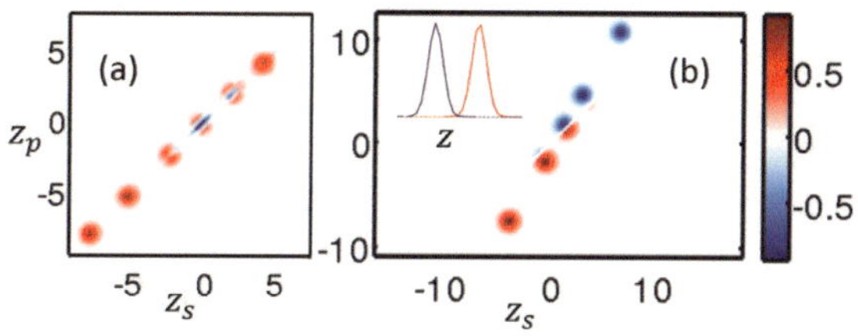

Figure 8. Evolution of the wave function $\varnothing_{ps}$ for (a) the same propagating speeds $v_p = v_s = 1$ and delay and (b) different speeds $v_p > v_s$ and different delays [26].

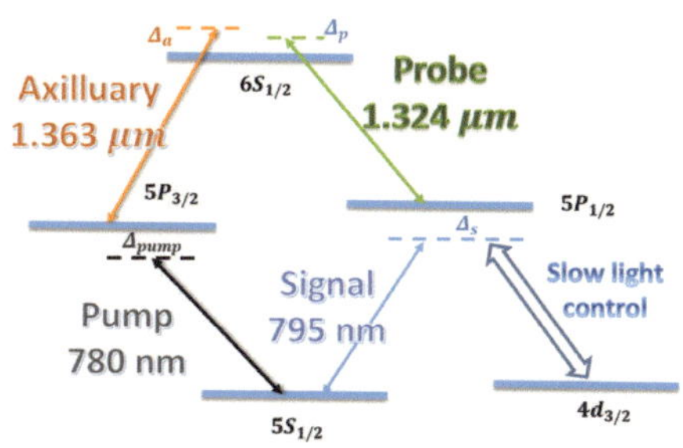

Figure 9. Configuration for four-wave mixing realized in Rb atomic vapor in hollow waveguides. The signal field is slowed via EIT by a strong coupling between levels of $5P_{1/2}$ and $4d_{3/2}$.

5.2. QND measurement with single emitters

Alternatively, Witthaut et al. proposed another scheme for QND measurement of single photons by using a single V-type emitter coupling to a 1D waveguide [30]. The configuration is depicted in **Figure 10**.

A V-type three-level emitter strongly couples to one end of semi-infinite waveguide. The signal photon drives the transition between $|g\rangle$ and $|e\rangle$. The coupling to the waveguide causes an external decay rate, Γ, of state $|e\rangle$. The metastable state $|s\rangle$ is decoupled from the waveguide. The emitter is initially prepared in a superposition state of $\alpha|g\rangle + \beta|e\rangle$ with $\beta = \sqrt{1-\alpha^2}$. The reflection amplitude of a single-photon input is given by

$$t_\Delta = \frac{\Delta + i(\gamma - \Gamma)}{\Delta + i(\gamma + \Gamma)}, \tag{8}$$

with Δ is the detuning between the carrier frequency and the transition frequency between $|g\rangle$ and $|e\rangle$.

A passing resonant photon then introduces a phase shift if and only if the emitter is in state $|g\rangle$. The transmission amplitude is given by $t_\Delta = (\gamma - \Gamma)/(\gamma + \Gamma)$ for this on resonance input. When $\Gamma \gg \gamma$, a π phase shift is imprinted on the photon. Then another classical control pulse is applied to invert the state to $-\beta|g\rangle + \alpha|e\rangle$. The complete procedure thus realizes the mapping

$$1\ signal\ photon: |g\rangle \to (\beta^2 + t_\Delta \alpha^2)|g\rangle + \alpha\beta(1 - t_\Delta)|s\rangle, \tag{9}$$

$$0\ signal\ photon: |g\rangle \to |g\rangle. \tag{10}$$

Measuring the phase shift imprinted on an incident classical laser pulse can measure the state of emitter. The emitter in $|s\rangle$ unambiguously reveals the presence of a signal single photon. This scheme is very unclear. They did not discuss how the phase of classical laser field can be shifted by an observable amount. It is also unclear how the single photon changes the state of emitter to be measured.

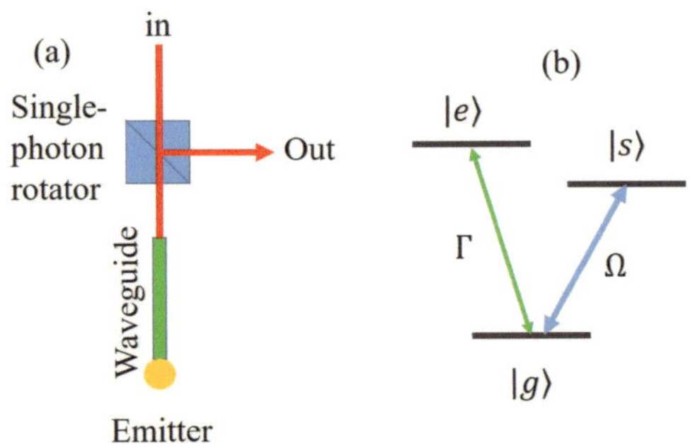

Figure 10. (a) Sketch of potential experimental setup for QND measurement of a single photon. The single-photon circulator is used to separate the input and output. (b) Level diagram of the emitter [30].

http://dx.doi.org/10.5772/intechopen.72871

For simplicity, set $v_a = v_p = 1$. Without loss of generality, a Gaussian input, $\varnothing_{ps}(t=0; z_p, z_s) = \left(1/\sqrt{\pi\tau_p\tau_s}\right)e^{-\left(z_p - z_{p,0}\right)^2/2\tau_p^2}e^{-\left(z_s - z_{s,0}\right)^2/2\tau_s^2}$ is applied, where $z_{p,0}$ and $z_{s,0}$ are the group delays of the probe and signal wavefunctions, respectively.

6. A possible bright future

QND measurement opens a door for precise measurement and versatile applications in photon-based quantum information processing. In principle, QND measurement enables repeated measurement of photon number, n, of a light beam. Because QND measurement does not disturb the photon number of light, it allows one to measure the photon number many times. This can surpass the standard quantum limit bounded by the "shot-noise" and allows to measure light with ultrahigh sensitivity. QND measurement down to the single photon level further enables potential application in quantum information processing. Remarkably, when a single signal photon can induce a π phase shift to another probe photon, the scheme for QND measurement essentially has the potential to implement a quantum controlled-phase gate between these two photonic modes. This kind of gate is a universal quantum gate for quantum computation. Another important application is to squeeze light via QND measurement. Although QND measurement has been well studied theoretically and has been realized in experiments, it is still questioned in its interpretation [31]. Monroe comments that photons can be independently generated once a signal photon is detected via absorption. He claims that the concept of QND measurement is confusing and should be demolished. However, his comments are also questionable. Squeezing light through QND measurement cannot be realized by simply generating photons according to the detection events. In summary, the concept of QND measurement applied to photons promises of great applications in quantum measurement. The progress approaching the single photon level may provide a simple router for implementing quantum information processing [32] or even quantum telescope [33].

Author details

Keyu Xia

Address all correspondence to: keyu.xia@nju.edu.cn

Nanjing University, Nanjing, China

References

[1] Wheeler JA, Zurek WH, editors. Quantum Theory and Measurement. Princeton: Princeton University Press; 1983

[2] Braginski VB, Vorontsov YI. Quantum-mechanical limitations in macroscopic experiments and modern experimental technique. Soviet Physics Uspekhi. 1975;**17**:644-650

[3] Thorne KS, Drever RWP, Caves CM, Zimmermann M, Sandberg VD. Quantum nondemolition measurements of Harmonic oscillators. Physical Review Letters. 1978;**40**(11):667-671. DOI: 10.1103/PhysRevLett.40.667

[4] Unruh WG. Quantum nondemolition and gravity-wave detection. Physical Review D. 1979;**19**(10):2888-2896. DOI: 10.1103/PhysRevD.19.2888

[5] Caves CM, Thorne KS, Drever RWP, Sandberg VD, Zimmermann M. On the measurement of a weak classical force coupled to a quantum-mechanical oscillator. I. Issues of principle. Reviews of Modern Physics. 1980;**52**(2):341-392. DOI: 10.1103/RevModPhys.52.341

[6] Semenov AD, Gol'tsman GN, Korneev AA. Quantum detection by current carrying superconducting film. Physica C. 2001;**351**:349-356

[7] Gol'tsman GN, Okunev O, Chulkova G, Lipatov A, Semenov A, Smirnov K, Voronov B, Dzardanov A, Williams C, Sobolewski R. Picosecond superconducting single-photon optical detector. Applied Physics Letters. 2001;**79**(6):705-707

[8] Marsili F, Verma VB, Stern JA, Harrington S, Lita AE, Gerrits T, Vayshenker I, Baek B, Shaw MD, Mirin RP, Nam SW. Detecting single infrared photons with 93% system efficiency. Nature Photonics. 2013;**7**(3):210-214. DOI: 10.1038/NPHOTON.2013.13

[9] Hofherr M, Rall D, Ilin KS, Semenov A, Gippius N, Huebers H-W, Siegel M. Superconducting nanowire single-photon detectors: quantum efficiency vs. film thickness. In: 9th European Conference on Applied Superconductivity (EUCAS 09); IOP Publishing; 2010. p. 012017. DOI: 10.1088/1742-6596/234/1/012017

[10] Imoto N, Haus HA, Yamamoto Y. Quantum nondemolition measurement of the photon number via the optical Kerr effect. Physical Review A. 1985;**32**(4):2287-2292. DOI: 10.1103/PhysRevA.32.2287

[11] Grangier P, Levenson JA, Poizat J-P. Quantum non-demolition measurements in optics. Nature. 1998;**396**(6711):537-542. DOI: 10.1038/25059

[12] Levenson MD, Shelby RM, Reid M, Walls DF. Quantum nondemolition detection of optical quadrature amplitudes. Physical Review Letters. 1986;**57**(20):2473-2476. DOI: 10.1103/PhysRevLett.57.2473

[13] Venkataraman V, Saha K, Gaeta AL. Phase modulation at the few-photon level for weak-nonlinearity-based quantum computing. Nature Photonics. 2012;**7**(2):138-141. DOI: 10.1038/NPHOTON.2012.283

[14] Tiarks D, Schmidt S, Rempe G, Duerr S. Optical pi phase shift created with a single-photon pulse. Science Advances. 2016;**2**:e1600036. DOI: 10.1126/sciadv.1600036

[15] Liu Z-Y, Chen Y-H, Chen Y-C, Lo H-Y, Tsai P-J, Yu IA, Chen Y-C, Chen Y-F. Large cross-phase modulations at the few-photon level. Physical Review Letters. 2016;**117**(20):203601. DOI: 10.1103/PhysRevLett.117.203601

[16] Gea-Banacloche J. Impossibility of large phase shifts via the giant Kerr effect with single-photon wave packets. Physical Review A. 2010;**81**(4):043823. DOI: 10.1103/PhysRevA.81.043823

[17] Shapiro JH, Razavi M. Continuous-time cross-phase modulation and quantum computation. New Journal of Physics. 2007;**9**:16. DOI: 10.1088/1367-2630/9/1/016

[18] Shapiro JH. Single-photon Kerr nonlinearities do not help quantum computation. Physical Review A. 2006;**73**(6):062305. DOI: 10.1103/PhysRevA.73.062305

[19] Dove J, Chudzicki C, Shapiro JH. Phase-noise limitations on single-photon cross-phase modulation with differing group velocities. Physical Review A. 2014;**90**(6):062314. DOI: 10.1103/PhysRevA.90.062314

[20] Johnson BR, Reed MD, Houck AA, Schuster DI, Bishop LS, Ginossar E, Gambetta JM, DiCarlo L, Frunzio L, Girvin SM, Schoelkopf RJ. Quantum non-demolition detection of single microwave photons in a circuit. Nature Physics. 2010;**6**(9):663-667. DOI: 10.1038/NPHYS1710

[21] Brune M, Bernu J, Guerlin C, Deleglise S, Sayrin C, Gleyzes S, Kuhr S, Dotsenko I, Raimond JM, Haroche S. Process tomography of field damping and measurement of Fock state lifetimes by quantum nondemotion photon counting in a cavity. Physical Review Letters. 2008;**101**(24):240402. DOI: 10.1103/PhysRevLett.101.240402

[22] Schuster DI, Houck AA, Schreier JA, Wallraff A, Gambetta JM, Blais A, Frunzio L, Majer J, Johnson B, Devoret MH, Girvin SM, Schoelkopf RJ. Resolving photon number states in a superconducting circuit. Nature. 2007;**445**(7127):515-518. DOI: 10.1038/nature0546

[23] Brune M, Haroche S, Lefevre V, Raimond JM, Zagury N. Quantum nondemolition measurement of small photon numbers by Rydberg-atom phase-sensitive detection. Physical Review Letters. 1990;**65**(8):976-979. DOI: 10.1103/PhysRevLett.65.976

[24] Nogues G, Rauschenbeutel A, Osnaghi S, Brune M, Raimond JM, Haroche S. Seeing a single photon without destroying it. Nature. 1999;**400**(6741):239-242. DOI: 10.1038/22275

[25] Reiserer A, Ritter S, Rempe G. Nondestructive detection of an optical photon. Science. 2013;**342**(6164):1349-1351. DOI: 10.1126/science.1246164

[26] Xia K, Johnsson M, Knight PL, Twamley J. Cavity-free scheme for nondestructive detection of a single optical photon. Physical Review Letters. 2016;**116**(2):023601. DOI: 10.1103/PhysRevLett.116.023601

[27] Shen J-T, Fan S. Theory of single-photon transport in a single-mode waveguide. I. Coupling to a cavity containing a two-level atom. Physical Review A. 2009;**79**(2):023837. DOI: 10.1103/PhysRevA.79.023837

[28] Xia K, Lu G, Lin G, Cheng Y, Niu Y, Gong S, Twamley J. Reversible nonmagnetic single-photon isolation using unbalanced quantum coupling. Physical Review A. 2014;**90**(4):043802. DOI: 10.1103/PhysRevA.90.043802

[29] Schmidt H, Hawkins AR. Electromagnetically induced transparency in alkali atoms integrated on a semiconductor chip. Applied Physics Letters. 2005;**86**(3):032106. DOI: 10.1063/1.1853528

[30] Witthaut D, Lukin MD, Sorensen AS. Photon sorters and QND detectors using single photon emitters. EPL. 2012;**97**(5):50007. DOI: 10.1209/0295-5075/97/50007

[31] Monroe C. Demolishing quantum nondemolition. Physics Today. 2011;**64**(1):8. DOI: 10.1063/1.3541926

[32] Osborne IS. Quantum optics: To catch a photon nondestructively. Science. 2016;**351**(6273):572. DOI: 10.1126/science.351.6273.572-b

[33] Kurek AR, Pieta T, Stebel T, Pollo A, Popowicz A. Quantum telescope: Feasibility and constraints. Optics Letters. 2016;**41**(6):1094-1097. DOI: 10.1364/OL.41.001094

Photon Counting for Studying Faint Astronomical Variable Signals in Optical Band

Filippo Ambrosino and Franco Meddi

Additional information is available at the end of the chapter

http://dx.doi.org/10.5772/intechopen.71072

Abstract

Although physics of neutron stars as pulsars together with their emission mechanisms leave discussions open, such objects represent the best targets to be deeply investigated by photon counting through the high-speed photometry technique. In this scenario, the capability of devices based on the silicon photomultiplier technology allows detecting single photons with remarkable time resolutions (few tens of nanoseconds). Whenever performing (optical) ground-based observations of variable sources, time of arrivals of incoming photons must be corrected because Earth's reference frame system is not inertial. Time corrections provide time of arrivals to be moved to the Solar System Barycentre inertial reference frame. If the pulsar belongs to a binary system, further corrective terms, due to the orbital motion of the companion star, have to be taken into account. In this chapter, we report experimental results obtained from observations performed on two different variable sources, the isolated Crab pulsar and Hz Her/Her X-1 binary system, with a very fast custom astronomical photometer.

Keywords: optical fast photometry, time of arrivals, variable astronomical sources, SiPM technology, data analysis

1. Introduction

Pulsars are highly magnetised, rotating neutron stars (NSs) (i.e. stars at the final state of their evolution) with an associated typical mass of about 1.4 $M_{\odot}$ (solar masses) as stated by the Chandrasekhar limit[1]. These objects emit electromagnetic radiation, detected as periodic

[1]In 1930, Chandrasekhar demonstrated that it was impossible for a white dwarf star (a stellar core remnant composed mostly of electron-degenerate matter) to be stable if its mass is greater than 1.4 $M_{\odot}$. If such a star does not completely burn its thermonuclear fuel, then this limiting mass may be slightly larger. A star that ends its nuclear-burning lifetime with a mass greater than the Chandrasekhar limit must become either a neutron star or a black hole.

signals (i.e. pulses), preferentially in radio band (especially if they are isolated). In some cases, energies associated to these pulsations are spread all over the electromagnetic spectrum. Being possible to find them either isolated or bounded in binary systems, different emission mechanisms take place. In this chapter, we only focus on optical ground-based observations performed on such sources.

Since the Earth's reference frame system is not inertial, the clock does not tick at a constant rate. It is a differential quantity due to both the Earth's rotation and revolution around the Sun. Hence, timing of periodic signals constitutes a fundamental problem that must be taken into account when performing data analysis in order to obtain corrected measurements (i.e. spin periods, orbital parameters, rotational light curves). For this reason, the Solar System Barycentre (SSB) is chosen as the origin of a new reference frame system considered as inertial in a very good approximation. Several corrective timing factors have to be included to perform a complete analysis of isolated pulsars. The equation converting photon ToAs from the terrestrial system to the SSB can be written as follows:

$$t_{SSB} = t_{obs} + t_{clk} + \Delta R_{\odot} + \Delta E_{\odot} - \Delta S_{\odot} - \Delta DM \tag{1}$$

where t_{obs} is the observed ToAs of photons, t_{clk} is the set of clock corrections through which universal time coordinated (UTC) is converted to Barycentric Dynamical Time (BDT) passing through terrestrial time (TT), $\Delta R_{\odot}$ is the Rømer delay, $\Delta E_{\odot}$ is the Einstein delay, $\Delta S_{\odot}$ is the Shapiro delay, and ΔDM is the delay due to the dispersion measure [1]. All these corrective terms are presented and explained in Section 2.

Pulsars can be also found to be bounded in binary systems, either low-mass X-ray binary (LMXB) or high mass X-ray binary (HMXB) depending on the mass of the companion star. In the first case, the mass of the latter does not exceed 1 $M_{\odot}$, while in the second case masses involved are larger than 5 $M_{\odot}$ [2]. One more class has been recently introduced for such systems called intermediate mass X-ray binary (IMXB), in which range of masses varies within the range $(2-3)\ M_{\odot}$.

The equation that permits to compute the true emission time must be completed with the following term:

$$t_{em} = t_{obs} - \frac{d}{c} - \frac{Z(t_{em})}{c} \tag{2}$$

where t_{obs} is ToAs of observed photons, d is the distance to the system, c is the speed of light, and $Z(t_{em})$ is the distance between the so-called line of nodes and the pulsar itself (see Section 2 for details). The term d/c can be neglected being a constant.

In this scenario, a high-speed photometer working in optical band (ranged from 320 nm up to 900 nm), called silicon fast astronomical photometer (SiFAP) [3, 4, 5] was conceived and realised at 'Sapienza University of Rome'. It is comprised of two channels: the first one is dedicated to study the science target, while the second one is devoted to monitor a reference star in the field of view (FoV). For both the two channels, a custom electronic chain capable to integrate the signal coming from the source in time windows down to 20 μs has been developed. Concerning the channel observing the target, an independent electronic system was added in order to tag the ToA of each single incoming photon with a time resolution of 25 ns.

A global positioning system (GPS) unit provides a reference time marker through the 1 pulse per second (1 PPS) signal, linked to UTC, with 25 ns resolution at 50% of the rising edge of the pulse itself. This signal drives two light-emitting diodes (LEDs) to have two optical markers superimposed on the data; the first one is at the beginning and the second one is at the end of the acquisition.

SiFAP was successfully mounted at 3.58 m Telescopio Nazionale Galileo (TNG) and 1.52 m Cassini telescope. Exciting results were obtained for the isolated Crab pulsar at TNG and Hz Her/Her X-1 binary system using Cassini telescope, respectively. By using two different data analysis approaches, Fourier and epoch folding search (EFS), a very good agreement between the observed spin period of the Crab pulsar and the one expected from ephemeris provided by Jodrell Bank Observatory (JBO) was obtained. The rotational light curve, showing the expected shape, was reconstructed for the Crab pulsar. Concerning Hz Her/Her X-1 system, its orbital period was found to be compatible with that extrapolated from the available literature, and the associated (simplified) orbital light curve was also successfully reconstructed.

2. The timing problem for isolated pulsars

2.1. Clock corrections

As stated earlier, the clock does not tick at a constant rate in the Earth's reference frame system suffering from the effects due to both the Earth's rotation and its revolution around the Sun. Time reference system is thus required to be constant. This can be achieved by considering some time conversions.

The first one is used to convert UTC into International Atomic Time (TAI). In this time domain, 1 s is defined as the time required for a ^{133}Cs atom at the ground state to complete exactly 9 192 631 770 oscillations [6]. It is given by:

$$TAI = UTC + N_{ls} + 10\ s \tag{3}$$

where N_{ls} is the number of leap seconds. If the difference between UTC and UT1 (Universal Time, also known as astronomical time or solar time, referred to the Earth's rotation) reaches 0.6 s, leap seconds are added in order to maintain this difference below 0.9 s. Until now (August, 2017), the number of leap seconds is 27, although 10 s more have to be considered because of historical reasons [7].

The second one in turn converts TAI into geocentric TT, using the following equation:

$$TT = TAI + 32.184\ s \tag{4}$$

where 32.184 s is an offset arising from historical issues.

The last time conversion term consists and allows TT to be converted into BDT (i.e. the time one would have when photons are collected from the SSB). An approximate formula, composed of 791 coefficients, developed by Fairhead and Breatagnon in 1990 [8], permits to convert TT into BDT taking into account the effects due to the time dilation. Such a formula can be expressed as:

$$BDT \approx TT + \sum_{i=1}^{478} A_i \sin\left(\omega_{A_i} T + \phi_{A_i}\right) + T \sum_{i=1}^{205} B_i \sin\left(\omega_{B_i} T + \phi_{B_i}\right) + \\ + T^2 \sum_{i=1}^{85} C_i \sin\left(\omega_{C_i} T + \phi_{C_i}\right) + T^3 \sum_{i=1}^{20} D_i \sin\left(\omega_{D_i} T + \phi_{D_i}\right) + T^4 \sum_{i=1}^{3} E_i \sin\left(\omega_{E_i} T + \phi_{E_i}\right) \tag{5}$$

where A_i, B_i, C_i, D_i, and E_i are coefficients expressed in microsecond, ω_{A_i}, ω_{B_i}, ω_{C_i}, ω_{D_i}, and ω_{E_i} are angular velocities expressed in rad/10^3 years. ϕ_{A_i}, ϕ_{B_i}, ϕ_{C_i}, ϕ_{D_i}, and ϕ_{E_i} are angular phases expressed in radians, and T is the number of Julian centuries since January 1, 2000 (i.e. the beginning of the J2000 epoch). T can be written as:

$$T = \frac{JD - 2\,451\,545}{36\,525} \tag{6}$$

where JD (in TT scale) is the time expressed in Julian Date[2] [8]. An accuracy of ~1 ns is achieved computing BDT through Eq. (5). More details describing the computation of the 791 coefficients can be found in Ref. [8].

2.2. The Rømer delay: A geometrical correction

The Danish astronomer Rømer was the first to study the geometrical correction caused by the variation of the path between the telescope collecting data and the SSB because of the Earth's rotation and movement around the Sun. Taking into account these two effects, the classical light-travel time is affected by a delay, called Rømer delay.

According to **Figure 1**, it is possible to compute the vector $\boldsymbol{r}_{B-O}$ pointing from the SSB towards the observatory site as the difference $\boldsymbol{r}_{B-O} = \boldsymbol{r}_{E-O} - \boldsymbol{r}_{B-E}$ [9] if the vector $\boldsymbol{r}_{B-E}$ pointing from the SSB towards the Geocentre and the vector $\boldsymbol{r}_{E-O}$ pointing from Geocentre towards observatory site are known. In addition to the geometrical correction, parallax effects should be included whenever observing targets close to the solar system. In our case, paths of incoming photons can be considered as parallel (i.e. not affected by the parallax) in the whole solar system in a good approximation because we are studying point sources located at infinity. Hence, considering such an approximation, the equation for the geometric time correction (i.e. the Rømer delay, $\Delta R_{\odot}$) can be expressed as:

[2]Julian dates are simply a continuous count of days and fractions since noon Universal Time on January 1, 4713 BC (on the Julian calendar). In order to increase the number of decimal digits modified Julian date (MJD) is often used.MJD corresponds to JD: 2400000.5.

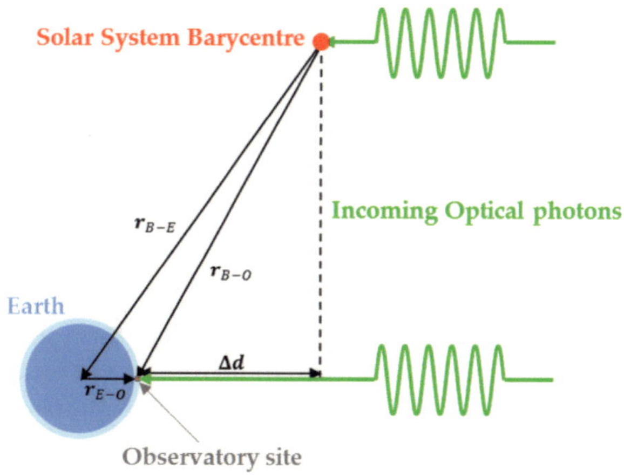

Figure 1. Schematic representation of the Rømer delay.

$$\Delta R_{\odot} = \frac{\boldsymbol{r}_{B-O} \cdot \widehat{\boldsymbol{n}}}{c} \tag{7}$$

where $\widehat{\boldsymbol{n}}$ is the unitary vector for the incoming photons and c is the speed of light.

2.3. The Einstein and Shapiro delays: Relativistic corrections

The time-varying gravitational potential and the Doppler shifts experienced by the observatory clock cause the so-called Einstein delay ($\Delta E_{\odot}$). This delay is the combined effect of two different terms: the gravitational redshift[3] and the time dilation due to motions of the Earth and other bodies.

The first effect can be corrected through Eq. (5), considering the clock to tick at the SSB. The second effect can be computed and corrected by using ephemeris (position and velocity) of the observatory site. The analytic equation describing the Einstein delay is given by:

$$\Delta E_{\odot} = \frac{\boldsymbol{r}_{E-O} \cdot \boldsymbol{v}_{Earth}}{c^2} \tag{8}$$

where $\boldsymbol{v}_{Earth}$ is the Earth's velocity with respect to the SSB.

When the light coming from a distant source travels close to massive objects, its path is not straight anymore but curved because of their gravitational fields. Because a curved line is longer than a straight path, the light will need more time to complete its travelling. The Shapiro delay ($\Delta S_{\odot}$) caused by the space–time curvature around massive objects can be thus written as:

[3]The gravitational redshift (or the Einstein shift) is the process by which electromagnetic radiation originating from a source that is in a gravitational field is reduced in frequency (or redshifted) when observed in a region at a larger gravitational potential [10].

$$\Delta S_{\odot} = -\frac{2GM_{Body}}{c^3} \ln(1 + \cos\theta) \tag{9}$$

where G is the gravitational constant, M_{Body} is the mass of the current object of which the gravitational field is considered, and θ is the angle between the pulsar and the Earth as seen from the Sun [11].

2.4. The dispersion measure delay

The time delay due to the propagation of a radiation with a given frequency with respect to one of infinite frequencies along a path of length d from the pulsar to the Earth is called Dispersion Measure (ΔDM) delay. The Dispersion Measure is expressed as follows:

$$\Delta DM = \frac{1}{c}\int_0^d \left(1 + \frac{f_p^2}{2f^2}\right) dl - \frac{d}{c} = \mathfrak{D}\frac{DM}{f^2} \tag{10}$$

where f_p is the frequency of a partially ionised plasma crossed by a signal with a given frequency f, and $\mathfrak{D}$ is the Dispersion Constant. The Dispersion Constant is defined as:

$$\mathfrak{D} \equiv \frac{e^2}{2\pi m_e c} \tag{11}$$

where e is the fundamental charge, and m_e is mass of the electron.

DM is the so-called Dispersion Measure, computed through the following integral:

$$DM = \int_0^d n_e dl \tag{12}$$

where n_e is the Galactic electron density distribution.

According to Eq. (10), DM is equal to zero in X-ray band being proportional to the inverse of the square of the signal's frequency. Moreover, it can be neglected for observations performed in optical band, while it must be taken into account for those in radio band [1].

3. Pulsars in binary systems

As a pulsar is not isolated but bounded to a companion star in a binary system, further corrective terms must be considered together with those explained in Section 2. The time related to the emission of radiation towards the observer varies cyclically depending on the orbital phase in which the pulsar is. It is immediately clear that for binary systems Doppler effect (concerning only the motion of the pulsar along the line of sight) must be considered. Starting from the Kepler's laws, a binary orbit can be fully described by seven parameters, five of which provide to define both the shape and orientation of the orbit, and the remaining two determine how the considered body moves along its orbit.

According to **Figure 2**, for data analysis of binary systems, it is sufficient to consider the following:

- the semi-major axis (a), defined as $a = [q/(1+q)]a_{sep}$, where $q = M_2/M_1$ which is the masses of the companion star and compact object, respectively, and a_{sep} is their orbital separation;
- the eccentricity (e) of the orbit;
- the inclination angle (i) as the angle between the orbit plane and the plane of the sky;
- the argument of periastron (ω) orienting the ellipse in the orbital plane. It is defined as the angle measured from the ascending node to the periastron;
- the epoch (T_{asc}) at which the object is passing on the ascending node of the orbit. Sometimes this parameter is replaced by the epoch $\left(T_{\frac{\pi}{2}}\right)$ of the passage at the superior conjunction (shifted by an angle of 90° with respect to the ascending node);
- the mean anomaly (m) specifies the current position of the body at a given time t. It is expressed as $m = 2\pi(t - T_{asc})/P_{orb}$, where P_{orb} is the orbital period of the binary system.

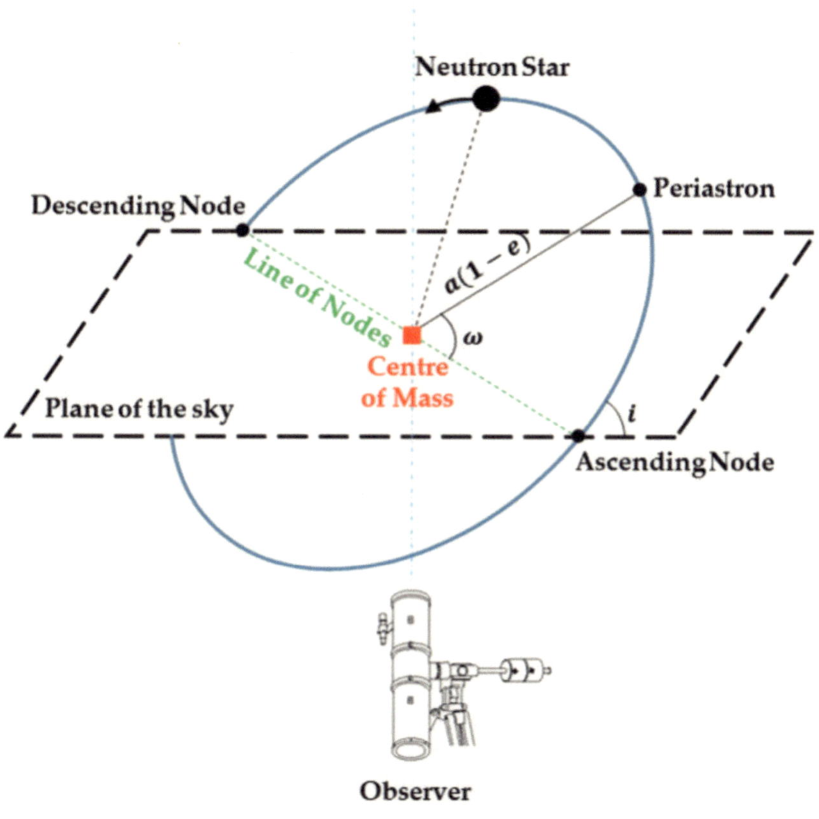

Figure 2. Orbital elements for a binary system. The line of nodes is the intersection between the orbital and sky planes.

The distance $Z(t)$ between the so-called line of nodes and the body (see **Figure 2**), when projecting the orbit onto the plane of the sky, can be expressed with a first-order approximation as:

$$Z(t) = a\ \sin i\left[\sin\left(m+\omega\right)+\frac{1}{2}e\sin\left(2m+\omega\right)-\frac{3}{2}e\sin\omega\right] \tag{13}$$

As stated earlier, $Z(t)$ is measured starting from the line of nodes, and it is considered as positive when the pulsar is farther than the line of nodes considering observer's reference direction. In this scenario, ToAs of observed photons are delayed with respect to emitted ones. Time of true emission can be computed through Eq. (2).

When a binary system is observed, the Doppler shift affecting the observed spin frequency (ν_{obs}) of the pulsar must be considered. Being the projected velocity of the pulsar $\dot{Z}(t)$, such an effect can be quantified as:

$$\nu_{obs}-\nu(t)=\nu(t)\frac{\dot{Z}(t)}{c}=u\frac{2\pi\nu(t)}{P_{orb}}\left[-\cos\left(m+\omega\right)-e\cos\left(2m+\omega\right)\right] \tag{14}$$

where $u = a \sin i/c$. Hence, if the orbital Doppler shift is not accounted for, it greatly limits the total amount of time over which pulsars can be observed and affects the data obtaining then a distorted folded pulse profile. Moreover, a maximum exposure time such that even a frequency residual $\delta\nu$ does not distort the folded pulse profile can be defined. For a given time bin Δt, the Doppler effect causes a variation on the frequency that can be evaluated as:

$$\delta\nu_{Dopp}\simeq\dot{\nu}_{Dopp}\ \Delta t\leq\ u\nu\left(\frac{2\pi}{P_{orb}}\right)^{2} \tag{15}$$

In order to satisfy this condition, we must have that:

$$\delta\nu_{Dopp}\ \leq\delta\nu_{min} \tag{16}$$

where $\delta\nu_{min} = 1/(nT_{exp})$, with n the number of phase bins which the folded spin period has been divided into, and T_{exp} is the total exposure time. In this way, Eq. (15) can be rewritten as:

$$\Delta t_{max}\leq\left(\frac{1}{nu\nu}\right)^{\frac{1}{2}}\frac{P_{orb}}{2\pi} \tag{17}$$

The knowledge of the relevant orbital parameters is surely needed to reconstruct corrected light curves. Starting from Eq. (2), and considering circular orbits (i.e. $e = 0$), the delay due to the Doppler effect on the true emission time of photons can be thus computed as:

$$t_{em} = t_{obs} - u\sin\left\{\frac{2\pi}{P_{orb}}\left[(t_{obs} - T_{asc}) - \frac{1}{2}\dot{P}_{orb}\frac{(t_{obs} - T_{asc})^2}{P_{orb}}\right]\right\} \tag{18}$$

$$t_{em} = t_{obs} - u\cos\left\{\frac{2\pi}{P_{orb}}\left[\left(t_{obs} - T_{\frac{\pi}{2}}\right) - \frac{1}{2}\dot{P}_{orb}\frac{\left(t_{obs} - T_{\frac{\pi}{2}}\right)^2}{P_{orb}}\right]\right\} \tag{19}$$

where both t_{obs} and T_{asc} $\left(T_{\frac{\pi}{2}}\right)$ must be in the same clock reference system (e.g. either UTC or BDT).

Orbital parameters that are mandatory to correct ToAs of photons coming from binary systems can be obtained in two ways. The first one consists in retrieving already existing ephemeris available in the literature. Unfortunately, problems could occur if tabulated parameters are too old and thus are needed to be updated.

In particular, in order to have more reasonable values of both epoch of ascending node and orbital period with its time derivative, if these two parameters are known at a specific reference time, they can be propagated [12] at the epoch of observation as:

$$T_{asc} = T^0_{asc} + NP^0_{orb} + \frac{1}{2}N^2P^0_{orb}\dot{P}^0_{orb} \tag{20}$$

$$T_{\frac{\pi}{2}} = T^0_{\frac{\pi}{2}} + NP^0_{orb} + \frac{1}{2}N^2P^0_{orb}\dot{P}^0_{orb} \tag{21}$$

$$P_{orb} = P^0_{orb} + \dot{P}^0_{orb}\Delta t \tag{22}$$

where T^0_{asc} $\left(T^0_{\frac{\pi}{2}}\right)$, P^0_{orb}, and $\dot{P}^0_{orb}$ are the epoch of ascending node (superior conjunction), the orbital period, and its first time derivative tabulated at a given reference time. Δt is the time interval spanning from T^0_{asc} $\left(T^0_{\frac{\pi}{2}}\right)$ up to the observation date, and N is the nearest integer of the ratio between Δt and P^0_{orb} [13]. The second way to get ephemeris is performing several observations in order to have many spin frequencies as a function of time.

Relativistic corrections (i.e. Einstein and Shapiro delays) due to the gravitational field of the pulsar's companion can be neglected for optical observations, not affecting the results. In particular, the Einstein delay is proportional to the eccentricity of the system and thus close to zero for small values of the eccentricity.

Although the Shapiro delay can contribute for tens of microseconds in the worst cases, it is not always possible to estimate its direct contribution. First, its contribution can be separated from that of the orbital motion only for nearly edge-on systems ($70^\circ < i \leq 90^\circ$) being dependent on the inclination angle.

Second, since the knowledge of both the mass of the companion star and that of the pulsar are sometimes unknown or known with a large relative error, time distortions could be even introduced when applying such a correction. Moreover, astrometric effects like parallax and proper motion (taken into account when considering coordinates of the object) can be also neglected because their contribution is much smaller than the effect caused by geometric issues (i.e. the Doppler shift).

4. SiFAP: high-speed photometry of variable sources

As already mentioned in Section 1, SiFAP was developed at the Department of Physics of 'Sapienza University of Rome' to study variable sources since 2009. During subsequent years, the instrument was improved to achieve both better photometric sensitivity and time resolution. The present version of SiFAP is comprised of two channels, Channel 0 observing the science target and Channel 2 monitoring a reference star in the FoV by using their own dedicated sensor. The sky background signal is acquired by the same sensors moving the telescope away from the two objects.

The top-level diagram of the architecture of SiFAP is shown in **Figure 3**. Detectors used, especially selected from the production batch for this application, are based on the Silicon Photo Multiplier (SiPM) technology. They are called multi pixel photon counters (MPPCs) provided by Hamamatsu Photonics[4].

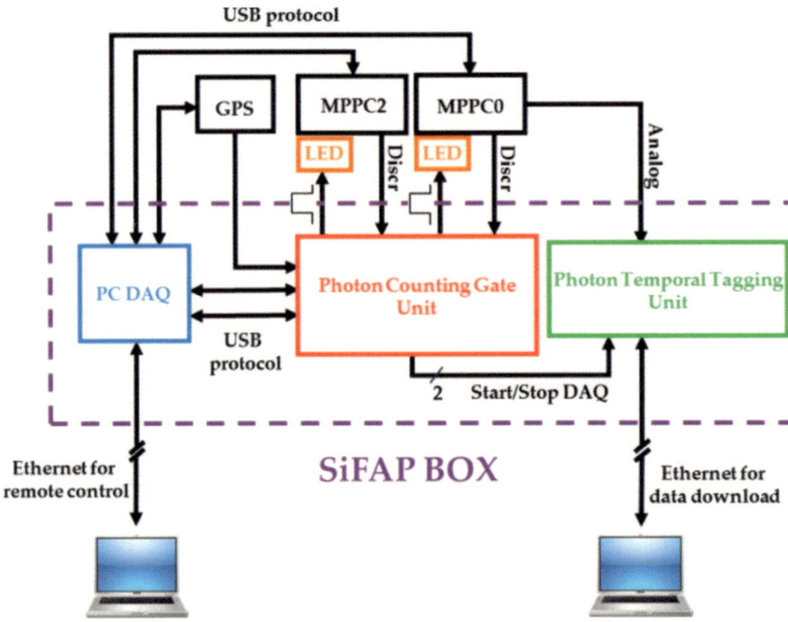

Figure 3. Top-level block diagram of the architecture of SiFAP.

[4]http://www.hamamatsu.com/us/en/index.html

Main characteristics of MPPCs (C13366 series) used by SiFAP are as follows:

- a double-stage Peltier cell cooling the sensor down to $-20°C$;
- an active area of $1.3 \cdot 1.3$ mm^2;
- a squared pixel size of 50 μm;
- a typical dark count rate of ~2500 cps (count per second);
- a photon detection efficiency (PDE) (the percentage of detected photons with respect to incident ones) covering the spectral range $(320-900)$ nm. It is peaked at a wavelength of 450 nm (blue) with a value of ~40%.

The whole system lies in a single unit, which contains two main custom blocks named photon counting gate (PCG) and photon temporal tagging (PTT), respectively. PCG is capable to collect data coming from the sensors in fixed integration time windows down to 20 μs, thanks to the handshake between a field-programmable gate array (FPGA) and a micro-processor (μP). PTT is an embedded system dedicated to observe only the science target (Channel 0) and developed to allow single photon tagging with a time resolution of 25 ns. The electronic chain of MPPC sensors provides counting photons with longer integration time windows within the range $(1-100)$ ms.

A further custom unit, named MS (Master Sync), was realised to both remotely control and synchronise PCG and PTT units. MS is composed of a dedicated micro-controller (μC) connected to a PC through a RS232 serial interface.

The GPS unit allows to link UTC to the 1PPS signal used to drive an electrical-optical transducer (i.e. LED) in order to have an optical temporal marker superimposed on the data at the beginning and the end of the acquisition. More detailed technical descriptions of SiFAP instrument are available in [3, 4, 5].

5. Data analysis techniques for variable signals

5.1. Discrete Fourier analysis

When observing pulsars searching for any regular pulsations (i.e. variations of flux) hidden into the signal, many numerical analysis techniques can be used. One of the most used methods is the Fourier analysis, which provides a transition from the time domain to the frequency domain. The continuous function that permits such a transition is the so-called Fourier transform, which is expressed as:

$$H(\nu_k) = \int_{-\infty}^{\infty} h(t)e^{2\pi i\nu_k t}\, dt \tag{23}$$

where $H(\nu_k)$ is the amplitude of the transformed signal $h(t)$, and ν_k is the sampling frequency.

For M data samples, the Fourier transform can be computed on M independent outputs only at a frequency value of:

$$\nu_k = \frac{k}{M\Delta t} \tag{24}$$

with $k = -M/2, \ldots, M/2$.

According to the Nyquist's theorem, the maximum achievable frequency (ν_{max}) is given by:

$$\nu_{max} = \frac{1}{2\Delta t} \tag{25}$$

so that frequencies higher than ν_{max} cannot be resolved. Because observational data are not continuous but sampled in discrete time intervals (Δt), the Fourier transform for M data samples can be rewritten in its discrete version as:

$$H(\nu_k) = \Delta t \sum_{l=0}^{M-1} h_l e^{\frac{2\pi ikl}{M}} \tag{26}$$

which represents the so-called discrete Fourier Transform (DFT). The accuracy ($\Delta\nu$) on frequency peaks achievable by such a method strongly depends on the total exposure time (T_{exp}) as:

$$\Delta\nu = \frac{1}{T_{exp}} \tag{27}$$

systematically affecting the estimates of frequencies. Unfortunately, since observations require the number of data samples to be very large in order to have both more statistics on faint signals and time accuracy, it would be appropriate using a fast algorithm (e.g. Cooley-Tukey) to perform the Fourier transform (FFT, Fast Fourier Transform).

The Fourier technique is particularly suitable in astronomy to detect periodicities. Being these periodic signals usually unknown and faint, their detection is realised by computing the power spectrum representing the frequency distribution of the squares of the Fourier transform coefficients.

In fact, the Parseval's theorem states that:

$$\sum_l h_l^2 = \frac{1}{M} \sum_k [H(\nu_k)]^2 \tag{28}$$

This means that the squared modulus of a function in the time domain is equal to the sum of the squares of its projection in the frequency domain. Hence, using the Parseval's theorem, the power spectrum can be written as:

$$P(\nu) = \frac{2}{N_\gamma} |H(\nu)|^2 \tag{29}$$

where N_γ is the total number of photons collected.

5.2. Epoch folding search

Unlike the Fourier analysis, EFS is a more refined method and is directly performed in the time domain. The standard approach to EFS consists in taking a dataset of a given total exposure time T_{exp} and defining a reasonable target period ($P^{\star}$), which can be determined by a preliminary FFT analysis although not always there are clear features in it, and anyway its accuracy on frequency peaks is limited by the signal time bin.

Starting from $P^{\star}$, a set of equispaced 'trial' periods (P_i) is created with a given time resolution. It is clear that $P_i \in [P_{min}, P_{max}]$ with $P_{min} < P^{\star} < P_{max}$. In turn, each trial period is divided into n time bins (often a phase is used instead of a time, making n phase bins be within the range from 0 up to 1). Hence, it is possible to map the whole dataset with such time (or phase) bins, producing a corresponding number of folded curves.

In this context, it can be easily verified that in absence of pulsations (or any secular trend), counts in each phase bin of folded curves produced by the set of trial periods are Poisson distributed with mean and variance best estimated by the mean number of counts per bin. Since the number of events in each phase bin is usually large, the number of counts (x_i) in the i-th bin is normally distributed with the mean equal to the variance $\mu_{exp} = \sigma^2_{exp}$. The statistic can be thus expressed as:

$$S = \sum_{i=1}^{n} \frac{\left(x_i - \mu_{exp}\right)^2}{\mu_{exp}} \tag{30}$$

where S represents a χ^2 distribution with $n-1$ DoF (Degrees of Freedom). Under the hypothesis of absence of pulsation, one expects that $S \approx n-1$. Hence, if S is much greater than its expected value, the statistics belongs no longer to a χ^2 distribution, showing that a nonuniform behaviour (periodic) is present in the acquired data.

The original procedure used for calculating S was provided by Leahy in 1983 [14] adopting the following parameters:

- $R = x_{tot}/T_{exp}$, where R is the total counting rate, x_{tot} is the total number of valid acquired events, and T_{exp} is the total exposure time;
- $R_i = x_i/T_i$, where R_i is the counting rate at the i-th bin, x_i is the number of counts in the i-th bin, and T_i is the time duration of i-th bin (it can differ bin-to-bin because of possible gaps in the acquisition);
- $\mu_{exp} = \mu_{\exp,i} = RT_i$, where μ_{exp} is the expected count rate which cannot be the same for each bin as T_i varies;
- $\sigma_i^2 = R/T_i$, where σ_i^2 is the variance of i-th bin.

Starting from Eq. (30), and after having defined these parameters, S can be expressed as:

$$\sum_{i=1}^{n}\frac{(R_iT_i - RT_i)^2}{RT_i} = \sum_{i=1}^{n}\frac{(R_i - R)^2}{\sigma_i^2} \tag{31}$$

A first threshold on the accuracy of the spin period of a pulsar can be estimated by guessing how much two periods must differ in order to provide two different profiles within statistical fluctuations. Another relation is based on the assumption that a shift of one phase bin when folding the light curve to a pulse profile with n phase bins has a significant influence of the pulse shape.

This assumption leads to write:

$$\delta P_{min} = \frac{P^2}{2nT_{exp}} \tag{32}$$

According to Monte Carlo simulations done by Leahy himself in 1987 [15], the accuracy on the best folding period found by EFS can be expressed as:

$$\sigma_P = \frac{P^2}{2T_{exp}}\left[\frac{S_{max}}{n-1} - 1\right]^{-0.63} \tag{33}$$

where S_{max} is the maximum value of S (corresponding to the best folding period).

6. Observational results

In this section, we report results of data analysis of two different pulsar signals acquired with SiFAP at 3.58 m TNG[5] (Telescopio Nazionale Galileo, Observatorio Astronomico Roque de Los Muchachos, La Palma, Canary Islands, Spain) and 1.52 m Cassini telescope[6] of Bologna Astronomical Observatory (Loiano, Bologna, Italy).

6.1. SiFAP at TNG: the Crab pulsar

The first science case we report here is the well-known isolated Crab pulsar (PSR B0531 + 21). The Crab pulsar is a relatively young NS, which is in the central region of Crab Nebula[7] (see **Figure 4**), the remnant of supernova SN 1054, and it was the first pulsar to be connected with a supernova remnant [16]. Isolated pulsars are in general associated to radio emissions, but some of them (like Crab pulsar) show also optical counterparts.

This source was chosen as a benchmark because of being one of the most widely studied variable sources in every band of the electromagnetic spectrum. Moreover, this object provides a very high signal-to-noise (S/N) ratio, allowing safe detection of the pulsed signal, even without performing post-processing data analysis if the primary mirror of the telescope is

---

[5] http://www.tng.iac.es/

[6] http://davide2.bo.astro.it/loiano/

[7] http://chandra.harvard.edu/photo/2002/0052/0052_xray_opt.jpg

Figure 4. Top: X-ray and optical combined images of Crab Nebula. The emitted radiation is well visible through magnetic poles. Bottom: optical image of Crab Nebula taken using Cassini telescope. The pulsar is in the central region of the Nebula and is indicated by two orthogonal segments.

enough large (3 m class or larger). In addition, it is very easy to obtain a large amount of data, which allow making comparisons among different data observations for such an object. Main properties of Crab pulsar are summarised in **Table 1**:

It is worth to emphasise that the spin period of the pulsar slows down by about 38 ns per day due to large amounts of energy carried away in the pulsar wind, although the second-time derivative of the spin period must be kept in mind to be not constant. Such an effect is thought to be due to rotational instabilities (glitches) of NS, which lead to a strong evidence for the existence of a fluid component inside it. During a glitch, in fact, a small sudden increase in rotation rate is observed in the sudden early arrival of pulses [1]. The light curve of the Crab pulsar is reported in **Figure 5** shows its typical double-peaked shape.

The first taller and thinner peak is believed to be due to the intrinsic emission of the pulsar, while the second one, lower and larger, should be caused by the re-emission of the first peak by the surrounding Nebula. Despite the Crab pulsar has a faint Optical counterpart, it is very bright in X-ray band, and the flux density and spectrum are known to be constant, with the

Crab pulsar		
Constellation		*Taurus*
Right ascension [Epoch J2000]	α	$05^h34^m31^s.97232$
Declination [Epoch J2000]	δ	$+22^\circ00'52''.069$
Apparent magnitude	V_{mag}	16.5 mag
Spin period @ (02/15/2014)	P_s	0.033692938448829(12) s
Spin period first derivative	$\dot{P}_s$	$4.2(1) \cdot 10^{-13}$ s/s
Distance	d	2.2(5) kpc

1σ uncertainties affecting the last digit(s) are presented within parentheses

Table 1. Main properties of Crab pulsar.

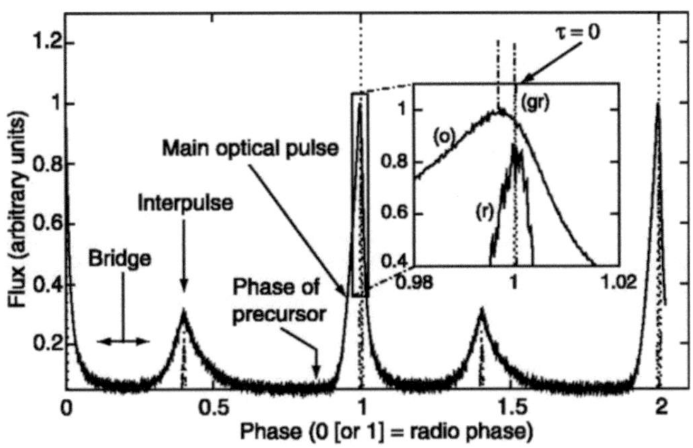

Figure 5. The Crab pulse profile showing the optical light curve (o) in the range (600 – 750) nm, the average radio light curve at 1380 MHz (r), and a single giant pulse at 1357.5 MHz (gr). Two periods are shown for clarity. Phase 0 corresponds to the arrival at the SSB of the peak radio pulse. The optical light curve was divided into 5000 phase bins. Image taken from http://www.ing.iac.es/PR/SH/SH2003/triffid.html.

exception of the pulsar itself. The pulsar provides a strong periodic signal that is usually used to check the timing of X-ray detectors.

In X-ray astronomy, this source is sometimes used as a flux density calibrator. In fact, crab and millicrab units were introduced. In particular, a millicrab corresponds to a flux density of about $2.4 \cdot 10^{11}$ erg/s/cm^2 ($2.4 \cdot 10^{14}$ W/m^2) within the range from 2 to 10 keV, for a 'crab-like' X-ray spectrum. Very few X-ray sources ever exceed one crab in brightness.

SiFAP observed the Crab pulsar at TNG on 26-02-2014 [17] for about 2400 s. After having performed both Fourier and EFS techniques, we reconstructed its rotational (spinning) light curve, shown in **Figure 6**.

We obtained two slightly different spin periods for the Crab pulsar. FFT analysis provided a spin period equal to ($0.0336927957 \pm 0.0000000014$) s, while the one obtained with EFS was ($0.0336929420 \pm 0.0000000050$) s, causing a phase shift as well. These two results had to be compared to the one extrapolated from JBO[8] database at the same observation date. The difference between the Crab pulsar spin period estimated from JBO and ours computed using FFT is ~140 ns. Such a difference was reduced to ~3 ns in the case of EFS, demonstrating that this method is more refined and robust.

It is worth to emphasise that several previous observations were performed using the Cassini telescope in order to optimise both the observational and analysis strategies.

6.2. SiFAP at Cassini telescope: Hz her/her X-1 binary system

Hz Her/Her X-1, also known as 4 U 1656 + 35, is a moderately strong X-ray binary source first studied by the Uhuru satellite. Such a system is classified as Intermediate Mass X-ray Binary (IMXB) being composed of a pulsar (Hercules X-1, spinning with a period of about 1.24 s) accreting mass from an A7 star (Hz Herculis) [18].

Hz Herculis was discovered in 1936, and classified as a variable star, while its collapsed companion was identified in 1972, thanks to the X-ray emission of the latter. Orbital parameters obtained from X-ray observations (i.e. X-ray ephemeris) and main features of Hz Her/Her X-1 [12, 19] are summarised in **Table 2**.

The possibility to observe periodic occultations between the two stars was favoured by the inclination angle, which is close to 90° (edge-on systems). Hz Herculis is included in the eclipse variable stars because its optical variability is due to the darkening by the collapsed companion (Her X-1). The magnitude variation of Hz Herculis in B-band is reported in **Figure 7**.

A strong modulation of about 35 d in X-ray intensity, which is believed to be due to the occultation of NS by the accretion disk, is also shown by such a system. This variation has a characteristic shape, with two maxima called main-on and short-on as shown in **Figure 8**.

---

[8] http://www.jb.man.ac.uk/pulsar/crab/crab2.txt

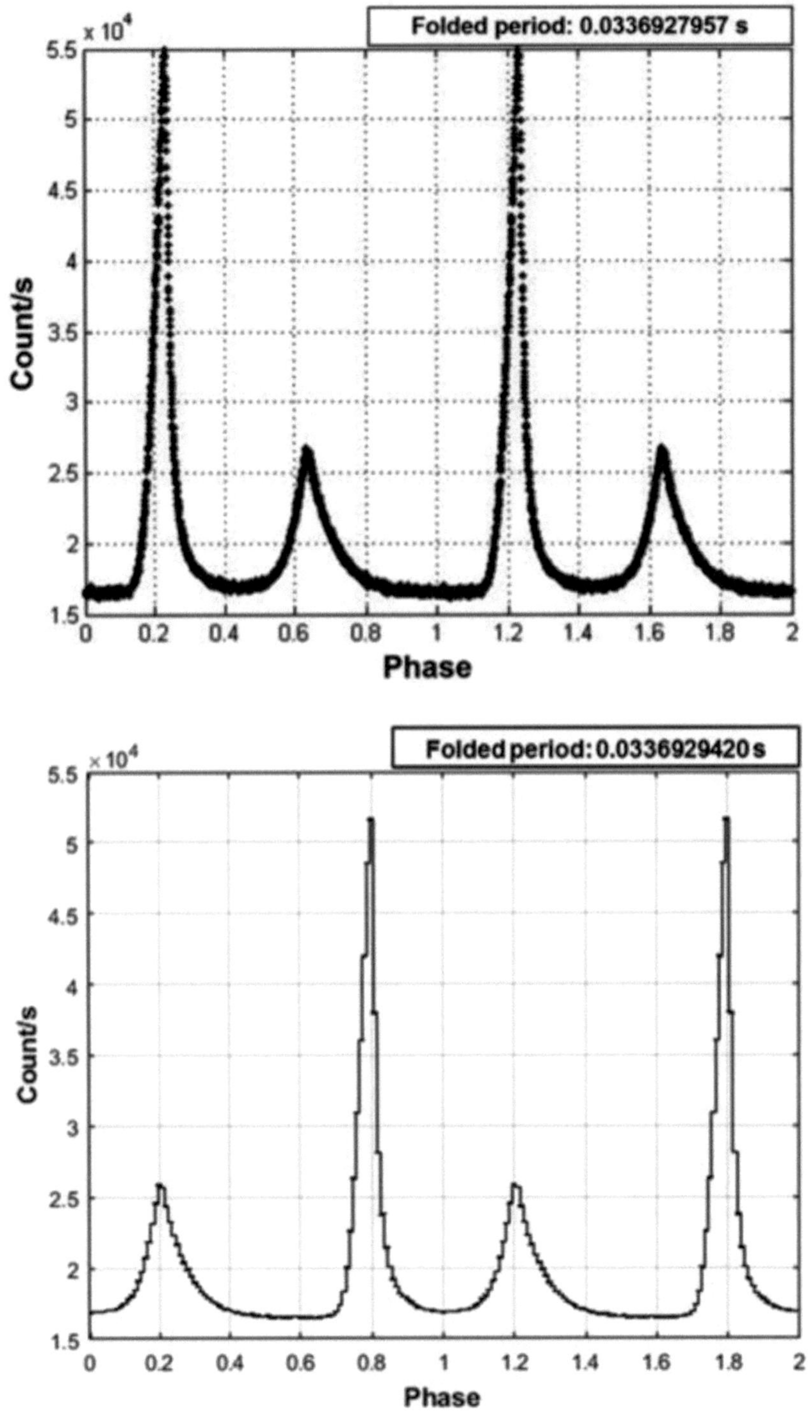

Figure 6. Top: Crab pulsar light curve obtained from data collected with SiFAP by using the Fourier approach. Bottom: Crab pulsar light curve obtained from data collected with SiFAP by using the EFS approach. The phase shift is due to the slight difference between the two computed spin periods.

Binary system		
Right Ascension [Epoch J2000]	α	$16^h57^m50^s.5$
Declination [Epoch J2000]	δ	$+35^\circ20'52''$
Distance	d	6.6(4) kpc
Orbital period	P_{orb}	1.700167202(1) d
Orbital period first derivative	$\dot{P}_{orb}$	$-1.33(7)\cdot10^{-8}$ d/yr
Projected orbital radius	$a\sin i/c$	13.1831(4) lt-s
Superior conjunction	$T_{\frac{\pi}{2}}$	54345.558195(80) MJD
Inclination angle	i	$83(4)^\circ$
Eccentricity	e	$<1.3\cdot10^{-4}$
Her X-1		
Mass	M_{X-1}	1.5(3) $M_\odot$
Spin period	P_s	1.237744750(60) s
Spin period first Derivative	$\dot{P}_s$	$-1.778(56)\cdot10^{-13}$ s/s
Magnetic field	B	$4.1\cdot10^{12}G$
X-ray luminosity	L_x	$3\cdot10^{37}$ erg/s
Hz Her		
Mass	M_{Hz}	2.3(3) $M_\odot$
Radius	R_{Hz}	4.2(2) $R_\odot$
Apparent magnitude	V_{mag}	13.2 – 14.7 mag

1σ uncertainties affecting the last digit(s) are presented within parentheses

Table 2. Orbital parameters and main properties of Hz Her/Her X-1 binary system.

The main-on, during which the maximum X-ray intensity can be observed, is followed by a 2–3 d lasting rapid drop of intensity called turn-off. The second maximum (called short-on or short-state) is about 78 d long, starts around $\phi_{35}=0.6$, where ϕ_{35} represents the 35 d phase.

The flux during the short-state has a maximum level of (10–20)% of the main-on intensity. In addition, it is worth to be noted that the optical signal is very much fainter than the X-ray one, with a pulsed fraction down to 0.1%.

Her X-1 generates a well-distinguishable X-ray pulse due to its rotation. Such a high-energetic radiation heats Hz Herculis atmosphere affecting its luminosity with a periodic cadence and making the spin signal to be detected in the optical band, thanks to thermal re-emission. It is also important to stress that the spin period is not monotonic but shows an evidence for spinning-up, in contrast with the monotonic spinning-down of the Crab pulsar. Such behaviour could be explained because of the inflow of matter towards NS, causing acceleration on its rotation.

SiFAP observed Hz Her/Her X-1 from August 25–28, 2016 at Cassini telescope. Four observations lasting about 2.5 h each were performed. Having only four data acquisitions (i.e. four

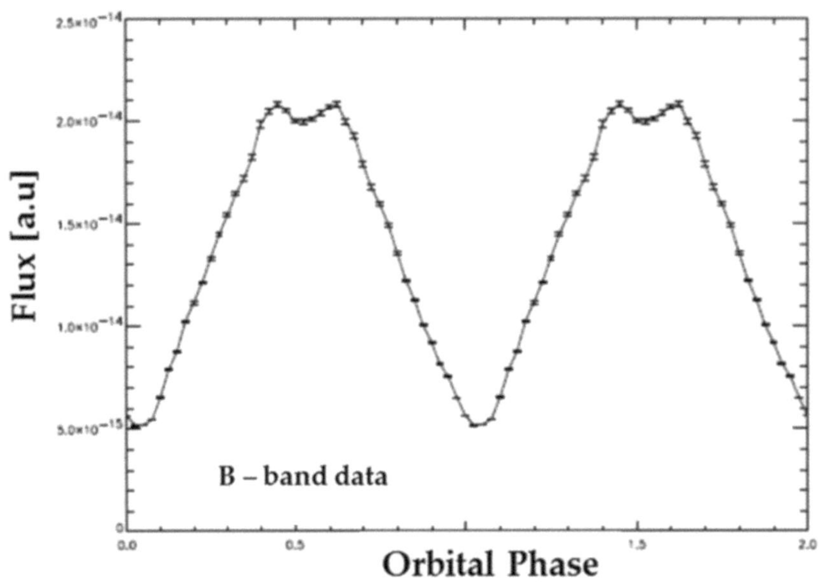

Figure 7. B-band orbital light curve of Hz Herculis ([19] and reference therein).

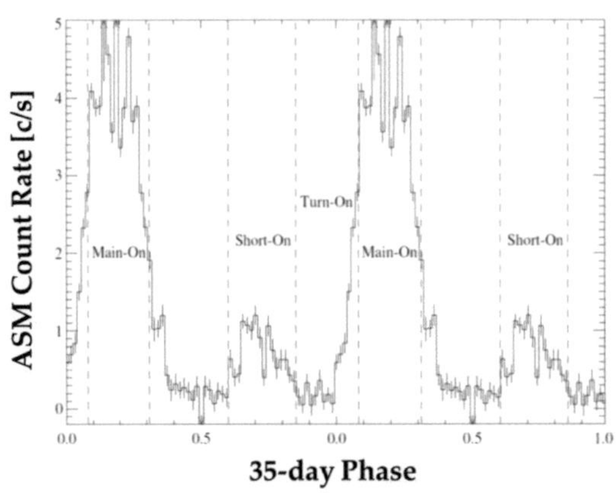

Figure 8. Light curve of the variability over 35 d, due to an accretion disk surrounding Her X-1 ([19] and reference therein).

data points and thus poor statistics) available, it was not possible to perform neither FFT nor EFS analysis.

In a first approximation, the expected orbital light curve of Hz Herculis, illustrated in **Figure 7**, can be considered to be well described by a sinusoid, although its shape is more complex. The orbital light curve reconstructed by analysing all data belonging to the four runs is reported in **Figure 9**. The plot shows four data points representing the instrumental magnitudes computed by averaging the total count of each observation as a function of time expressed in MJD unit.

The value of the orbital period obtained by sinusoidally fitting data was found to be equal to (1.70017 ± 0.00058) d, in an optimal agreement with that expected one, differing by only $3.6 \cdot 10^{-6}$ d. It is worth considering that this result could not be directly compared with that represented in **Figure 7** because observations done with SiFAP were performed without using any filter. The resulting plot thus shows a profile integrated all over the optical band, the region of the electromagnetic spectrum where PDE of MPPC spans.

We also tried to reconstruct the rotational light curve without any success. This was due to two main factors. The first one concerned the poor knowledge about optical ephemeris of the system. It is known that the optical counterpart of Hz Her/Her X-1 system arises from the reprocessing of X-ray radiation. Therefore, X-ray ephemeris reported in **Table 2** was not suitable to correct ToAs for the orbital motion of the system. In this way, we were not able to merge efficiently acquired data in order to increase the statistics and thus the S/N ratio. In fact, putting together observations which have not been previously corrected for the Doppler effect would smear pulsations out. This happens because spin periods (frequencies) are varying during the orbital period.

The second constraint was due to the physical condition of the binary system itself. Unfortunately, we estimated that our observations were performed when the system was in a not optimal orbital phase for being observed. Moreover, the periodic precession of the disk (35 d modulation) played an unfavourable role because the system was in the state with a very low intensity.

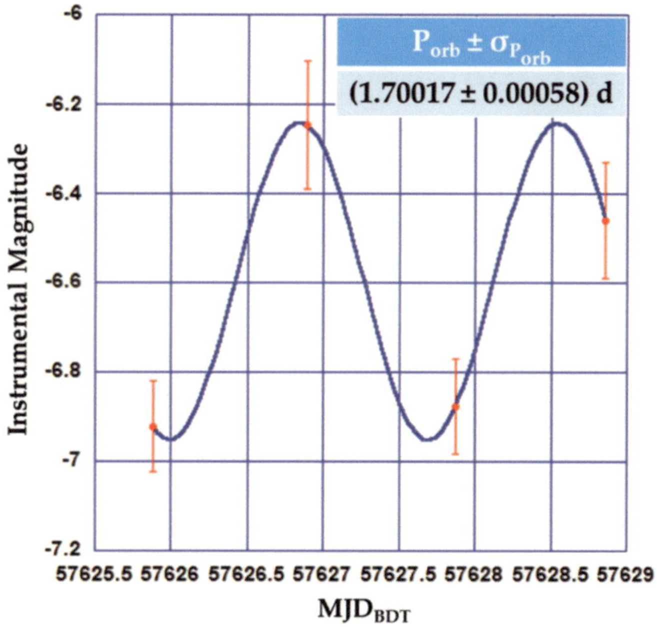

Figure 9. Light curve of the orbital period of Hz Her. Four data points represent the instrumental magnitude computed by averaging the total count of each observation as a function of time.

7. Conclusions

In this chapter, we showed how astronomical variable sources could be studied by photon counting through the (optical) high-speed photometry technique with ground-based telescopes and instrumentation. The most difficult effort to obtain correct results is analysing data taking into account several perturbation factors such as the Earth's not inertial reference frame system and orbital motion of binary systems. Although all these effects could deteriorate measurements, anyway very good and promising results can be obtained by applying above-described time corrections.

In this scenario, the SiFAP instrument gave its contribution with interesting results, despite more specific and deeper knowledge on this kind of variable sources can be achieved, thanks to multiwavelength (even simultaneous) measurements. In fact, the opportunity to observe such targets in more than one electromagnetic band is very useful to understand both the physics and emission mechanisms and their possible correlation.

Acknowledgements

The authors acknowledge the research project Protocol C26A15YCJ4 funded by the Department of Physics of the 'Sapienza University of Rome' for the financial support.

Moreover, the authors would remind that results obtained by SiFAP on the Crab pulsar are based on observations made with the Italian 3.58 m Telescopio Nazionale Galileo (TNG) operated on the island of La Palma by the Fundación Galileo Galilei of the INAF (Istituto Nazionale di Astrofisica) at the Spanish Observatorio del Roque de los Muchachos of the Instituto de Astrofisica de Canarias.

In addition, the authors would also like to thank the 1.52 m Cassini telescope, which is run by INAF-Osservatorio Astronomico di Bologna at Loiano (Bologna), supported measurements on the binary system Hz Her/Her X-1 performed with SiFAP.

Author details

Filippo Ambrosino[1,2]* and Franco Meddi[2]

*Address all correspondence to: filippo.ambrosino@roma1.infn.it

1 INAF-IAPS, Rome, Italy

2 Sapienza University of Rome, Rome, Italy

References

[1] Lorimer DR, Kramer M. Handbook of Pulsar Astronomy. 4th ed. Cambridge University Press; 2004. p. 50, 51, 85-86, 206-208. ISBN 978-0-521-82823-9

[2] Liu QZ, van Paradijs J, van den Heuvel EPJ. A catalogue of low-mass X-ray binaries in the galaxy, LMC, and SMC (fourth edition). Astronomy and Astrophysics. 2007;**469**(2):807-810. DOI: 10.1051/0004-6361:20077303

[3] Meddi F, Ambrosino F, Nesci R, Rossi C, Sclavi S, Bruni I, Ruggieri A, Sestito S. A new fast silicon photomultiplier photometer. Publications of the Astronomical Society of the Pacific. 2012;**124**(915):448-453. DOI: 10.1086/665925

[4] Ambrosino F, Meddi F, Nesci R, Rossi C, Sclavi S, Bruni I. SiFAP: A simple sub-millisecond astronomical photometer. Journal of Astronomical Instrumentation. 2013;**2**(1):1350006. DOI: 10.1142/S2251171713500062

[5] Ambrosino F, Cretaro P, Meddi F, Rossi C, Sclavi S, Bruni I. The latest version of SiFAP: Beyond microsecond time scale photometry of variable objects. Journal of Astronomical Instrumentation. 2016;**5**(3):1650005-1651267. DOI: 10.1142/S2251171716500057

[6] Huang T-Y, Xu B-X, Zhu J, Zhang H. The concepts of international atomic time (TAI) and terrestrial dynamic time (TDT). Astronomy and Astrophysics. 1989;**220**(1-2):329-334

[7] McCarthy DD, Hackman C, Nelson RA. The physical basis of the leap second. The Astronomical Journal. 2008;**136**(5):1906-1908. DOI: 10.1088/0004-6256/136/5/1906

[8] Fairhead L, Bretagnon P. An analytical formula for the time transformation TB-TT. Astronomy and Astrophysics. 1990;**229**(1):240-247

[9] Lundholm, M.; Mikhalev, V.; Nilsson, P.. [dissertation]. Time Correction for the PoGOLite Project: 2011. 53 pp. Available from: http://www.diva-portal.org/smash/get/diva2:571972/FULLTEXT02

[10] Wilhelm K, Dwivedi BN. On the gravitational redshift. New Astronomy. 2014;**31**:8-13. DOI: 10.1016/j.newast.2014.01.012

[11] Ambrosino F. GUIDA: A graphical user Interface for optical data analysis of isolated pulsars. Publications of the Astronomical Society of the Pacific. 2015;**127**(955):931-939. DOI: 10.1086/683014

[12] Staubert R, Klochkov D, Wilms J. Updating the orbital ephemeris of Hercules X-1; rate of decay and eccentricity of the orbit. Astronomy and Astrophysics. 2009;**500**(2):883-889. DOI: 10.1051/0004-6361/200911690

[13] Stelzer B, Staubert R, Wilms J, Geckeler RD, Gruber D, Rothschild R. Evolution of the orbital period of her X-1: Determination of a new ephemeris using RXTE data. In: The Fourth Compton Symposium; AIP Conference Proceedings. 1997. pp. 753-757. DOI: 10.1063/1.54147

[14] Leahy DA, Elsner RF, Weisskopf MC. On searches for periodic pulsed emission - the Rayleigh test compared to epoch folding. The Astrophysical Journal. 1983;**272**(Sept. 1):256-258. DOI: 10.1086/161288

[15] Leahy DA. Searches for pulsed emission: Improved determination of period and amplitude from epoch folding for sinusoidal signals. Astronomy and Astrophysics. 1987;**180**(1-2):275-277

[16] Zeilik M, Gregory S. Introductory Astronomy and Astrophysics. 4th ed. Saunders College Publishing; 1998. p. 369. ISBN 0-03-006228-4

[17] Ambrosino F, Meddi F, Rossi C, Sclavi S, Nesci R, Bruni I, Ghedina A, Riverol L, Di Fabrizio L. Fast multichannel astronomical photometer based on silicon photo multipliers mounted at the Telescopio Nazionale Galileo. In: Ground-based and Airborne Instrumentation for Astronomy V; Proceedings of the SPIE. 2014;**9147**(91478R):10. DOI: 10.1117/12.2064649

[18] Reynolds AP, Quaintrell H, Still MD, Roche P, Chakrabarty D, Levine SE. A new mass estimate for Hercules X-1. Monthly Notices of the Royal Astronomical Society. 1997;**288**(1): 43-52. DOI: 10.1093/mnras/288.1.43

[19] Available from: http://d-nb.info/971486271/34

Chapter 5

Computational Methods for Photon-Counting and Photon-Processing Detectors

Luca Caucci, Yijun Ding and Harrison H. Barrett

Additional information is available at the end of the chapter

http://dx.doi.org/10.5772/intechopen.72151

Abstract

We present computational methods for attribute estimation of photon-counting and photon-processing detectors. We define a photon-processing detector as any imaging device that uses maximum-likelihood methods to estimate photon attributes, such as position, direction of propagation and energy. Estimated attributes are then stored at full precision in the memory of a computer. Accurate estimation of a large number of attributes for each collected photon does require considerable computational power. We show how mass-produced graphics processing units (GPUs) are viable parallel computing solutions capable of meeting the required computing needs of photon-counting and photon-processing detectors, while keeping overall costs affordable.

Keywords: photon-processing detectors, maximum-likelihood estimation, GPU, parallel processing, gamma-ray photons, charged particles

1. Introduction

In broad terms, detectors used in imaging can be classified into a small number of categories depending on their working principles. These categories include integrating detectors, pixelated photon-counting detectors as well as a new class of detectors, which we refer to as photon-processing detectors.

An integrating detector measures charges accumulated at each pixel location. These charges are induced by light impinging on the detector and are proportional to the average number of photons incident on each pixel. Dedicated circuitry reads out these changes and converts them to numbers roughly proportional to the charge accumulated at each pixel.

A photon-counting detector works by counting the number of photoelectric interactions observed during the exposure time. Count registers associated with each pixel are read at the end of the exposure time, thus making the output of photon-counting detectors a collection of pixel counts.

A photon-processing detector may use any existing detector technology to measure several "attributes" for each photon entering the detector. Attributes include the photon position, its direction of propagation and the amount of energy it deposited in the detector. This is accomplished by reengineering the detector design so that additional information can be extracted from raw unprocessed data. Important aspects of any photon-processing detector include the algorithm used to estimate photon attributes from raw data as well as how these attributes are represented and stored in the memory of a computer.

Particle-processing detectors are a variation on photon-processing detectors and are designed to detect charged particles, such as alpha and beta particles. Particle-processing detectors enable a new imaging technique—called charged-particle emission tomography (CPET)—which attains high-resolution 3D imaging in living organisms so long as accurate estimation of parameters for each charged particle is available.

This chapter is organized as follows. Section 2 provides an overview of detectors suitable for photon counting and photon processing. Maximum-likelihood estimation (MLE) and its properties are discussed in some detail in Section 3. The next section—Section 4—introduces graphics processing units (GPUs) and the compute unified device architecture (CUDA) programming model. Section 5 presents algorithms for photon-counting detectors, while Section 6 discusses photon- and particle-processing detectors and presents fast GPU-based algorithms for maximum-likelihood estimation of photon parameters. Finally, Section 7 summarizes this chapter and discusses possible applications of photon-processing detectors.

A portion of this chapter has been adapted from Y. Ding, *"Charged-Particle Emission Tomography"* [1].

2. Detectors for photon counting and photon processing

2.1. Gamma-ray cameras

Gamma-ray cameras are used in nuclear medicine to image gamma-ray photons emitted by radioactive elements. The first gamma-ray camera was developed by Hal Oscar Anger in 1957 [2]. Anger's original design, often referred to as an "Anger camera," is still widely used today. A diagram of an Anger camera is provided in **Figure 1**.

An Anger camera includes a scintillation crystal, a light guide and an array of photomultiplier tubes (PMTs). When a gamma-ray photon interacts with the scintillation crystal, a burst of visible-light photons is produced. Some of these photons propagate through the crystal and the light guide and enter one or more PMTs. When a photon enters a PMT and interacts with it, a measurable electrical signal in the form of a narrow current pulse is produced. This pulse is transmitted to amplifying electronics, so that it can be analyzed. A transimpedance amplifier converts the current pulse to voltage. A shaping amplifier further amplifies the signal and reshapes it, making it broader and smoother. A broad signal is easier to sample via an analog-to-digital converter. The output of the analog-to-digital converter can be scaled to obtain an

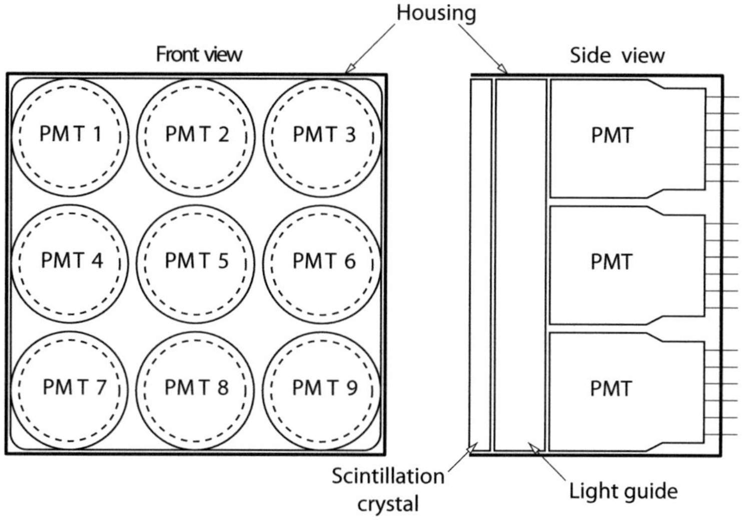

Figure 1. Diagram of a typical gamma-ray camera (adapted from [3]).

integer number representing the number of photons entering the PMT. Digitized samples collected from each of the K PMTs are then scanned for events. Detected events are stored in the memory of a computer in the form of scaled PMT samples $g_1, \ldots, g_K$.

Detailed analysis of the physical processes that take place inside the scintillation crystal and each PMT allows us to derive a statistical model for the scaled PMT samples $g_1, \ldots, g_K$ produced by a gamma-ray camera with K PMTs. Because of noise, PMT samples $g_1, \ldots, g_K$ can be thought of random variables. If we normalize each PMT signal by the gain of the PMTs and we ignore the noise in the gain, random variables $g_1, \ldots, g_K$ can be shown to be conditionally independent and to follow Poisson statistics with means, respectively, $\overline{g}_1(\boldsymbol{R}, E), \ldots, \overline{g}_K(\boldsymbol{R}, E)$ [4]. Thus, we can write:

$$\mathrm{pr}(g_1, \ldots, g_K \mid \boldsymbol{R}, E) = \prod_{k=1}^{K} \frac{[\overline{g}_k(\boldsymbol{R}, E)]^{g_k} \exp\left[-\overline{g}_k(\boldsymbol{R}, E)\right]}{g_k!}. \qquad (1)$$

Functions $\overline{g}_1(\boldsymbol{R}, E), \ldots, \overline{g}_K(\boldsymbol{R}, E)$ are called mean detector response functions (MDRFs), and they describe the mean detector response upon detection of a gamma-ray photon with energy E at location $\boldsymbol{R}$.

2.2. Semiconductor detectors for charged particles

Semiconductor pixelated detectors can be used to measure position and energy of charged particles, including alpha and beta particles. One possible detector configuration consists of a layer of semiconductor material (which we refer to as the "active volume"), a set of anodes placed on one side of the detector's active volume, and some data-processing circuitry (such as application-specific integrated circuits or ASICs) that measures the anode signals and converts them into digital signals.

When a charged particle enters the detector's active volume and deposits some of its energy, electron-hole pairs are produced along the particle's track. The electrons and holes drift in opposite directions under an electric field applied throughout the detector's active volume. This process is accompanied by the production of electrical charges, which are collected by electrodes on one side of the detector's active volume. These charges are then converted to digital signals (e.g., number of electron-hole pairs produced) and are either sent to a computer or accumulated in count registers to form an image.

An example of a semiconductor pixelated detector for alpha and beta particle is the Medipix2 sensor (**Figure 2**) developed at CERN [5]. The Medipix2 sensor features an array of 256 × 256 square pixels of size 55 μm. The counter in each pixel of a Medipix2 sensor can record the duration of an event that is above a threshold, from which the energy collected at each pixel and the particle's residual energy can be measured.

A statistical model for the data produced by a semiconductor detector for charged particles (such as the Medipix2 sensor) must take into account the so-called charge sharing effect [6] as well as many variables, including particle's position $\mathbf{R}$ and energy E, its angle of incidence (denoted as the unit vector $\vec{s}$) and bias voltage V_{bias} applied across the semiconductor. Some recent results for the Medipix2 sensor have been reported in [1, 7]. When a highly energetic particle enters the detector, a large number of charges will be collected at its electrodes. In such a case, the statistics of pixel outputs $g_1, \ldots, g_M$ (where M denotes the number of detector pixels) conditioned on $\mathbf{R}$, E, $\vec{s}$ and V_{bias} approach Gaussian statistics, and we can write:

$$\mathrm{pr}(g_1, \ldots, g_M | \mathbf{R}, E, \vec{s}, V_{bias}) = \prod_{m=1}^{M} \frac{1}{\sqrt{2\pi\sigma_m^2\left(\mathbf{R}, E, \vec{s}, V_{bias}\right)}} \exp\left[-\frac{\left(g_m - \overline{g}_m\left(\mathbf{R}, E, \vec{s}, V_{bias}\right)\right)^2}{2\sigma_m^2\left(\mathbf{R}, E, \vec{s}, V_{bias}\right)} \right], \tag{2}$$

in which $\overline{g}_m\left(\mathbf{R}, E, \vec{s}, V_{bias}\right)$ is the mean of the mth pixel and $\sigma_m\left(\mathbf{R}, E, \vec{s}, V_{bias}\right)$ is the standard deviation of g_m.

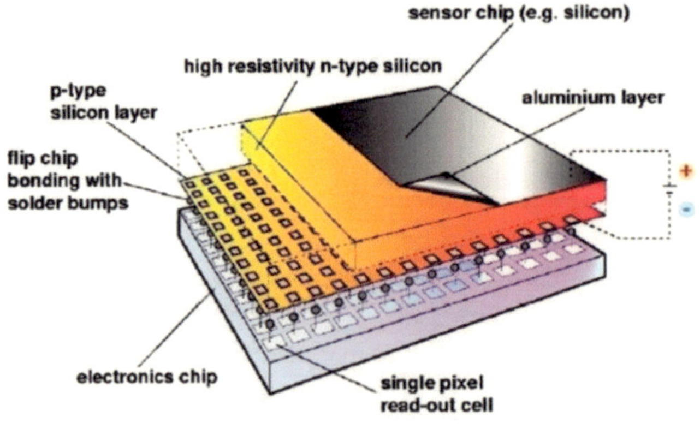

Figure 2. Diagram of the Medipix2 chip sensor (https://medipix.web.cern.ch).

2.3. Intensified charge-coupled detectors

A charge-coupled detector (CCD) is a semiconductor device that produces a pixelated image by converting incoming photons into electric charges, which are then stored at each pixel location. These charges are induced by photons with energy exceeding the semiconductor bandgap. The most general form for the mean output $\overline{g}_m$ (calculated by imaging the same object over and over again) is [8]:

$$\overline{g}_m = \int_{det} d^2R \int_0^{\infty} dE \int_{hemi} d\Omega \int_0^T dt\, \eta_m\left(\mathbf{R}, E, \vec{s}\right) L\left(\mathbf{R}, E, \vec{s}, t\right), \tag{3}$$

in which m varies from 1 to the total number of pixels M, $\eta_m\left(\mathbf{R}, E, \vec{s}\right)$ is the quantum efficiency at pixel m, point $\mathbf{R}$ on detector face, photon energy E and along direction $\vec{s}$. The function $L\left(\mathbf{R}, E, \vec{s}, t\right)$ is the spectral photon radiance at point $\mathbf{R}$ for photon energy$E$, time t and along direction $\vec{s}$ [8, 9]. In Eq. (3), the spatial extent of the detector was denoted as "det" and "$\int_{\text{hemi}} d\Omega$" means integration over all the possible directions $\vec{s}$ incident on the detector. Finally, integration over the time variable t starts at time $t = 0$ and ends at time $t = T$.

An intensified charge-coupled detector (ICCD) uses an image intensifier (such as a microchannel plate (MCP)) to amplify scintillation light before imaging it onto a CCD sensor. The image intensifier provides optical gain (in the range from 10^5 to 10^6 or more) so that low-energy optical photons (emitted, e.g., upon interaction of a charged particle with a scintillator) can be imaged with practically any CCD sensor. Lenses, usually placed between the image intensifier and the CCD sensor, reimage the image intensifier's output window on the CCD sensor. Examples of intensified charge-coupled detectors include the iQID sensor developed at the University of Arizona by Brian W. Miller [10].

A proper statistical model for an intensified charge-coupled detector must consider both the statistics of the output produced by the image intensifier and the statistics of the data produced by the CCD sensor. To find a model for the image intensifier, we begin by noticing that each point in the CCD sensor can be propagated back through the lenses all the way to the entrance face of the image intensifier. Therefore, we can consider the number of photons p_m impinging on the image intensifier at locations that fall within pixel m on the CCD sensor. Under broad conditions, we can show that p_m obeys Poisson statistics and we denote the mean of p_m as $\overline{p}_m$. For large enough $\overline{p}_m$, the statistics of p_m are approximatively Gaussian.

A general expression that relates $\overline{p}_m$ to the sensor output $\overline{g}_m$ takes the form:

$$\overline{g}_m = \overline{A}\overline{p}_m, \tag{4}$$

in which $\overline{A}$ denotes the mean of the image intensifier amplification (gain) A . The variance, σ_m^2, of $\overline{g}_m$ is related to $\overline{p}_m$ and the statistics of A as follows [7]:

$$\sigma_m^2 = \overline{p}_m\left(\sigma_A^2 + \overline{A}^2\right) + \sigma_{\text{read}}^2, \tag{5}$$

in which σ_A^2 is the variance of the amplification A and σ_{read}^2 denotes the variance of the sensor's readout noise. If the blur introduced by the image intensifier and optics located between the image intensifier and the CCD sensor is smaller than the size of a sensor pixel, then output g_m is independent on $g_{m'}$ for any $m' \neq m$. If we further assume that the amplification A and the readout noise also obey Gaussian statistics, we can write [1, 7]:

$$\text{pr}(g_1, \dots, g_M) = \prod_{m=1}^{M} \frac{1}{\sqrt{2\,\pi\,\sigma_m^2}} \exp\left[-\frac{\left(g_m - \overline{g}_m\right)^2}{2\,\sigma_m^2}\right]. \tag{6}$$

3. Maximum-likelihood estimation

Maximum-likelihood estimation (MLE) is a statistical method that uses observed noisy data to estimate model parameters. For a good historical treatment of the concept of maximum-likelihood estimation, the interested reader can consult [11]. Given a set of observed data and an underlying model (which depends on some unknown parameters), MLE calculates the values of the parameters that better explain the observed data. The observed data that are used for maximum-likelihood estimation are realizations of random variables. Thus, parameters we estimate from these data are realizations of random variables as well.

Maximum-likelihood estimation can, in principle, be used with all the detectors discussed above. For example, we show how maximum-likelihood estimation is used to estimate position of interaction from PMT data, and we discuss an efficient parallel algorithm for it. Moreover and as we argue in Section 6, maximum-likelihood estimation is the estimation method of choice for photon-processing detectors.

3.1. Mathematical description

Let us denote the parameters we want to estimate as the vector $\boldsymbol{\theta}$. The model itself is characterized by a probability density function (PDF), denoted as $\text{pr}(\boldsymbol{x}|\boldsymbol{\theta})$. We use the vector $\boldsymbol{x}$ to refer to the *complete data*, while we denote the *incomplete data* as $\boldsymbol{y}$ [3, 8]. We stress that we do not directly observe $\boldsymbol{x}$, but only indirectly and through the vector $\boldsymbol{y}$. Vectors $\boldsymbol{x}$ and $\boldsymbol{y}$ are statistically related via the PDF $\text{pr}(\boldsymbol{y}|\boldsymbol{x})$. Probability density functions $\text{pr}(\boldsymbol{x}|\boldsymbol{\theta})$ and $\text{pr}(\boldsymbol{y}|\boldsymbol{x})$ allow us to write

$$\text{pr}(\boldsymbol{y}|\boldsymbol{\theta}) = \int pr(\boldsymbol{y}|\boldsymbol{x})\, pr(\boldsymbol{x}|\boldsymbol{\theta})\, d\boldsymbol{x}, \tag{7}$$

in which $\text{pr}(\boldsymbol{y}|\boldsymbol{\theta})$ is the PDF of the observed data $\boldsymbol{y}$ given the parameter $\boldsymbol{\theta}$. Eq. (7) above takes into account two separate "mechanisms" that, when concatenated, produce a sample $\boldsymbol{y}$ from the value of $\boldsymbol{\theta}$. The first mechanism produces the complete data $\boldsymbol{x}$ according to $\text{pr}(\boldsymbol{x}|\boldsymbol{\theta})$, while the second mechanism samples $\text{pr}(\boldsymbol{y}|\boldsymbol{x})$ to produce the incomplete data $\boldsymbol{y}$.

MLE solves the estimation problem by finding the vector $\boldsymbol{\theta}$ that maximized the *likelihood* $\text{L}(\boldsymbol{\theta};\boldsymbol{y})$ for observed data $\boldsymbol{y}$. Mathematically, this concept is formalized as:

$$\widehat{\boldsymbol{\theta}}_{ML} = \text{argmax}_{\boldsymbol{\theta}}\ \text{pr}(\boldsymbol{y}|\boldsymbol{\theta}) = \text{argmax}_{\boldsymbol{\theta}}\ \text{L}(\boldsymbol{\theta};\boldsymbol{y}) \tag{8}$$

in which the "hat" symbol denotes an estimated quantity, and we have defined the likelihood as:

$$\mathrm{L}(\boldsymbol{\theta}; \boldsymbol{y}) = \mathrm{pr}(\boldsymbol{y}|\boldsymbol{\theta}). \tag{9}$$

Likelihood $\mathrm{L}(\boldsymbol{\theta}; \boldsymbol{y})$ has to be interpreted as a function of $\boldsymbol{\theta}$ for fixed (measured) $\boldsymbol{y}$. In Eq. (8), we used "$\mathrm{argmax}_{\boldsymbol{\theta}}\, \mathrm{L}(\boldsymbol{\theta}; \boldsymbol{y})$" to denote the value of $\boldsymbol{\theta}$ that maximizes the likelihood. Because $\boldsymbol{y}$ is the result of a noisy measurement, the actual value of $\boldsymbol{y}$ in Eq. (8) will change if the measurement is repeated. In other words, $\boldsymbol{y}$ is a random quantity, and this implies that the ML estimate $\widehat{\boldsymbol{\theta}}_{\mathrm{ML}}$ is random as well.

Alternatively, $\widehat{\boldsymbol{\theta}}_{\mathrm{ML}}$ can be calculated by rewriting Eq. (8) as

$$\widehat{\boldsymbol{\theta}}_{ML} = \mathrm{argmax}_{\boldsymbol{\theta}} \ln\left[\mathrm{pr}(\boldsymbol{y}|\boldsymbol{\theta})\right] = \mathrm{argmax}_{\boldsymbol{\theta}}\, \ell(\boldsymbol{\theta}; \boldsymbol{y}), \tag{10}$$

in which we have introduced the log-likelihood [8]

$$\ell(\boldsymbol{\theta}; \boldsymbol{y}) = \ln\left[\mathrm{L}(\boldsymbol{\theta}; \boldsymbol{y})\right]. \tag{11}$$

Because the logarithm is a strictly monotonic function, the expression in Eq. (10) is equivalent to the one in Eq. (8). Often, the log-likelihood $\ell(\boldsymbol{\theta}; \boldsymbol{y})$ is numerically easier to calculate with a computer than the likelihood $\mathrm{L}(\boldsymbol{\theta}; \boldsymbol{y})$.

3.2. Properties of ML estimates

Maximum-likelihood estimates have many desirable properties. Some of these properties are summarized below.

- **Asymptotic efficiency.** If $\boldsymbol{y}$ represents a set of repeated independent and identically distributed samples $y_1, \ldots y_M$, asymptotic efficiency of MLE implies that, as M increases, the variance of each component of $\widehat{\boldsymbol{\theta}}_{\mathrm{ML}}$ converges to the smallest possible value, which is given by the *Cramér-Rao lower bound* [12, 13].

- **Functional invariance.** Assume the ML estimate of a parameter vector $\boldsymbol{\theta}$ is $\widehat{\boldsymbol{\theta}}_{\mathrm{ML}}$ and consider a function $u(\boldsymbol{\theta})$ of $\boldsymbol{\theta}$. We can identify $u(\boldsymbol{\theta})$ with the parameter vector $\boldsymbol{\mu}$, and we can consider a maximum-likelihood estimate $\widehat{\boldsymbol{\mu}}_{\mathrm{ML}}$ of $\boldsymbol{\mu}$. Then [14]

$$\widehat{\boldsymbol{\mu}}_{\mathrm{ML}} = u\left(\widehat{\boldsymbol{\theta}}_{\mathrm{ML}}\right). \tag{12}$$

  This equation shows that the property of being a maximum-likelihood estimate is preserved if we consider a function of the maximum-likelihood estimate itself.

- **Sufficiency.** Any quantity $T(y_1, \ldots, y_M)$ calculated from samples $y_1, \ldots, y_M$ and used to estimate an unknown parameter vector $\boldsymbol{\theta}$ is said to be a *sufficient statistic* for $y_1, \ldots, y_M$ if no other quantity that can be calculated from the same samples would provide additional information regarding the value of the parameter vector $\boldsymbol{\theta}$. In simple terms, a sufficient statistic is a function of the samples $y_1, \ldots, y_M$ that "compresses" them without losing any information about $\boldsymbol{\theta}$. Sufficiency for a maximum-likelihood estimate $\widehat{\boldsymbol{\theta}}_{\mathrm{ML}}$ can be stated by saying that $\widehat{\boldsymbol{\theta}}_{\mathrm{ML}}$ is a function of a sufficient statistic for $\boldsymbol{\theta}$ [15].

- **Consistency.** Consistency of an estimator regards the behavior of the estimator as the sample size M increases. Consider the case in which $\boldsymbol{y}$ is a set of repeated independent and identically distributed samples $y_1, \ldots, y_M$. It is possible to show that, when the range of the elements of $\boldsymbol{y} = (y_1, \ldots, y_M)$ does not dependent on the parameter vector $\boldsymbol{\theta}$, there exists a maximum-likelihood estimate $\widehat{\boldsymbol{\theta}}_{\mathrm{ML}}$ that, as M increases, converges in probability to the true value of the parameter vector. A consistent maximum-likelihood estimate is unique [16].

- **Asymptotic normality.** Because the ML estimate $\widehat{\boldsymbol{\theta}}_{\mathrm{ML}}$ of $\boldsymbol{\theta}$ is a random variable, it makes sense to consider its probability density function. As the sample size M increases, the probability density function of $\widehat{\boldsymbol{\theta}}_{\mathrm{ML}}$ converges to the probability density function of a normally distributed random variable with mean equal to the true value of the parameter we want to estimate and covariance matrix equal to the inverse of the *Fisher information matrix* [17].

4. Graphics processing units and CUDA

Driven by the insatiable demand for real-time rendering in gaming and entertainment, graphics processing units (GPUs) have become highly parallel devices capable of running general-purpose code. Newer products that offer an ever-increasing amount of computational power are constantly introduced in the market at very competitive prices.

Programming languages have been developed to harness the parallel capabilities of GPU devices. The most widespread language for GPU programming is called compute unified device architecture (CUDA), which was introduced in 2006 by NVIDIA. Due to its similarity to C, CUDA has rapidly become the *de facto* programming language for GPUs.

4.1. The CUDA programming model

In CUDA, the GPU is usually referred to as the *device* and the computer that hosts it is referred to as the *host*. Many GPU devices can be installed in the same host, and it is not uncommon to have systems with more than one GPU device. Each GPU device has its own memory, which we refer to as *device memory*. In CUDA, it is also common to refer to the memory installed in the host as *host memory*. CUDA provides library functions to allocate blocks of memory in device memory and to transfer blocks of data from host memory to device memory and vice versa. As shown in **Figure 3**, a typical GPU device includes some GPU cores (ranging in number from a few hundreds to a few thousands) and some control logic.

Programmers access the parallel capabilities of a CUDA-enabled device by writing *kernels*, which are pieces of code that look very similar to regular C functions. In CUDA, a kernel is run in parallel on many different GPU cores. A kernel in execution is referred to as a *thread*. Threads are grouped into blocks, which can be 1D, 2D or 3D, and blocks are grouped into a grid. Grids can be 1D, 2D or 3D. The size and dimensionality of blocks and grids are decided

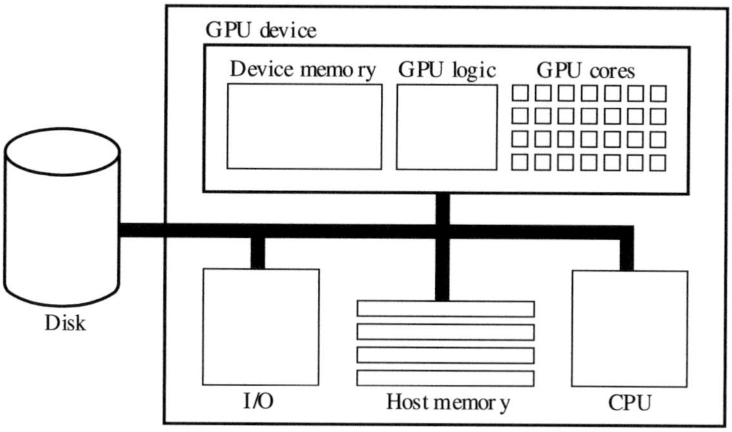

Figure 3. Diagram of a computer equipped with a GPU device (adapted from [3]).

by the programmer via an *execution configuration,* which is also used to call a kernel and instruct the GPU hardware to execute threads.

In a GPU device, thread scheduling is extremely efficient and it is performed by the hardware and without the intervention of the programmer. To improve performance, the hardware also suspends execution for threads that are waiting for completion of memory transfers between device memory and GPU registers. When that happens, the hardware selects for execution threads that already have data to be processed. The programmer is typically unaware of what threads are running at any given time, nor does he know what kernel instruction is being executed by a specific thread. In other words, the programmer cannot rely on any particular scheduling order of GPU threads. There are, however, situations in which it is necessary to ensure that a block of threads has reached a certain instruction in a kernel before all the threads in the block can continue. In CUDA, this is accomplished via *synchronization barriers*.

Synchronization barriers are often used when threads have to exchange data with each other via shared variables. Without any synchronization mechanism, a thread will not be able to know if the content of a shared variable has already been written by a cooperating thread. Synchronization barriers solve this problem by suspending thread execution until *all* the threads in the same block have reached a synchronization barrier. In CUDA, synchronization barriers are allowed only among the threads in the same block.

GPU devices are equipped with different memory spaces. This includes *global memory* (which is used to share data between the host and the device) as well as *shared memory*. While global memory is rather slow and physically separated from the GPU cores, shared memory is much faster and it is built on the same chip as the GPU cores. Threads use shared memory to efficiently share data among them.

Another type of memory space available in a GPU device is *texture memory*. As the name suggests, texture memory has been designed to speed up and facilitate 3D rendering in computer games and computer-generated scenes. This is the reason why texture memory supports unique features including hardware-based on-the-fly interpolation of texture data.

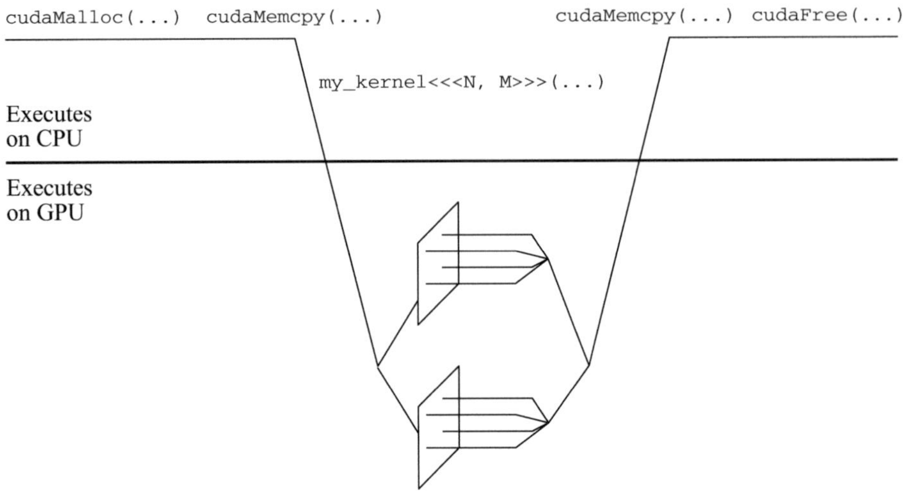

Figure 4. Workflow of a CUDA application (adapted from [3]).

4.2. Workflow of a CUDA application

The basic steps that are needed to execute a kernel are summarized in **Figure 4**.

In a typically CUDA application, one or more blocks of device memory are allocated by the host via the **cudaMalloc(...)** CUDA library function. The host then copies input data from host memory via one or more **cudaMemcpy(...)** function calls. Kernel execution is started with a call of the form **my_kernel < <<N, M>> > (...)**, in which **my_kernel** is the name of the kernel, **N** is the grid size and **M** is the block size. Parameters, such as pointers to device memory, are passed to the kernel as parameters enclosed in parentheses. Once *all* the threads have finished executing, the control returns to the CPU. Results can be copied from device memory to host memory via one or more calls to **cudaMemcpy(...)**. Finally, device memory that was previously allocated is released via the **cudaFree(...)** call.

The CUDA environment automatically defines read-only built-in variables that can only be used in a kernel. These variables include **blockIdx**, **blockDim** and **threadIdx**. The variable **threadIdx** enumerates threads within each block in ascending order and starting from 0. Similarly, **blockIdx** enumerates blocks within the grid. The size of each block is contained in the variable **blockDim**. Built-in variables are used by the programmer to calculate which element(s) of the input array(s) a thread has to work on, or where in device memory a result has to be stored.

5. Algorithms for photon-counting detectors

To make our discussion more concrete, we begin this section by considering a GPU algorithm for maximum-likelihood estimation (MLE) of position of interaction of gamma-ray photons in an Anger camera. We then comment on ways to adapt our algorithm to other cases, including photon-counting and photon-processing detectors (Section 6).

We showed in Section 2.1 that digitized PMT signals $g_1, \dots, g_K$ obey Poisson statistics and we denoted the means of $g_1, \dots, g_K$ as $\overline{g}_1(\boldsymbol{R}, E), \dots, \overline{g}_K(\boldsymbol{R}, E)$, respectively. If photon energy E is known, the likelihood for the estimation of position $\boldsymbol{R}$ under the assumption of Poisson noise is written as:

$$\mathrm{L}(\boldsymbol{R}; g_1, \dots, g_K, E) = \prod_{k=1}^{K} \frac{[\overline{g}_k(\boldsymbol{R}, E)]^{g_k} \exp\left[-\overline{g}_k(\boldsymbol{R}, E)\right]}{g_k!}. \tag{13}$$

Functions $\overline{g}_1(\boldsymbol{R}, E), \dots, \overline{g}_K(\boldsymbol{R}, E)$ are called mean detector response functions (MDRFs) and they can be either measured, derived analytically or estimated via simulation codes. Using Eq. (13), an ML estimate $\widehat{\boldsymbol{R}}_{\mathrm{ML}} = (\widehat{x}_{\mathrm{ML}}, \widehat{y}_{\mathrm{ML}})$ of $\boldsymbol{R} = (x, y)$ can be found as:

$$\widehat{\boldsymbol{R}}_{\mathrm{ML}} = \mathrm{argmax}_{\boldsymbol{R}} \mathrm{L}(\boldsymbol{R}; g_1, \dots, g_K, E). \tag{14}$$

Equivalently, we can consider the logarithm of $\mathrm{L}(\boldsymbol{R}; g_1, \dots, g_K, E)$ in the maximization step and write:

$$\widehat{\boldsymbol{R}}_{\mathrm{ML}} = \mathrm{argmax}_{\boldsymbol{R}} \sum_{k=1}^{K} \left\{ g_k \log\left[\overline{g}_k(\boldsymbol{R}, E)\right] - \overline{g}_k(\boldsymbol{R}, E) \right\}, \tag{15}$$

in which we omitted the $\log(g_k!)$ term as it does not depend on $\boldsymbol{R}$ and, therefore, it will not affect the estimate $\widehat{\boldsymbol{R}}_{\mathrm{ML}}$.

The algorithm we present here uses the fact that, for fixed $g_1, \dots, g_K$ and E, the log-likelihood $\ell(\boldsymbol{R}; g_1, \dots, g_K, E) = \log \mathrm{L}(\boldsymbol{R}; g_1, \dots, g_K, E)$ is a smooth function of $\boldsymbol{R}$. Hence, maximum-likelihood estimate $\widehat{\boldsymbol{R}}_{\mathrm{ML}}$ can be searched for in an iterative fashion by first evaluating $\ell(\boldsymbol{R}; g_1, \dots, g_K, E)$ over a coarse grid of S_x-by-S_y points that uniformly spans the whole detector space. The point of the grid that maximizes $\ell(\boldsymbol{R}; g_1, \dots, g_K, E)$ is used in the next iteration as the center of a new grid smaller than the previous one by a factor $\alpha > 1$. This process is repeated M times. We refer this algorithm as the *contracting grid algorithm* [18, 19].

Figure 5 shows pseudocode for a possible GPU implementation of the contracting grid algorithm. We used superscripts to make it clear on which memory space a given variable is stored. Variables with no superscript will denote either numerical constants (such as the contracting factor α) or local variables, typically stored in GPU registers.

The algorithm of **Figure 5** assumes that an array of R PMT sample vectors $g_0, \dots g_{R-1}$ is available. These data are stored in device memory and we decided to use in our GPU implementation a grid of size $R \times 1 \times 1$ with 2D blocks of size $S_x \times S_y$. This thread hierarchy follows naturally from the data we have to process and how we process them. In fact, the block index is used to index one of the $g_0, \dots g_{R-1}$ vectors, while the 2D thread index is used to identify a point of the contracting grid (of size $S_x \times S_y$).

$$
\begin{array}{l}
\text{function 2D-ML}(\mathbf{g}_0^{[\text{global}]},\ldots,\mathbf{g}_{R-1}^{[\text{global}]})\\
\quad i \leftarrow \texttt{threadIdx.x}\\
\quad j \leftarrow \texttt{threadIdx.y}\\
\quad \text{if } (i=0)\wedge(j=0) \text{ then}\\
\qquad r \leftarrow \texttt{blockIdx.x}\\
\qquad \mathbf{g}^{[\text{shared}]} \leftarrow \mathbf{g}_r^{[\text{global}]}\\
\qquad x_\star^{[\text{shared}]} \leftarrow (a+b)/2\\
\qquad y_\star^{[\text{shared}]} \leftarrow (c+d)/2\\
\qquad \Delta_x^{[\text{shared}]} \leftarrow (b-a)/S_x\\
\qquad \Delta_y^{[\text{shared}]} \leftarrow (d-c)/S_y\\
\quad \text{end if}\\
\quad \texttt{\_\_syncthreads}\\
\quad \text{for } m=0,\ldots,M-1 \text{ do}\\
\qquad x \leftarrow x_\star^{[\text{shared}]} + [i-(S_x-1)/2]\cdot\Delta_x^{[\text{shared}]}\\
\qquad y \leftarrow y_\star^{[\text{shared}]} + [j-(S_y-1)/2]\cdot\Delta_y^{[\text{shared}]}\\
\qquad \ell_{i,j}^{[\text{shared}]} \leftarrow 0\\
\qquad \text{for } k=0,\ldots,K-1 \text{ do}\\
\qquad\quad \bar{g}_k \leftarrow \texttt{tex2DLayered}(\bar{\mathbf{g}}^{[\text{texture}]},x,y,k)\\
\qquad\quad \text{if } (g_k^{[\text{shared}]}\neq 0)\vee(\bar{g}_k\neq 0) \text{ then}\\
\qquad\qquad \ell_{i,j}^{[\text{shared}]} \leftarrow \ell_{i,j}^{[\text{shared}]} + g_k^{[\text{shared}]}\cdot\log(\bar{g}_k)-\bar{g}_k\\
\qquad\quad \text{end if}\\
\qquad \text{end for}\\
\qquad \texttt{\_\_syncthreads}\\
\qquad \text{if } (i=0)\wedge(j=0) \text{ then}\\
\qquad\quad \ell_{\max} \leftarrow -\infty\\
\qquad\quad \text{for } i_{\text{test}}=0,\ldots,S_x-1 \text{ do}\\
\qquad\qquad \text{for } j_{\text{test}}=0,\ldots,S_y-1 \text{ do}\\
\qquad\qquad\quad \text{if } \ell_{\max} < \ell_{i_{\text{test}},j_{\text{test}}}^{[\text{shared}]} \text{ then}\\
\qquad\qquad\qquad \ell_{\max} \leftarrow \ell_{i_{\text{test}},j_{\text{test}}}^{[\text{shared}]}\\
\qquad\qquad\qquad i_{\max}^{[\text{shared}]} \leftarrow i_{\text{test}}\\
\qquad\qquad\qquad j_{\max}^{[\text{shared}]} \leftarrow j_{\text{test}}\\
\qquad\qquad\quad \text{end if}\\
\qquad\qquad \text{end for}\\
\qquad\quad \text{end for}\\
\qquad\quad \Delta_x^{[\text{shared}]} \leftarrow \Delta_x^{[\text{shared}]}/\alpha\\
\qquad\quad \Delta_y^{[\text{shared}]} \leftarrow \Delta_y^{[\text{shared}]}/\alpha\\
\qquad \text{end if}\\
\qquad \texttt{\_\_syncthreads}\\
\qquad \text{if } (i=i_{\max}^{[\text{shared}]})\wedge(j=j_{\max}^{[\text{shared}]}) \text{ then}\\
\qquad\quad x_\star^{[\text{shared}]} \leftarrow x\\
\qquad\quad y_\star^{[\text{shared}]} \leftarrow y\\
\qquad \text{end if}\\
\qquad \texttt{\_\_syncthreads}\\
\quad \text{end for}\\
\quad \text{if } (i=0)\wedge(j=0) \text{ then}\\
\qquad r \leftarrow \texttt{blockIdx.x}\\
\qquad \hat{x}_r^{[\text{global}]} \leftarrow x_\star^{[\text{shared}]}\\
\qquad \hat{y}_r^{[\text{global}]} \leftarrow y_\star^{[\text{shared}]}\\
\quad \text{end if}\\
\text{end function}
\end{array}
$$

Figure 5. GPU pseudocode for ML estimation via a contracting-grid search algorithm.

Our GPU implementation uses shared memory to either store data that are used multiple times during thread execution (this would be the case, e.g., of PMT data vector g_r) or to share common variables among all the threads in the same block. Each thread in a block calculates the likelihood $\ell(x, y; g^{[\text{shared}]})$ for one of the points in the contacting grid and shares the value of the likelihood among all the threads in the same block.

MDRF data (previously estimated via simulation codes [4]) are stored in a 2D *layered* texture and used during the calculation of the log-likelihood $\ell(x, y; g^{[\text{shared}]})$ (denoted as $\ell_{i,j}^{[\text{shared}]}$ in the pseudocode). MDRF data are transparently interpolated by the hardware during texture fetching. Moreover, we set texture boundary conditions so that, should the point (x, y) fall outside the detector's entrance face, $\overline{g}_k(x, y)$ would evaluate to 0. Physically, this can be interpreted as no PMT signals being produced for a gamma-ray "interaction" outside the detector's entrance face.

Besides code speed and clarity, a layered texture makes our code extremely flexible. By changing S_x, S_y and/or α, it is possible to change the size of the contracting grid or its contracting factor to find the desired trade-off between speed and estimation accuracy.

5.1. Comments and applications to photon counting

The algorithm we discussed above was specifically designed for gamma-ray data and it uses calibration data in the form of mean detector response functions (MDRFs). The output of the algorithm is a list of positions in the form $\{(\widehat{x}_0, \widehat{y}_0), \dots, (\widehat{x}_{R-1}, \widehat{y}_{R-1})\}$. This list can directly be fed to an algorithm for list-mode image reconstruction. Implementation details and results are reported in [20]. Common practice, however, is to *bin* the list-mode data and count the number of points $(\widehat{x}_r, \widehat{y}_r)$ that fall within each bin. As we argue in [21], one drawback of this step is that it introduces some error, as all the points within each bin are represented with a single point location.

The algorithm we presented in **Figure 5** is one example of a contracting grid algorithm for maximum-likelihood estimation. The main assumption we made was that the likelihood $\mathrm{L}(\boldsymbol{g} \,|\, \boldsymbol{\theta})$ (or its logarithm) is a smooth function of $\boldsymbol{\theta}$, the vector of parameters we want to estimate. This is true for many estimation problems. Therefore, the algorithm of **Figure 5** provides a general pattern for the implementation of maximum-likelihood estimation on a GPU device.

6. Photon-processing detectors and algorithms

For each photon-absorption event in a detector, there are many parameters we can consider. These parameters include photon position $\boldsymbol{R}$ with respect to a plane or a reference point, direction of propagation $\vec{s}$ and energy E the photon deposited in the detector. Depending on the application (e.g., single-photon emission computed tomography for 4D angiography or coincidence detection in positron emission tomography), we might also need to consider the time t the photon impinged on the detector. Finally, some imaging techniques (such as two-photon quantum imaging) do require measurements of quantum mechanical parameters, one example being quantum spin.

6.1. Mathematical description

We refer to a set of photon parameters as an *attribute vector,* and we denote it as $\boldsymbol{A}$. Hence, depending on the application, an attribute vector might have five or more components. We denote the number of components of $\boldsymbol{A}$ as N. Because of noise, it is not possible to estimate exactly the components of $\boldsymbol{A}$ and we use the notation $\widehat{\boldsymbol{A}}$ to denote an estimated attribute vector.

We define a *photon-processing* detector as any imaging device that [9]:

- uses a gain mechanism (such as an image intensifier) to obtain multiple measurements (e.g., multiple pixel values) for each absorbed photon;
- uses these measurements to perform maximum-likelihood estimation of photon attribute vector $\widehat{\boldsymbol{A}}_j$, for$j = 1, \ldots, J$;
- stores the estimated attributes at full precision as a list $\widehat{\mathcal{A}} = \left\{\widehat{\boldsymbol{A}}_1, \ldots, \widehat{\boldsymbol{A}}_J\right\}$ and without performing any binning.

Photon-processing detectors are fundamentally different than photon-counting detectors. While photon-counting detectors only consider photon position and record the number of photons (or charged particles) that fall within each bin over a predetermined amount of time, photon-processing detectors use maximum-likelihood to estimate a wide range of attributes and retain all the estimated information at full precision as the list $\widehat{\mathcal{A}} = \left\{\widehat{\boldsymbol{A}}_1, \ldots, \widehat{\boldsymbol{A}}_J\right\}$.

The full information from a photon-processing detector is retained if we simply store the N estimated attributes for each of J photons as the list $\widehat{\mathcal{A}}$, but an equivalent construction as a random point process in attribute space offers new theoretical insights. From $\widehat{\mathcal{A}}$, we introduce this point process as [3, 9]

$$u(\boldsymbol{A}) = \sum_{j=1}^{J} \delta\left(\boldsymbol{A} - \widehat{\boldsymbol{A}}_j\right), \tag{16}$$

where $\delta(\ldots)$ is the N-dimensional Dirac delta function. The mean of the point process $u(\boldsymbol{A})$ is obtained by averaging over the statistics of each of the attribute vectors $\widehat{\boldsymbol{A}}_1, \ldots, \widehat{\boldsymbol{A}}_J$ and then over J itself for a given object $\boldsymbol{f}$. This calculation gives a function $\overline{u}(\boldsymbol{A} \mid \boldsymbol{f})$ of $\boldsymbol{A}$ for fixed $\boldsymbol{f}$. This function can be regarded as a vector $\overline{\boldsymbol{u}}(\boldsymbol{f})$ in the Hilbert space $\mathbb{L}_2(\mathbb{R}^N)$, which is the vector space of square-integrable functions of N real variables. As shown in [22, 23], we can introduce the linear operator $\mathcal{L}$ that maps the object $\boldsymbol{f}$ (belonging to the infinite-dimensional Hilbert space $\mathbb{L}_2(\mathbb{R}^3)$) to $\overline{\boldsymbol{u}}(\boldsymbol{f})$. In symbols,

$$\overline{\boldsymbol{u}}(\boldsymbol{f}) = \mathcal{L}\boldsymbol{f}. \tag{17}$$

A similar expression but for photon-counting detectors is:

$$\overline{\boldsymbol{g}}(\boldsymbol{f}) = \mathcal{H}\boldsymbol{f}, \tag{18}$$

in which the vector $\overline{g}(f)$ is the mean over many realizations of the vector g. Vector $\overline{g}(f)$ belongs to the Euclidian vector space $\mathbb{E}^M$, which is the space of all M-dimensional vectors. We refer to $\mathcal{H}$ as a continuous-to-discrete operator [8] as it maps the function $f(\boldsymbol{r})$ of continuous variable $\boldsymbol{r}$ to a discrete vector $\overline{g}(f)$ with M components. On the other hands, $\mathcal{L}$ is a continuous-to-continuous operator as it maps $f(\boldsymbol{r})$ to the function $\overline{u}(\boldsymbol{A} \mid f)$ of continuous variable $\boldsymbol{A}$ [8].

The key difference between $\mathcal{H}$ and $\mathcal{L}$ is that $\mathcal{H}$ must necessarily have a nontrivial null space, as it maps vectors in an infinite-dimensional Hilbert space to vectors in a finite-dimensional vector space. This means that for any imaging system that produces photon-counting data, there exist nonzero objects f_{null} that, on average, do not produce any data. Equivalently, we can say that there exist two objects f_1 and f_2 with $f_1 \neq f_2$ for which $\mathcal{H}f_1 = \mathcal{H}f_2$. A continuous-to-continuous operator—such as the operator $\mathcal{L}$ defined above—maps an object in an infinite-dimensional Hilbert space to another infinite-dimensional Hilbert space. Therefore, the same dimensionality analysis we considered for $\mathcal{H}$ does not apply to $\mathcal{L}$. In fact, $\mathcal{L}$ might allow a lower dimensional null space than $\mathcal{H}$ for the same imaging system [21].

6.2. Relationship to radiometry and the Boltzmann transport equation

The word "radiometry" refers to a set of techniques used in optics to describe and calculate the distribution of light. An important concept used in radiometry is that of radiance, denoted as $L(\boldsymbol{r}, \vec{s})$, which is a function that describes the radiant flux in an optical system as a function of three-dimensional spatial position $\boldsymbol{r}$ and direction $\vec{s}$. Since the radiant flux is measured in Watts, the units of $L(\boldsymbol{r}, \vec{s})$ are Watts per square meter per steradian, or W/ (m$^2\cdot$ ster) [8]. From the radiance, other important radiometric quantities can be calculated. This includes the irradiance (power per unit area), radiant intensity (power per unit solid angle) and radiant flux (power).

Spectral dependence can be introduced in the basic definition of radiance by considering radiance per unit wavelength λ, which we denote as $L_\lambda(\boldsymbol{r}, \vec{s}, \lambda)$. The units of $L_\lambda(\boldsymbol{r}, \vec{s}, \lambda)$ are W/ (m$^2\cdot$ ster$\cdot$ nm), provided that the units of wavelength are nanometers (nm). Spectral radiance can also be measured in photon units by first expressing wavelength in terms of energy ($E = hc/\lambda$, in which h is Planck's constant and c is the speed of light) and then by dividing $L_\lambda(\boldsymbol{r}, \vec{s}, \lambda)$ by the energy of a photon. We denote this new quantity as $L_{p,E}(\boldsymbol{r}, \vec{s}, E)$, and its units are (photons/s)/(m$^2\cdot$ ster). Finally, we can consider a time-dependent spectral photon radiance, and we denote this function as $L_{p,E}(\boldsymbol{r}, \vec{s}, E, t)$.

In radiometry, the Boltzmann transport equation (BTE) allows to calculate $L_{p,E}(\boldsymbol{r}, \vec{s}, E, t)$ at any point inside an optical system by taking into account absorption, emission, scattering and propagation of light. In its most general form, the BTE is written as [8]:

$$\frac{\partial L_{p,E}}{\partial t} = \left[\frac{\partial L_{p,E}}{\partial t}\right]_{\text{abs}} + \left[\frac{\partial L_{p,E}}{\partial t}\right]_{\text{emiss}} + \left[\frac{\partial L_{p,E}}{\partial t}\right]_{\text{sc}} + \left[\frac{\partial L_{p,E}}{\partial t}\right]_{\text{prop}}. \tag{19}$$

Each term on the right-hand side can be worked out explicitly [8, 9] to obtain:

$$\frac{\partial L_{p,E}}{\partial t} = -c_m \mu_{\text{tot}} L_{p,E} + c_m \Xi_{p,E} + \mathcal{K} L_{p,E} - c_m \vec{s} \cdot \nabla L_{p,E}, \tag{20}$$

where c_m is the speed of light in the medium, μ_{tot} is the total attenuation coefficient (with contributions from both absorption and scattering processes), $\mathcal{K}$ is an integral operator describing the angular and energy dependence of the scattering, and $\Xi_{p,E}$ describes any light source. In general, the function $\Xi_{p,E}$ depends on $\boldsymbol{r}$, $\vec{s}$, E and t. If the light source is isotropic (i.e., $\Xi_{p,E}\left(\boldsymbol{r}, \vec{s}, E, t\right)$ does not depend on $\vec{s}$) and independent of time, we can write

$$\Xi_{p,E}\left(\boldsymbol{r}, \vec{s}, E, t\right) = \frac{1}{4\pi} f(\boldsymbol{r}, E), \tag{21}$$

where the 4π term (units: ster) accounts for integration of a solid angle over a sphere. Under these hypotheses, a steady-state solution to Eq. (20) is found by setting the partial derivative $\partial L_{p,E}/\partial t$ to zero. The result is

$$c_m \mu_{tot} L_{p,E} - \mathcal{K} L_{p,E} + c_m \vec{s} \cdot \nabla L_{p,E} = \frac{c_m}{4\pi} f, \tag{22}$$

which can be further rewritten in operator form as

$$\frac{4\pi}{c_m} \mathcal{B}\, \mathbf{L}_{p,E} = \boldsymbol{f}, \tag{23}$$

provided that

$$\mathcal{B} = \mathrm{c_m} \mu_{\text{tot}} - \mathcal{K} + \mathrm{c_m} \vec{\mathrm{s}} \cdot \nabla. \tag{24}$$

We refer to $\mathcal{B}$ as the Boltzmann operator. If we insert Eq. (23) into Eq. (17), we get

$$\overline{\boldsymbol{u}}(\boldsymbol{f}) = \frac{4\pi}{c_m} \mathcal{L}\, \mathcal{B}\, \mathbf{L}_{p,E}, \tag{25}$$

which describes a practical way to obtain the function $\overline{\boldsymbol{u}}(\boldsymbol{f})$ from knowledge of the radiance function $L_{p,E}\left(\boldsymbol{r}, \vec{s}, E\right)$ inside an optical system and the Boltzmann operator $\mathcal{B}$.

6.3. Particle-processing detectors

An example of a particle-processing detector for beta particles is shown in **Figure 6**. This detector includes two layers of ultrathin phosphor foils separated by an air gap, an image intensifier, a high numerical aperture lens system and a light sensor [1, 7].

In **Figure 6**, an incoming beta particle interacts with a layer of phosphor (just a few microns thick) at location $\boldsymbol{r}_1 = (x_1, y_1)$ where it deposits some of its energy, which the layer of phosphor gives off as a flash of visible light. The particle further propagates and interacts at $\boldsymbol{r}_2 = (x_2, y_2)$

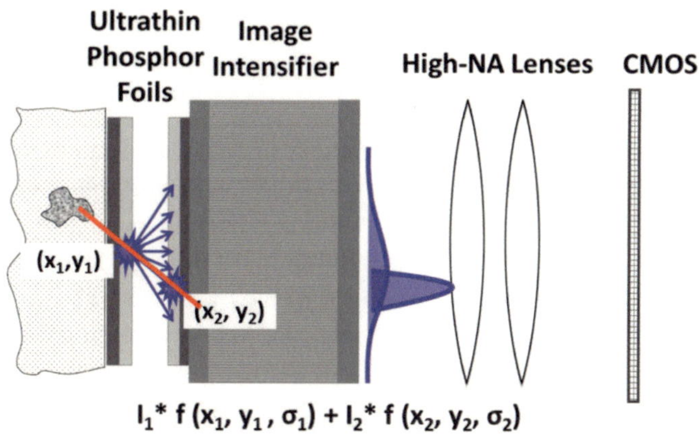

Figure 6. A particle-processing detector for beta particle (adapted from [1]).

with a second phosphor layer, thus producing a flash of visible light here as well. Light flashes produced at each layer get amplified by an image intensifier and imaged onto a sensor. The flash of light generated at the first layer spreads out considerably as it propagates through the air gap, thus resulting in a much broader signal on the sensor than that corresponding to the flash of light generated at the second layer. This is used to determine at which layer each flash of light was generated. The direction $\vec{s}$ of the particle and its position at either phosphor foil can be estimated from the two interaction positions and the distance between the two foils (assumed known). If the particle's residual energy E is of interest, the second phosphor foil can just be replaced by a thick scintillator, so that the particle is stopped.

A CUDA algorithm for maximum-likelihood estimation of position and particle direction for the setup of **Figure 6** has been developed by Garrett Hinton at the Center for Gamma-Ray Imaging, University of Arizona. Following the same approach outlined in Section 5, the algorithm uses a four-dimensional contracting grid to simultaneously estimate location $\boldsymbol{r}_1 = (x_1, y_1)$ and $\boldsymbol{r}_2 = (x_2, y_2)$ from an image frame collected with an ultrafast CMOS camera. Experimental setup and preliminary results have been presented in [1, 7].

7. Summary and applications

This chapter provided a general overview of detector technology and algorithms for photon counting and photon processing. We started by describing detectors suitable for photon counting and photon processing. Statistical models for the data produced by these detectors were presented. We then introduced maximum-likelihood estimation and discussed its properties. We emphasized that a maximum-likelihood estimate is any parameter that maximizes the likelihood function given the detector output, and we pointed out that the likelihood function is the probability density function of the detector output conditioned on the parameter being estimated.

We then commented on graphics processing units (GPUs) and the CUDA programming environment. Through CUDA-like pseudocode, we provided a maximum-likelihood algorithm for estimation of position of interaction for gamma-ray cameras. This algorithm used a contracting-grid approach to find a maximum of the likelihood function. Our approach heavily relied upon GPU textures to quickly retrieve calibration data. The same approach is applicable to many estimation problems.

Photon-processing detectors were introduced and defined as any system that collects multiple measurements to perform maximum-likelihood estimation of multiple event parameters (such as position, direction and energy), which are stored as a list and in full precision in the memory of a computer. The same data can also be represented as a point process, and we introduced a linear operator that maps the object being imaged to the mean of this point process. We used a dimensionality analysis to describe the advantages of photon-processing detectors over photon-counting detectors.

Particle-processing detectors are a variation of photon-processing detector. As an emerging technology, particle-processing detectors will find applications in many fields, one of them being medical imaging. In a new technique, called charged-particle emission tomography (CPET), particle-processing detectors are being evaluated for 3D *in vivo* imaging with alpha and beta particles [1, 7]. Like photon-processing detectors, particle-processing detectors convey a larger amount of information than conventional detectors for charged particles. This enables high-resolution 3D reconstruction of the distribution of radionuclides emitting charged particles without the need to kill an animal to image a collection of thinly sliced tissue sections.

Drug development will take advantage of CPET to determine drug pharmacokinetics, 3D transduction across cell membranes and targeting to tissues of interest. In the development of internal radioimmunotherapy, CPET imaging can be used to collect data on the 3D heterogeneous distributions of targeting molecules and in the estimation of delivered radiation dose. Finally, CPET will likely become a valuable technique in the emerging fields of personalized medicine and theranostics, in which diagnostics and therapy are combined in an attempt to avoid the "one-size-fits-all" approach to treatment that is often successful for some patients but not for others.

Acknowledgements

This chapter has been supported by National Institutes of Health (Grants R01 EB000803 and P41 EB002035).

Author details

Luca Caucci[1]*, Yijun Ding[2] and Harrison H. Barrett[1,3]

*Address all correspondence to: caucci@email.arizona.edu

1 Department of Medical Imaging, University of Arizona, Tucson, AZ, United States

2 Department of Radiation Oncology, University of Colorado Denver, Denver, United States

3 College of Optical Sciences, University of Arizona, Tucson, AZ, United States

References

[1] Yijun Ding. Charged-Particle Emission Tomography [dissertation]. Tucson, AZ; 2016.

[2] Anger HO. Scintillation Camera. Review of Scientific Instruments. 1958;**29**(1):27-33

[3] Luca Caucci. Task Performance with List-Mode Data [dissertation]. Tucson, AZ; 2012

[4] Hunter WCJ, Barrett HH, Furenlid LR. Calibration method for ML estimation of 3D interaction position in a thick gamma-ray detector. IEEE Transactions on Nuclear Science. 2009;**56**(1):189-196

[5] Llopart X, Campbell M, Dinapoli R, San Segundo D, Pernigotti E. Medipix2: A 64-k pixel readout Chip with 55-μm. IEEE Transactions on Nuclear Science. 2002;**49**(5):2279-2283

[6] Bouchami J, Gutiérrez A, Houdayer A, Jakůbek J, Lebel C, Leroy C, Macana J, Martin JP, Platkevič M, Pospíši S, Teyssierl C. Study of the charge sharing in silicon pixel detector by means of heavy ionizing particles interacting with a Medipix2 device. Nuclear Instruments and Methods in Physics Research Section A: Accelerators, Spectrometers, Detectors and Associated Equipment. 2011;**633**(Supplement 1):S117-S120

[7] Ding Y, Caucci L, Barrett HH. Charged-particle emission tomography. Medical Physics. 2017;**44**(6):2478-2489

[8] Harrison H. Barrett, Kyle J. Myers. Foundations of Image Science. Hoboken, NJ: Wiley-Interscience; 2004

[9] Caucci L, Myers KJ, Barrett HH. Radiance and photon noise: Imaging in geometrical optics, physical optics, quantum optics and radiology. Optical Engineering. 2016;**55**(1):013102

[10] Miller BW, Gregory SJ, Fuller ES, Barrett HH, Barber HB, Furenlid LR. The iQID camera: An ionizing-radiation quantum imaging detector. Nuclear Instruments and Methods in Physics Research Section A: Accelerators, Spectrometers, Detectors and Associated Equipment. 2014;**767**:146-152

[11] Aldrich J. R. A. Fisher and the making of maximum likelihood 1912–1922. Statistical Science. 1997;**12**(3):162-176

[12] Cramér H. Mathematical Methods of Statistics. Princeton, NJ: Princeton University Press; 1956

[13] Rao CR. Information and the accuracy attainable in the estimation of statistical parameters. Bulletin of the Calcutta Mathematical Society. 1945;**37**:81-89

[14] Zehna PW. Invariance of maximum likelihood estimators. Annals of Mathematical Statistics. 1966;**37**(3):744

[15] Moore DS. Maximum likelihood and sufficient statistics. The American Mathematical Monthly. 1971;**78**(1):50-52

[16] Huzurbazar VS. The likelihood equation, consistency and the maxima of the likelihood function. Annals of Human Genetics. 1947;**14**(1):185-200

[17] Fisher RA. Theory of statistical estimation. Mathematical Proceedings of the Cambridge Philosophical Society. 1925;**22**(5):700-725

[18] Furenlid LR, Hesterman JY, Barrett HH. Real-time data acquisition and maximum-likelihood estimation for gamma cameras. In: 14th IEEE-NPSS Real Time Conference; 4–10 June 2005. Stockholm, Sweden; 2005. p. 498-501

[19] Hesterman JY, Caucci L, Kupinski MA, Barrett HH, Furenlid LR. Maximum-likelihood estimation with a contracting-grid search algorithm. IEEE Transactions on Nuclear Science. 2010;**57**(3):1077-1084

[20] Luca Caucci, William C. J. Hunter, Lars R. Furenlid, Harrison H. Barrett. List-mode MLEM Image Reconstruction from 3D ML Position Estimates. In: IEEE Nuclear Science Symposium Conference Record (NSS/MIC), 30 Oct–6 Nov. 2010. Knoxville, TN, USA; 2010. p. 2643-2647

[21] Luca Caucci, Abhinav K. Jha, Lars R. Furenlid, Eric W. Clarkson, Matthew A. Kupinski, Harrison H. Barrett. Image Science with Photon-Processing Detectors. In: IEEE Nuclear Science Symposium and Medical Imaging Conference (NSS/MIC), 27 Oct–2 Nov, 2013. Seoul, South Korea; 2013. p. 1-7

[22] Andre Lehovich. List-mode SPECT Reconstruction Using Empirical Likelihood [dissertation]. Tucson, AZ; 2005

[23] Caucci L, Barrett HH. Objective assessment of image quality. V. Photon-counting detectors and list-mode data. Journal of the Optical Society of America A. 2012;**29**(6):1003-1016

Chapter 6

Laser-Induced Fluorescence of Hydroxyl (OH) Radical in Cold Atmospheric Discharges

Jan Voráč, Pavel Dvořák and Martina Mrkvičková

Additional information is available at the end of the chapter

http://dx.doi.org/10.5772/intechopen.72274

Abstract

The application of laser-induced fluorescence (LIF) to measurement of absolute concentration of hydroxyl radicals in cold atmospheric discharges is described. Though only the case of OH is presented, the method can be directly applied to other molecules as well. Starting from the rate equations for the LIF process, the main formulas for two- and multi-level excitation scheme are derived. It is also shown how to use partially saturated LIF in practice, enhancing the signal-to-noise ratio. Practical tips for automating the data evaluation are given, allowing processing large data sets, particularly suitable for planar measurements. Gas temperature estimation from fluorescence on different rotational absorption lines is shown as an attractive method for obtaining temperature maps with high spatial resolution. The important aspects of calibration are discussed, particularly the overlap of the laser line with the selected absorption line and the measurement of the Rayleigh scattering for sensitivity calibration, together with the common sources of errors. The application of OH(A, $v' = 0 \leftarrow$ X, $v'' = 0$) excitation scheme to the effluent of atmospheric pressure plasma jet ignited in argon and of OH(A, $v' = 1 \leftarrow$ X, $v'' = 0$) to the plasma of coplanar surface barrier discharge in air and in water vapor is shown.

Keywords: laser-induced fluorescence, atmospheric pressure plasma, hydroxyl, concentration, spectroscopy

1. Introduction

Even though the hydroxyl (OH) in laboratory discharges readily emits radiation, which may be the source of many interesting information [1], there is no known direct and reliable method of determining the total OH radical density from the emission spectra. Indeed, the radicals in the ground electronic states are usually the most abundant and their effect on the plasma chemistry is dominant, but they are not observable by simple emission spectroscopy. Absorption-based spectroscopic techniques allow access directly to the ground electronic states. In absorption spectroscopy, the decrease in the light intensity after passing through the probed volume is observed. The information obtained by absorption spectroscopy is thus line-integrated. On the other hand, the evaluation of the measurement is quite straightforward and absorption measurements could be considered less vulnerable to errors compared to fluorescence measurements. The sensitivity of absorption spectroscopy for single pass may be insufficient for some laboratory discharges and multi-pass cells need to be constructed, complicating the experiment and the data evaluation.

Laser-induced fluorescence (LIF) is an interesting combination of absorption and emission spectroscopy. It is naturally more complicated than either of these methods alone, both experimentally and theoretically. In return, it offers higher sensitivity compared to single-pass absorption measurements, good spatial resolution given by intersection of the laser beam, and the optical path of observation. To obtain the absolute concentration, many things need to be taken into account and there are many partial measurements that need to be done and evaluated. In this chapter, we show how the OH fluorescence measurements in atmospheric pressure discharges can be performed, including tricks that can be applied for conditions with low ratios of wanted-to-unwanted signals. We shall also show what information may be extracted from the partial steps that are necessary for the concentration determination.

Hydroxyl is a diatomic radical consisting of an oxygen and a hydrogen atom. It contains an unpaired electron, which makes it a highly reactive species—its lifetime in an open atmosphere is typically less than a second [2]. It can bind to a polymer surface, forming a functional group and increasing its wettability [3]. It has been proved that introducing OH radicals into mixtures containing volatile organic compounds leads to removal of these pollutants, for example, [4]. It has also been found that the OH radical plays an important role in the response of immune system to infection [5–7]. OH radical radiation on the $A^2\Sigma^+ \rightarrow X^2\Pi$ electronic transition with the strongest band in the near UV at 306 nm is also readily present in nearly all laboratory discharges, as the water impurities are notoriously difficult to avoid and the water molecule is easily dissociated, forming an OH radical. Part of the dissociation products is created directly in the excited $A^2\Sigma$ electronic state. This is responsible for a considerable amount of UV radiation also in plasma jets used for plasma medicine, introducing another mean of affecting both the patient's tissue and the eventual invading microorganisms.

Consequently, OH has been intensively spectroscopically studied. The fundamental work of Dieke and Crosswhite [8] offered a good description of the quantum states with satisfactorily precise values necessary for the energy levels calculation. Luque and Crosley later added calculations for transition probabilities [9]. Together with improved calculations of energy levels, this has already been sufficient for precise simulations of the molecular spectrum as available in their free software LIFBASE [10].

2. Structure of OH radical and its spectrum

In this section, we briefly summarize the structure of the OH radical focusing on how it affects its spectrum. For details, see the comprehensive book of Herzberg and Spinks [11].

http://dx.doi.org/10.5772/intechopen.72274

Symbol	Meaning
J	Total angular momentum
N (K in [11])	Total angular momentum apart from the electronic spin
S	Angular momentum of electronic spin

Table 1. List of quantum numbers used in this chapter and their meaning.

The two most relevant electronic states for OH spectroscopy are the ground state X $^2\Pi$ and the first excited state A $^2\Sigma^+$, in further text we call them briefly X and A. They are approximately 4 eV apart. This energy gap corresponds to photon wavelengths around 300 nm. Both mentioned electronic states have resulting spin 1/2 and are thus doublets.

The ground electronic state is a Π state, that is, the resulting length Λ of the projection of the electronic orbital angular momentum to the internuclear axis is 1. For lower rotational states, the dominant fine-structure-splitting mechanism is the spin-orbit coupling. However, for

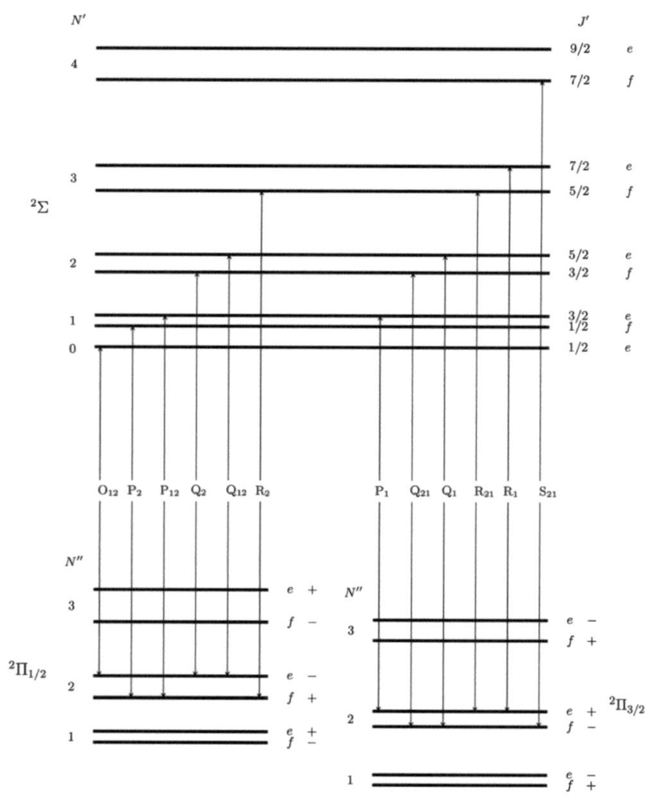

Figure 1. Diagram of $^2\Sigma \leftrightarrow {}^2\Pi$ transitions. The J values are omitted for the $^2\Pi$ state. It would be $J = N - 1/2$ for the $^2\Pi_{1/2}$ state and $J = N + 1/2$ for the $^2\Pi_{3/2}$. The lines are labeled according to Hund's case (b), that is, P for $\Delta N = -1$, Q for $\Delta N = 0$ and R for $\Delta N = 1$. Note that the Π states do not have $N = 0$.

higher rotational states the spin-rotational coupling starts to gain importance and for $J \geq 8.5$ starts to dominate. The ground electronic state is thus an intermediate state between Hund's cases (a) and (b). Consequently, neither the N nor the J quantum number is considered "good," but both are "almost good" and can be used to label the rotational states depending on the preference.[1] The relation between the two is $J = N \pm 1/2$ for the fine-structure components, usually labeled 1 and 2, respectively. See **Table 1** for summary of the quantum numbers. The Λ-doubling effect is also present. Effectively, there are always four different states with the same vibrational and rotational excitation—two orientations of the electronic spin (spin-orbit and spin-rotational splitting) and two orientations of the projection of total orbital momentum of electrons to the internuclear axis. Due to symmetry selection rules (only $+ \leftrightarrow -$ transitions are allowed), only one of the two Λ-doublet components is available for the photon absorption. This is important to take into account when calculating the radical concentration from the LIF measurements. The ± parity is derived from the (anti)symmetry of the wavefunction with respect to mirroring by an arbitrary plane containing the internuclear axis. Because this property switches with the rotational quantum number, a notation e, f was introduced as $\binom{e}{f} = \pm(-1)^{J-1/2}$, see **Figure 1**.

The first excited state A is a Σ state, that is, $\Lambda = 0$. Σ electronic states are best described by Hund's case (b), the good quantum number for rotational state is N. There is no spin-orbit coupling and no Λ-doubling effect. The fine-structure splitting happens solely due to spin-rotational interaction. In this case, the energy spacing of the doublet levels is given by $\gamma(N + 1/2)$, that is, increases linearly with rotational quantum number. The energy splitting of the doublet components is smaller compared to the X $^2\Pi$ state.

The structure of the OH energy levels is sketched in **Figures 1–3**. The A $^2\Sigma$ state is affected by the spin-rotation interaction which causes a splitting of each rotational state into two sublevels. Similarly, the X $^2\Pi$ state is affected dominantly by the spin-orbit interaction. Each of these two effects causes doubling of each expected rotational line, so we have four lines for $\Delta J = -1$, four lines for $\Delta J = 0$, and four other lines for $\Delta J = +1$. This makes 12 lines altogether. The Λ-doubling of the $^2\Pi$ state, on the other hand, introduces no further splitting of the observed rotational lines. This is because of the $+ \leftrightarrow -$ parity selection rule. Because of the Λ-doubling, the Q(N'') lines have different lower state than P(N'') or R(N'') lines. Here, we have used the standard spectroscopic abbreviation: single prime (N') is used to describe the upper state of a transition (here exclusively the electronic state A) and double prime (N'') describes the lower state (here exclusively the X electronic state).

[1]In this text, we adopt the newer notation with J for the quantum number describing the total angular momentum (apart from the nuclear spin that will be neglected in the whole chapter), N for the total angular momentum *apart from the electronic spin*, R for pure rotation (seldom used), and S for angular momentum of electronic spin, as in [12, 13]. In Herzberg's book, the same quantum numbers are called J, K, N, and S, respectively.

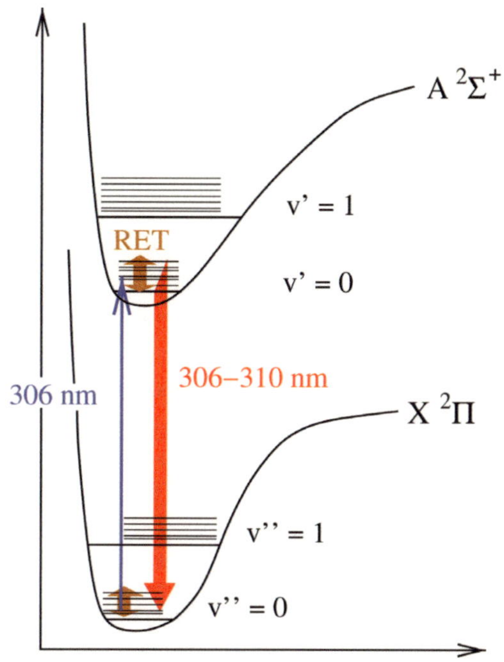

Figure 2. Fluorescence schema of the excitation to the $v' = 0$ state of OH radicals.

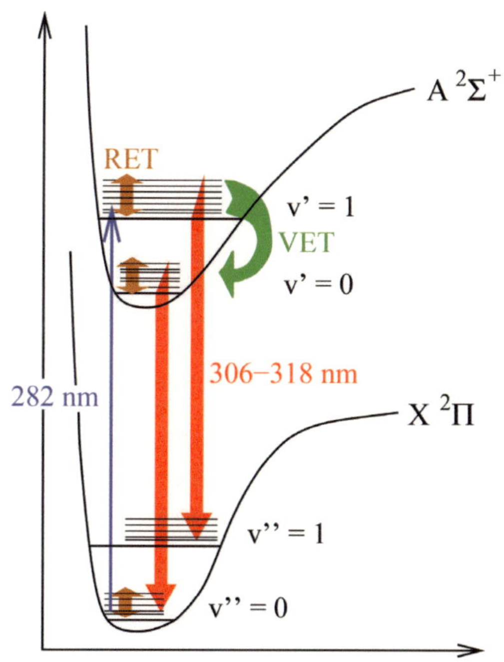

Figure 3. Fluorescence schema of the excitation to the $v' = 1$ state of OH radicals.

3. Excitation schema

Although only one specific rotational level can be excited by laser with narrow spectral linewidth, collisional processes can change both vibrational and rotational state of the excited molecular species. As a result, the fluorescence spectrum of OH radicals usually consists of numerous rotational lines that are merged to several vibrational bands. The composition of fluorescence spectrum depends markedly on particular vibrational state that is excited by laser:

1. The simplest situation is realized when OH radicals are excited to the ground vibrational state $A^2\Sigma^+(v' = 0)$ [14]. When no thermal excitation to higher vibrational states occurs, the only intensive fluorescence radiation is the 0 – 0 vibrational band located between 306 and 314 nm. Other vibrational bands originating from the ground vibrational state are weak due to their low Franck-Condon factors. Since excitation and fluorescence wavelengths coincide, the fluorescence signal cannot be separated from scattered laser radiation spectrally. Consequently, this excitation schema can be used for LIF experiments only when temporal separation can be used, that is, when the lifetime of the excited OH state is longer than the laser pulse duration and the fluorescence signal can be detected after the end of the laser pulse.

2. The spectral separation is simply achieved when OH radicals are excited to the first vibrationally excited state $A^2\Sigma^+(v' = 1)$ [15–18]. In this case, excitation wavelengths around 282 nm are used. Fluorescence vibrational bands 1 – 1 (312–320 nm) and 0 – 0 (306–314 nm) are usually used for detection, which can be separated from scattered excitation radiation by spectral filters. The 0 – 0 vibrational band arises due to the collisional vibrational energy transfer (VET) and its intensity depends on the ratio between the VET rate and the total deexcitation rate of the $(v' = 1)$ state.

3. Alternatively, higher vibrational states can be excited as well. In these cases, the kinetics of excited states is even more complicated. Excitation to the $A^2\Sigma^+(v' = 3)$ state is sometimes used, since this state undergoes rapid predissociation, which reduces the dependence of excitation state lifetime on collisional quenching [19]. On the other hand, the dissociation reduces the fluorescence signal, and the influence of collisional processes on the lifetime of excited OH radicals is not entirely avoided, especially at high pressure. That is why we further concentrate only on excitation to the $v' = 0$ and $v' = 1$ state.

The vibrational and rotational distribution of the excited state should be taken into account during processing of measured data, since rovibrational levels differ in fluorescence quantum yield and wavelengths. Consequently, the sensitivity of the method depends on the rovibrational distribution. The intensity of the measured fluorescence signal can be expressed as the sum of contributions from all spectral lines generated by the excited OH radical

$$M_f = \iiint_V \frac{\Omega}{4\pi} \int_0^{\infty} \sum_{v'v''} \sum_{J'J''} \sum_{\alpha'\alpha''} F\left(\lambda^{v'J'\alpha'}_{v''J''\alpha''}\right) D\left(\lambda^{v'J'\alpha'}_{v''J''\alpha''}\right) A^{v'J'\alpha'}_{v''J''\alpha''} N^{v'J'\alpha'}\, \mathrm{d}t\mathrm{d}V \tag{1}$$

with the following meaning of used symbols: Ω is the solid angle in which the fluorescence is collected by the detector, F is the transmittance of optics in front of the detector (usually the transmittance of an interference filter), D is the detector sensitivity, A is the emission coefficient for the particular line, and N is the concentration of excited OH radical in the particular level described by vibrational number v', total angular momentum number J', and doublet component α'. (The doublet component can have two values, usually labeled as 1 and 2.) λ is the wavelength of the particular line. The signal is integrated over the whole detection volume and the whole time of the detection process. The absolute detector sensitivity is usually not known and needs to be calibrated. However, the relative dependence of the sensitivity is simply achievable. Therefore, we describe the detector sensitivity as a product $D = D_a \cdot d$ of a known function of wavelength $d(\lambda)$ and an unknown constant D_a.

Concentration of all excited levels involved in the fluorescence process can be calculated by means of a set of tens kinetic equations. Luckily, at atmospheric pressure the rotational equilibrium of each particular vibronic $A^2\Sigma^+$ state is often reached quickly, which can be used for a considerable simplification of Eq. (1). In this simplification, we describe the relative population of each rotational level by the Boltzmann factor

$$f_{v'J'\alpha'} = \frac{(2J'+1)\exp\left(-\frac{E_{v'J'\alpha'}}{kT}\right)}{\sum_{J'\alpha'}(2J'+1)\exp\left(-\frac{E_{v'J'\alpha'}}{kT}\right)} \tag{2}$$

and we assume that saturation effects are negligible. At these assumptions, the kinetic equations of the fluorescence process enable to express the fluorescence signal by

$$M_f = N_{Xi}\frac{\kappa B}{c}E_f\tau_1\left[\sum_{J'J''\alpha'\alpha''} F\left(\lambda^{v'=1J'\alpha'}_{v''=1J''\alpha''}\right)d\left(\lambda^{v'=1J'\alpha'}_{v''=1J''\alpha''}\right)A^{v'=1J'\alpha'}_{v''=1J''\alpha''}f_{v'=1J'\alpha'}(T) + \right.$$
$$\left. + V\tau_0\sum_{J'J''\alpha'\alpha''} F\left(\lambda^{v'=0J'\alpha'}_{v''=0J''\alpha''}\right)d\left(\lambda^{v'=0J'\alpha'}_{v''=0J''\alpha''}\right)A^{v'=0J'\alpha'}_{v''=0J''\alpha''}f_{v'=0J'\alpha'}(T)\right]\iiint_V D_a\frac{\Omega}{4\pi}s\,dV. \tag{3}$$

for excitation to the $v' = 1$ state and detection of $1-1$ and $0-0$ vibrational bands and by

$$M_f = N_{Xi}\frac{\kappa B}{c}E_f\tau_0\left[\sum_{J'J''\alpha'\alpha''} F\left(\lambda^{v'=0J'\alpha'}_{v''=0J''\alpha''}\right)d\left(\lambda^{v'=0J'\alpha'}_{v''=0J''\alpha''}\right)A^{v'=0J'\alpha'}_{v''=0J''\alpha''}f_{v'=0J'\alpha'}(T)\right]\iiint_V D_a\frac{\Omega}{4\pi}s\,dV. \tag{4}$$

for excitation to the $v' = 0$ state. N_{Xi} denotes the concentration of OH radicals in the particular rotational level of the ground vibronic state $X^2\Pi(v'' = 0)$ from which the excitation occurred, B is the absorption coefficient of the particular excitation transition, c is the speed of light and κ is the overlap term of the absorption and laser line [20], E_f is the mean energy of laser pulses during the measurement of the fluorescence, τ is the lifetime of the relevant vibronic state, and V is the $A^2\Sigma^+(v' = 1) \rightarrow A^2\Sigma^+(v' = 0)$ VET rate constant. The spatial laser beam profile s is proportional to the area density of laser energy and it is normalized to one when integrated in the plane perpendicular to the laser beam direction, that is, $\int\int_S s\,dS = 1$. The value of the integral $\int\int\int_V D_a \frac{\Omega}{4\pi} s\,dV$ can be calibrated by means of Rayleigh scattering, see Section 9.

Eqs. (3) or (4) can be used for the calculation of OH concentration in the ground state. Since N_{Xi} includes only OH radicals in one particular rotational level, it is necessary to calculate the concentration of all OH radicals in the ground vibronic state $X^2\Pi(v''=0)$ by means of the Boltzmann factor

$$N_X = N_{Xi}/f_{Xi} = N_{Xi}\frac{2\sum_j\left(2J_j+1\right)\exp\left(-\frac{E_j}{kT}\right)}{(2J_i+1)\exp\left(-\frac{E_i}{kT}\right)}. \quad (5)$$

Factor 2 in the numerator is caused by the Λ-doubling.

4. Instrumentation

For a typical state-of-art LIF experiment, an extensive instrumentation is necessary. An example of such setup for a point like or line-resolved measurement is shown in **Figure 4** and the necessary beam expansion optics for a planar LIF imaging is in **Figure 5**.

The core instrument of every LIF experiment is a laser system tunable in wavelength. The power source of such laser system is usually a powerful pulsed laser, typically Nd:YAG. Its frequency-doubled (532 nm) or -tripled (355 nm) output is used to pump a wavelength-tunable light source, typically a dye laser with a movable dispersive element in the cavity to control the resonance wavelength. The tuning range of dye lasers is limited by the available dyes, but can be greatly enhanced by using nonlinear frequency conversion. This way, visible light (typically in the range of 556–662 nm for the well-behaving family of rhodamine dyes) can be converted to as low as 200 nm.

Alternatively, solid-state optical parameter oscillator (OPO)-tunable lasers can be used, but despite the noticeable recent advances in this technology, dye lasers with frequency conversion

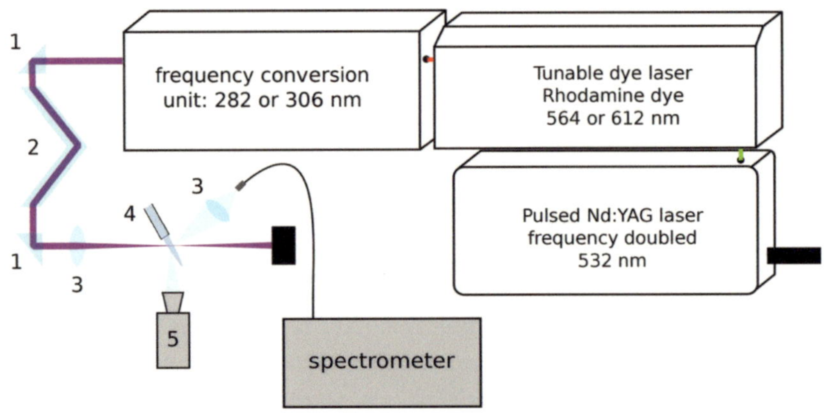

Figure 4. Instrumentation of an LIF experiment. 1 – Right-angle prism, 2 – Fresnel rhomb tandem to turn polarization (only, if necessary), 3 – Spherical positive UV lens, 4 – Investigated plasma source, 5 – ICCD camera synchronized with the laser.

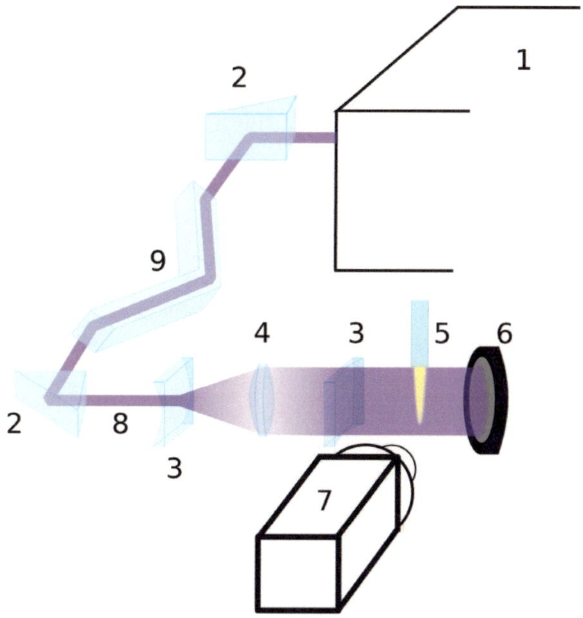

Figure 5. Instrumentation of a planar LIF experiment. 1 – Pulsed tunable laser, 2 – Right-angle prism, 3 – Cylindrical negative UV lens, 4 – Spherical positive UV lens, 5 – Investigated plasma source, 6 – Laser-power meter, 7 – ICCD camera synchronized with the laser, 8 – Laser beam, 9 – Fresnel rhomb tandem to turn polarization (if necessary). Reprinted from [14] ©IOP publishing. Reproduced with permission. All rights reserved.

extensions achieve spectrally narrower output, which is often advantageous for LIF experiments. Consequently, LIF using OPO lasers is in minority, though some achievements have been reported [21–23]. Another alternative may be a use of tunable excimer lasers, but this is limited to species that absorb in their relatively narrow tuning range [24]. Also, fluorescence measurements using tunable diode lasers (TDLs) should be mentioned. This technique offers the lowest spectral width of the output line among the mentioned methods, but as the tuning range is strongly limited, it is often insufficient to cover a satisfying number of rotational lines in the molecular spectra. Notable work on atomic LIF by TDL is, for example, [25].

The polarization of the laser output depends on the particular settings of the frequency conversion unit and it should be always cared for. The laser-induced fluorescence can be anisotropic [26]. Also, in most laboratory plasmas, the main scattering mechanism is Rayleigh scattering on the gas particles, causing possibly unwanted background signal. This mechanism does not take place in the direction of the light polarization and can thus be avoided by using horizontally polarized light for excitation. The laser polarization can be turned by 90° using a Fresnel rhomb tandem. A single Fresnel rhomb acts like a quarter wave plate, whereas a tandem of two Fresnel rhombs has an effect similar to half-wave plate, but working for broad-band radiation.

For fluorescence detection, any suitable light detector that can be synchronized with the laser pulses can be used. Photomultiplier tubes give directly temporally resolved result and are very sensitive for low-light detection, but using them for spatially resolved results is a difficult and often lengthy task. Nevertheless, notable results have been achieved by point-like measurements, scanning the area of interest [21, 27]. An intensified charge-coupled device (ICCD) camera offers an alternative, which gives directly two-dimensionally resolved results, basically a digital photograph, where the resolution is given mainly by the imaging optics. The modern ICCDs offer short exposure times, down to fractions of a nanosecond. The temporal development of fluorescence can then be obtained by taking a series of fluorescence images with varying delays of the exposure gate with respect to the laser pulse. When both spatial and temporal resolutions are desirable, the criterion for decision could be the amount of points necessary in time versus space.

The laser-induced fluorescence is often not the only light signal. If the investigated medium is an active discharge, it often radiates itself. If this background radiation is stable enough, it can be easily subtracted from the measured signal. Furthermore, the laser light is scattered on the particles of the investigated medium, either by Rayleigh scattering on atoms and molecules or by Mie scattering on larger particles like dust or droplets. The former may be eliminated or suppressed by using horizontally polarized light (see earlier text), but in general, the propagation of scattered laser light in the direction of the detector cannot be totally avoided. The laser and the fluorescence light can be separated either spectrally or temporally, depending on the particular experiment. If the fluorescence lifetime is long enough, the detection gate can be started after the end of the laser pulse, excluding the scattered laser light. Eq. (4), however, calculates the total fluorescence, including the photon emitted during the laser pulse. A correction for the fluorescence that was not acquired this way must be calculated. This is relatively easy, as the temporal development of the fluorescence is given by convolution of the laser pulse with a decay exponential, where the laser pulse temporal shape, the fluorescence lifetime, and the delay of the fluorescence detection gate are necessary inputs.

Another common strategy is the spectral separation. To achieve this, the laser wavelength must sufficiently differ from the wavelength of the fluorescence, often leading to more complicated excitation schemata, like the one in **Figure 3**. The most convenient way is a use of bandpass or edge filters in front of the detector. These usually still transmit a non-negligible amount of scattered laser light, which should be quantified by measuring with laser detuned from the absorption wavelength and then subtracted from the measured fluorescence signal. Some groups have used a grating monochromator instead [27]. This increases the complexity of the setup and decreases the sensitivity of fluorescence detection, with the possible benefit of obtaining spectrally resolved fluorescence. This is important for fundamental fluorescence research, but redundant for routine measurement of concentration.

4.1. The instruments used in our work

- Frequency-doubled pulsed Nd:YAG laser (Quanta-Ray PRO-270-30)
- Dye laser (Sirah-D-24-EG) with either a mixture of Rhodamine B + Rhodamine 101 for 612-nm radiation or Rhodamine 6G for 564 nm. In both cases is the output of the dye laser frequency doubled to achieve the wavelength of OH absorption.
- ICCD camera (PI-MAX 1024RB-25-FG43, 16-bit gray-scale resolution) with a UV-transmitting objective lens.

http://dx.doi.org/10.5772/intechopen.72274

5. Intensity of the measured fluorescence signal: parasitic effects

When investigating discharges via LIF, the excitation of chosen species usually is not the only consequence of the powerful laser radiation. In the following paragraphs, we would like to highlight the other parasitic effects of the laser-discharge interactions, which can affect the intensity of signal measured by the detector, namely:

- Scattering of laser on the surfaces of the discharge reactor
- Laser-induced fluorescence of used materials
- Laser-induced breakdown or other influences of the discharge
- Photodissociation of reactive species

The importance of laser-surface interactions varies greatly in different types of discharges. Special care should be taken in the case of surface discharges, where the laser beam passes the discharge in the proximity of the surface of electrodes or dielectrics. To avoid these phenomena, several tips can be realized. Where it is possible, the laser-surface interaction can be reduced by the proper choice of materials and discharge design. A smooth surface reduces laser scattering toward detector and the use of materials with low absorption in the OH-excitation wavelength range reduces parasitic fluorescence signals. In order to reduce contact of laser beam with solid surfaces, proper alignment of the beam is advisable, including the use of diaphragms and elimination of reflections to the discharge zone, for example, by means of windows tilted at Brewster angle.

Laser-induced breakdown was observed in dielectric barrier discharges (DBD) [28, 29]. When the laser beam hits the surface of dielectrics, it can cause a release of the electron charges trapped at the dielectrics. That can lead to ignition of the discharge at voltage lower than the regular ignition voltage threshold. It means that the laser pulse can behave as the trigger of the discharge. Moreover, during the measurement the laser can increase the intensity of the discharge and also the intensity of spontaneous emission of OH radicals. To avoid the laser-induced breakdown of AC discharges, it is helpful to synchronize the laser with the discharge and to perform the measurement in such a moment in the period of the supplied voltage when the electric field is not strong enough to allow the breakdown.

Fortunately, most of the parasitic effects are not strongly dependent on the laser wavelength. To determine their influence, we can measure the signal with laser wavelength slightly detuned from the OH absorption line. Then, the detector should register the same amount of the scattered laser, fluorescence of surfaces, and increase of signal caused by artificial breakdown. This signal can be subtracted from the OH fluorescence measurement.

In **Figure 6**, we present the example of strong parasitic effects in the case of measurement of OH radicals in diffuse coplanar dielectric barrier discharge, described in detail in [29]. The fluorescence signal was measured in different phases of the period of the supplied voltage. The dotted line shows the spontaneous emission of discharge measured with laser off, showing that the active discharge occurs mainly in the phase regions of 150–200° and 330–380°. The dash-dotted line is the signal measured with the laser wavelength tuned to the center of the

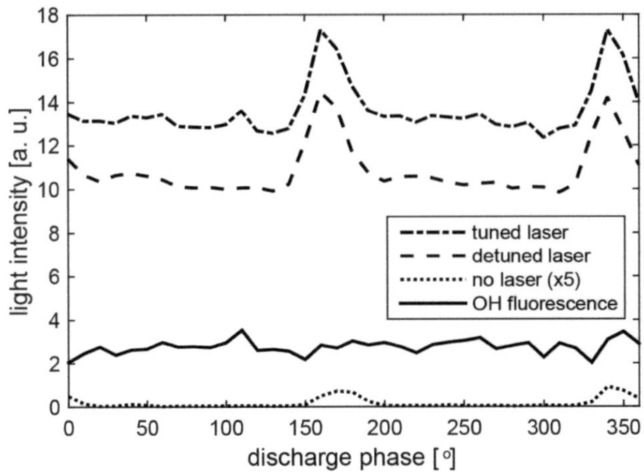

Figure 6. Signal strength vs. discharge phase. Dotted line was measured without laser radiation and is magnified 5×. Dash-dotted line is with the laser tuned to the OH absorption line, dashed line is with laser detuned. The solid line is the difference between the dash-dotted and the dashed.

absorption OH line. From that signal it may seem that the concentration of OH radicals is strongly increased during the active discharge. However, very similar evolution can be seen in the dashed line, which represents the measurement performed with laser wavelength detuned from OH absorption line. Since this signal does not include fluorescence of OH radicals induced directly by the laser, its temporary increase must be caused by the artificial breakdown of the discharge. Apart from this temporary increase the dashed line is permanently raised due to the scattered laser radiation and the fluorescence of the dielectric surface. The OH concentration should be calculated from the solid line, i.e. from the difference between the tuned and detuned measurement. Similar problematics of parasitic effects was studied also in [30] for two-photon absorption LIF of atomic hydrogen in a surface DBD.

Another effect which can cause artificial increase of measured signal is photodissociation. The laser photons needed for OH LIF measurement have energy sufficient for dissociation of, for example, ozone molecule. Oxygen atoms then can react with water vapor to form new OH radicals [31]. To cause fluorescence radiation, these artificially created radicals need one photon to be formed and second photon to be excited. Therefore, the photodissociation results in a quadratic increase in the dependence of the signal intensity on the energy of laser pulses. To avoid the effects of photodissociation, the shape of these dependences should be checked and the measurement should be taken in the region of energy where the dependence is not quicker than linear.

6. Intensity of the measured fluorescence signal: saturation effects

Another factor leading to deviation from linear relation of measured fluorescence signal and the exciting laser power is the effect of saturation. It appears with higher laser power when the depletion of the ground state by laser excitation, and the laser-stimulated emission processes are no longer negligible. Under such conditions, further increase of exciting laser power causes lower increase of fluorescence signal than expected. In the limit of very high laser power, the signal should be no longer dependent on the exciting laser power, leading to theoretically easier evaluation of the measurements [32]. However, in practice it is found that this approach is strongly influenced by spatial and temporal inhomogeneities in the laser beam—in other words, it is quite difficult to guarantee the full saturation conditions across the whole laser beam. This approach is nowadays not very popular and a vast majority of LIF publications works with linear regime. Remaining in the limits of linear regime may be disadvantageous for some experiments due to low fluorescence intensity. Working in partially saturated regime is a possible solution. The partial saturation for 0 – 0 excitation schema was thoroughly investigated in [29]. Let us summarize the main practical results.

Under the condition that rotational energy transfer (RET) in both electronic states is much faster than the spontaneous and collisional deexcitation of the laser-excited electronic state, the dependence of the time-integrated fluorescence intensity M_f on the exciting laser pulse energy E_L can be expressed as

$$M_f(E_f) = \alpha\left(E_f - \frac{p_1 E_f^2}{p_1 E_f + p_2}\left[\frac{e^{-(p_1 E_f + p_2)} - 1}{p_1 E_f + p_2} + 1\right]\right), \tag{6}$$

where

$$\alpha = N_{Xi}\frac{\kappa B}{c}\tau_0\left[\sum_{J'J''\alpha'\alpha''} F\left(\Lambda^{v'=0J'\alpha'}_{v''=0J''\alpha''}\right) d\left(\Lambda^{v'=0J'\alpha'}_{v''=0J''\alpha''}\right) A^{v'=0J'\alpha'}_{v''=0J''\alpha''} f_{v'=0J'\alpha'}(T)\right]\iiint_V D_a \frac{\Omega}{4\pi} s\,dV \tag{7}$$

describes the theoretical slope that would be observed in the linear case, compare with Eq. (4), and is called *fluorescence gain*. p_1 and p_2 are the parameters describing the saturation. In this form, the equation may be fitted to the measured dependence $M_f(E_f)$ by minimizing the sum of squared residuals with α, p_1, and p_2 as fit parameters. This formula was found to describe the dependence in the whole range from linear to fully saturated regime quite accurately, but was not practical for use with real-life data. The complexity of the problem caused the fits often to fail and the least-squares optimization was very slow. For this reason, the formula was substituted by a simpler one

$$M_f(E_L) = \frac{\alpha E_f}{1 + \beta E_f}, \tag{8}$$

where β is a parameter describing the degree of saturation, but has no direct physical meaning, as this formula is only an approximative one. The use of formula (8) can be justified by the fact that the second-order polynomial expansion matches with the second-order polynomial expansion of formula (6), and it has been shown that the results are within 10% deviation from the correct value of fluorescence gain α (from fitting Eq. (6)), if the following procedure is followed [29]:

1. First fit the whole range of measured data with the simplified Eq. (8), or its linearized form

$$\frac{1}{M_f(E_f)} = \frac{1}{\alpha E_f} + \frac{\beta}{\alpha}. \tag{9}$$

as the linear fits are very fast and robust. Take the resulting parameter estimates $\widehat{\alpha}_0$ and $\widehat{\beta}_0$.

2. Check the degree of saturation for the whole energy range. Exclude points with $\widehat{\beta}_0 E_f > 1$, reducing the data range such that the simplification (8, 9) can be used. Obviously, if you have no data left after this step, you should adjust the range of laser pulse energies during your next measurement accordingly.

3. Repeat the fit from step 1 with the reduced range. The resulting estimate $\widehat{\alpha}_1$ from this fit is now influenced by systematic error due to the use of the simplified equation by less than 10%.

The article [29] deals specifically with 0 – 0 excitation. This was motivated by the fact that the saturation effects were much stronger under conditions with low quenching and long lifetime of the laser-induced fluorescence, where this excitation scheme is advantageously used. For conditions with fast quenching, where the more complicated 0 – 1→1 – 1 + 0 – 0 scheme must be used, the saturation effects were usually weak.

7. Collisional processes

The role of collisional processes on quenching and redistribution of excited states was outlined in Section 3. At atmospheric pressure, collisional processes often present the fastest depopulation mechanism of excited states and they need to be taken into account during processing of measured data.

Collisional quenching of excited OH radicals depends on both the vibronic state of OH radical and the type of colliding species. Whereas in rare gases collisional quenching can be negligible when compared with radiative deexcitation (the radiative lifetime of OH(A) is around 700 ns), air at atmospheric pressure reduces the lifetime of excited OH radicals to nanoseconds or fractions of nanoseconds. Water vapor quenches excited OH even by an order of magnitude faster. Strong dependence of excited OH lifetime on the type of collisional partners enables to determine the gas composition, which will be shown in Section 10. On the other hand, this dependence complicates LIF evaluation, especially in environments with inhomogeneous gas composition or temperature.

Second collisional process is the vibrational energy transfer. Since VET noticeably affects the fluorescence spectra, it is important to take into account during LIF data processing especially if the detection sensitivity varies with wavelength. The influence of collisional partners on fluorescence spectra can be demonstrated by comparison of spectra measured in air and in water vapor when OH radicals were excited to the $v' = 1$ state: in air, most of excited OH radicals were transferred to the $v' = 0$ state prior to photon emission and the 0 – 0 band with head located at 306 nm dominated to the fluorescence spectrum. By contrast, the OH(A) lifetime in water vapor was so short that the OH radicals did not have time for VET; most of photons were emitted directly from the $v' = 1$ state and mainly radiation of the 1 – 1 band around 315 nm was observed [33].

When excitation to the $v' = 1$ state is realized, the two fluorescence bands 1 – 1 and 0 – 0 are usually detected. The 0 – 0 vibrational band that results from VET is delayed behind the 1 – 1 band that is activated directly by laser excitation. Similarly, as the temporal evolution of the 1 – 1 band intensity can be described by convolution of the excitation rate (i.e., the temporal profile of laser pulse) with the exponential decay of the $v' = 1$ state, the temporal evolution of the delayed 0 – 0 band can be calculated as a convolution of the temporal profile of the 1 – 1 band (which follows the $v' = 1$ state population) with the exponential decay of the $v' = 0$ state. This holds only when the VET rate constant does not vary during the fluorescence process, that is, when eventual changes of rotational distribution on the $v' = 1$ state do not influence the VET rate strongly.

Third collisional process is the rotational energy transfer, which tends to populate all rotational levels of vibrational states that appear in the fluorescence process. At atmospheric pressure, RET is sometimes so fast that the excited state is practically all the time in rotational equilibrium [20, 34]. In such a case, each vibrational state can be treated as only one effective level, which simplifies the LIF data processing considerably. When RET is not so fast, it is often sufficient to take out the one directly excited rotational level and to suppose that all other rotational levels are in rotational equilibrium [33, 35]. In general, when RET is not fast enough, rotational levels may not reach equilibrium during the OH^* lifetime and each vibrational state may be described by its own rotational temperature. Finally, the rotational distribution can be far from any thermalized state and population of each rotational level should be determined individually.

The collisional rate constants for $A^2\Sigma^+(v' = 0)$ and $(v' = 1)$ OH states can be found, for example, in [36–46] for quenching, in [39, 42–45] for VET, and in [45, 47, 48] for RET.

8. Spectral overlap of laser line and absorption line

The measured fluorescence signal depends also on the spectral profile of the laser beam and the absorption line of measured species. That is reflected by the quantity κ appearing in Eqs. (3) and (4)

$$\kappa = \int_{\nu} a(\nu)\, l(\nu)\, \mathrm{d}\nu, \tag{10}$$

where a is the spectral profile of the absorption line and l spectral profile of the laser, both normalized by $\int_{-\infty}^{\infty} a(\nu)\mathrm{d}\nu = \int_{-\infty}^{\infty} l(\nu)\mathrm{d}\nu = 1$.

The value of spectral overlap κ can be determined via measurement of the dependence of the fluorescence intensity on the laser wavelength, varied in the vicinity of the absorption line. Resulting spectral profile of measured signal is convolution of the profile of laser line and absorption line. When the laser line profile is known—in our case, the line is assumed to have a Gaussian profile with FWHM 0.4 pm, based on laser specifications from the vendor—the shape of absorption line can be obtained by deconvolution of the laser shape from the measured profile. However, numerical deconvolution is known to be highly sensitive to the noise of experimental data. Alternatively, the absorption line profile can be obtained by least-squares fitting. In that case, the absorption profile can be assumed to have a Voigt shape and the best fit between the convolution of the unknown Voigt profile with the known (or assumed) laser line and the measured profile of the excitation line is found. Then, the value of κ can be calculated using Eq. (10).

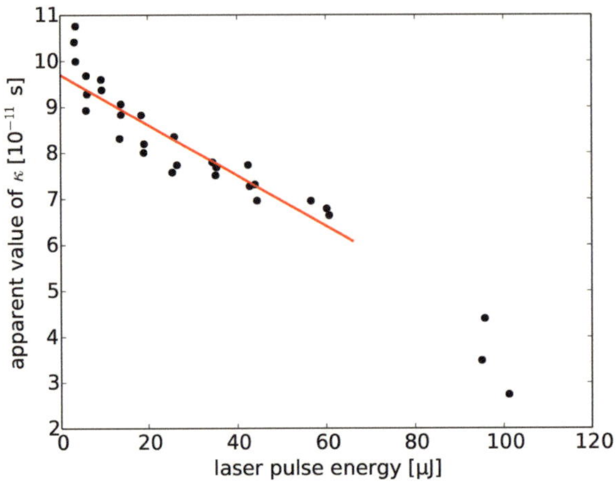

Figure 7. Apparent value of the spectral overlap κ versus the mean laser pulse energy during the measurement. The calculated value of κ is obviously affected by saturation effects. Therefore, its value extrapolated to the limit of zero laser pulse energy was used.

It should be noted that the obtained value of κ is dependent not only on temperature or pressure but is strongly affected also by saturation effects described in Section 6. Saturation lowers the fluorescence intensity mainly in the center of the absorption line, which leads to apparent broadening of the line. Therefore, for higher laser intensity the apparent value of κ is lower than in linear regime, as illustrated in **Figure 7**. For the evaluation of experiments, it is advised to carefully determine the value of κ not affected by saturation; otherwise, the significant systematic error can influence the results.

9. Calibration

The unknown integral $\int\int\int_V D_a \frac{\Omega}{4\pi} s\mathrm{d}V$ in Eqs. (3) and (4) usually needs to be calibrated. The constant D_a could be measured by means of a calibrated light source, but this procedure requires an independent determination of the laser-plasma interaction volume and does not take into account that the detection sensitivity may depend on the incidence angle of fluorescence radiation on the detection unit. Consequently, it is advantageous to calibrate the whole integral $\int\int\int_V D_a \frac{\Omega}{4\pi} s\mathrm{d}V$ directly, which can be achieved by measurement of Rayleigh scattering [49, 50] on a gas at known pressure and temperature.

The intensity of signal measured by Rayleigh scattering experiment can be expressed as

$$M_r = \frac{\mathrm{d}\sigma_r}{d\Omega} N_r \frac{E_r}{h\nu_r} d(\lambda_r) \int\int\int_V D_a \Omega s \mathrm{d}V + m_s E_r, \tag{11}$$

where $\mathrm{d}\sigma_r/d\Omega$ is the differential cross section for Rayleigh scattering, N_r is the concentration of atoms or molecules of the used gas, E_r is the mean energy of laser pulses during the collection of the scattering signal. ν_r and λ_r are the frequency and wavelength of laser radiation, respectively. $m_s E_r$ is a parasitic signal caused by laser scattering on surrounding objects or eventual dust particles in the gas. When this parasitic signal is eliminated, Eq. (11) enables to determine the value of the unknown integral.

Besides optimization of laser beam shape and discharge apparatus, there are two basic ways on how to eliminate the parasitic signal. The first way can be used when the discharge apparatus can be placed into a vacuum chamber. Since Rayleigh scattering intensity depends on gas pressure whereas scattering on surrounding objects not, it is easy to determine the m_s value from the dependence of measured scattering signal (M_r) on pressure and subtract the parasitic signal from measured data.

However, in some experiments vacuum cannot be used and Rayleigh scattering must be realized in a reactor that is open to ambient atmosphere. Besides laser scattering on reactor parts, dust particles may penetrate to the scattering volume and increase the parasitic signal. In this case, it is possible to use the anisotropy of Rayleigh scattering: Rayleigh scattering does not emit light to the direction parallel with laser polarization. Since scattering on surrounding objects and eventual dust particles does not depend markedly on laser polarization, it is possible to determine the parasitic signal intensity as the signal detected when laser polarization is turned by 90°.

Especially when dust particles can penetrate to the scattering volume, it is desirable to use suitable statistical processing of measured data. If sufficient number of scattering measurements is realized, it is possible to exclude deviating values (e.g., that differ from the mean value by more than triple of sample standard deviation) and/or to calculate the mean value of scattering signals from points between the first and third quartile.

10. Planar LIF in the effluent of a radio-frequency plasma jet

The plasma pencil is a radio-frequency (13.56 MHz)-driven plasma jet ignited in a fused silica capillary (inner diameter 2 mm) with a hollow-driving electrode encompassing the capillary [51, 52]. There is no grounded electrode. The plasma is ignited in argon gas flowing through the capillary and is thus divided from the electrode by a dielectric barrier of the capillary wall. The plasma pencil has been successfully used, for example, for cleaning of historical artifacts [53] or as an excitation source for atomic spectrometry [54]. The power values reported in this section are the power output of the RF generator, not the power absorbed in the plasma.

The argon gas was supplied by a series of pipes. It is known that water molecules adsorbed on the inner walls of pipes are slowly released into the gas phase [55]. The resulting humidity of the working gas is not necessarily unwanted as it allows the formation of OH radicals that play an important role in the plasma jet applications. For serious experiments, however, it should be kept constant and monitored. For this experiment, this was established by humidifying the gas artificially through thin membrane and waiting long enough for the humidity to stabilize. This was verified by measuring the humidity of the feed gas with a capacitive aluminum oxide moisture sensor (Panametrics, MMS 35).

In the effluent of an argon plasma jet, the lifetime of laser-excited OH is longer than the laser pulse and the simpler 0 – 0 excitation scheme may be used. To know the rotational temperature of OH(X) that is necessary for the absolute density determination, see Eq. (5), several rotational absorption lines were probed, namely $R_1(2,4,7)$. Whenever lower rotational temperature was expected (lower driving power or higher argon flow rates), also $R_1(5, 6)$ were probed for accuracy.

To obtain planar-resolved LIF images, the beam expansion optics was used, see **Figure 5**. The thickness of the laser sheet was notably smaller than the inner diameter of the jet capillary; the shown results are thus cross sections through a plane containing the axis of the effluent. The inhomogeneous laser beam profile was determined by acquiring an image of Rayleigh scattering on laboratory air. Prior to this, the detector was tested for the absence of vignetting or other obvious imaging errors. As the laser was propagating horizontally and no significant absorption losses were observed, only the vertical laser profile was of interest: $E(y)$. This is a necessary input for measurement of the fluorescence gain α, see Section 6 about the saturation effects. To obtain the fluorescence gain, a series of fluorescence images were captured, with simultaneously scanning and recording the laser pulse energy. This allows us to form a dependence $M_f(x, y) = f(E_f(y))$ for every pixel of the image and fit Eq. (9) to it. The python code for evaluation could look like this:

```
import numpy
import scipy.stats as st
laser_vertical_profile = numpy.load('laser_profile.npy') #vector
laser_vertical_profile /= numpy.sum(laser_vertical_profile)
#normalize

fluo_vs_pwr = numpy.load('fluo_vs_pwr.npy') #3D array - series of
ICCD images
                        #axes: 0-frame no. 1-vertical, 2-horizontal
laser_pulse_Es = numpy.load('laser_pulse_Es.npy') #vector

#usually, fluorescence images are accumulated over several laser
Pulses
#this is to get them on the same scale
#assumption: the laser pulse energy does not change much during
#acquiring a single image
ratio = len(laser_pulse_Es) // len(fluo_vs_pwr) #integer division
laser_Es = numpy.zeros(len(fluo_vs_pwr))
for i in range(len(fluo_vs_pwr)):
  #calculate mean energy of laser pulses measured during the
  #accumulation of each fluorescence image
  laser_Es[i] = numpy.mean(laser_pulse_Es[i*ratio:(i+1)*ratio])

#run fit of every pixel
#z denotes the number of the image zmax,
ymax, xmax = fluo_vs_pwr.shape
alphas = numpy.zeros((ymax, xmax))
alphas_errors = numpy.zeros((ymax, xmax))
for y in range(ymax):
```

```
#prepare the laser-pulse-energy vector for the y-position
en_y = numpy.zeros(zmax)
for frame in range(zmax):
    en_y[frame] = laser_Es[frame] * laser_vertical_profile[y]
for x in range(xmax):
    inv_lif = fluo_vs_pwr[:,y,x]**(-1)
    slope,intercept,r,p,stderr = st.linregress(en_y**(-1), inv_lif)
    alpha0 = 1/slope
    beta0 = intercept*alpha0
    #second-round fit with restricted range
    slope,intercept,r,p,stderr = st.linregress(en_y[(en_y*beta0)<1]**(-1),
                                               inv_lif[(en_y*beta0)<1])
    alpha1 = 1/slope
    beta1 = intercept*alpha1
    #keep the result in 2D array of alphas
    alphas[y,x] = alpha1
    alphas_errors[y,x] = stderr
```

Here, we have advantageously used the linearized Eq. (9) and the `linregress` function of `scipy.stats` library, which calculates the best fitting parameters analytically and is thus fast and robust.

Probing several absorption lines and calculating their fluorescence gain for each pixel allows estimating the rotational temperature with planar resolution, see the example in **Figures 8** and **9** and Eq. (5). Rotational temperature is often expected to agree with the kinetic temperature of the gas. The molecular fluorescence can be thus used for thermometry with very high spatial resolution, disturbing the measured plasma only weakly compared to most other thermometric

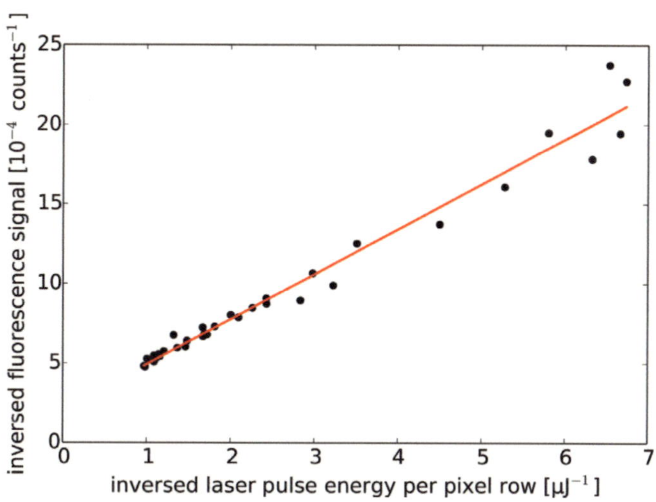

Figure 8. An example of the fluorescence gain determination. In the shown case, the fluorescence gain $\alpha = (3500 \pm 100)$ counts/μJ.

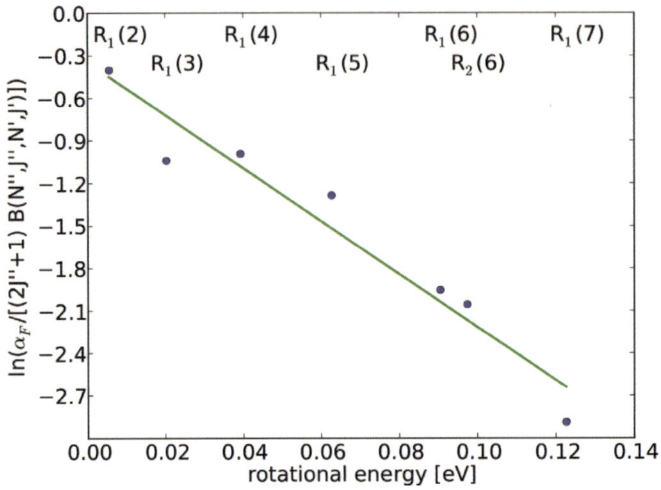

Figure 9. Fluorescence gains obtained from probing different absorption lines show Boltzmann distribution corresponding to rotational temperature $T = (620 \pm 9)$ K. J'' is the quantum number of the probed lower state and B is the absorption coefficient of the transition. Reprinted from [14] ©IOP publishing. Reproduced with permission. All rights reserved.

methods, but with the assumption that the investigated plasma source and the laser profile stay stable for the whole measurement time. An example of such temperature measurement is shown in **Figure 10**. A result of gas dynamics simulation is added for comparison. The simulation considers a simplified case without plasma, where hot argon is blown into room-temperature air. For further details, see [14].

To obtain the absolute concentration, the quantum gain of the fluorescence is also necessary, that is, the fraction of laser-excited molecules that got quenched must be quantified. In practice, the real fluorescence lifetime is measured and compared with the natural lifetime of the laser-excited state. As mentioned in Section 4, the fluorescence lifetime may be measured by taking a series of short-exposure fluorescence images with varying delays after the laser pulse. The gate width was set to 5 ns, which was sufficient for this task, as the OH fluorescence lifetime in argon is expected to be in the order of tens to hundreds of nanoseconds. The fluorescence decay for the 0 – 0 excitation scheme is expected to follow a single exponential, provided that the rotational energy transfer is much faster, which was satisfied in this case. Like the fluorescence gain parameter α, also the fluorescence lifetime τ is space-dependent. To evaluate the series of images and obtain a map of fluorescence lifetimes, a code similar to that for the fluorescence gain α was used. The theoretical function $M_f(t) = A\mathrm{e}^{-t/\tau}$ could in principal be linearized by taking a logarithm, but in practice, a constant offset was found to be non-negligible, so the model function was $M_f(t) = A\mathrm{e}^{-t/\tau} + C$, which cannot be linearized anymore. Nevertheless, the implementation of Levenberg–Marquardt algorithm available in `scipy` or `lmfit` libraries was found to work satisfactorily fast and robust, see an example fit in **Figure 11**.

The map of fluorescence lifetimes is not only necessary for calculating the map of OH concentration but also contains information about the local quenching rate. For simple cases where only two nonreactive gas environments with sufficiently different quenching rates meet, this can be used to estimate the mixing ratio at each position. For the resulting fluorescence lifetime τ holds.

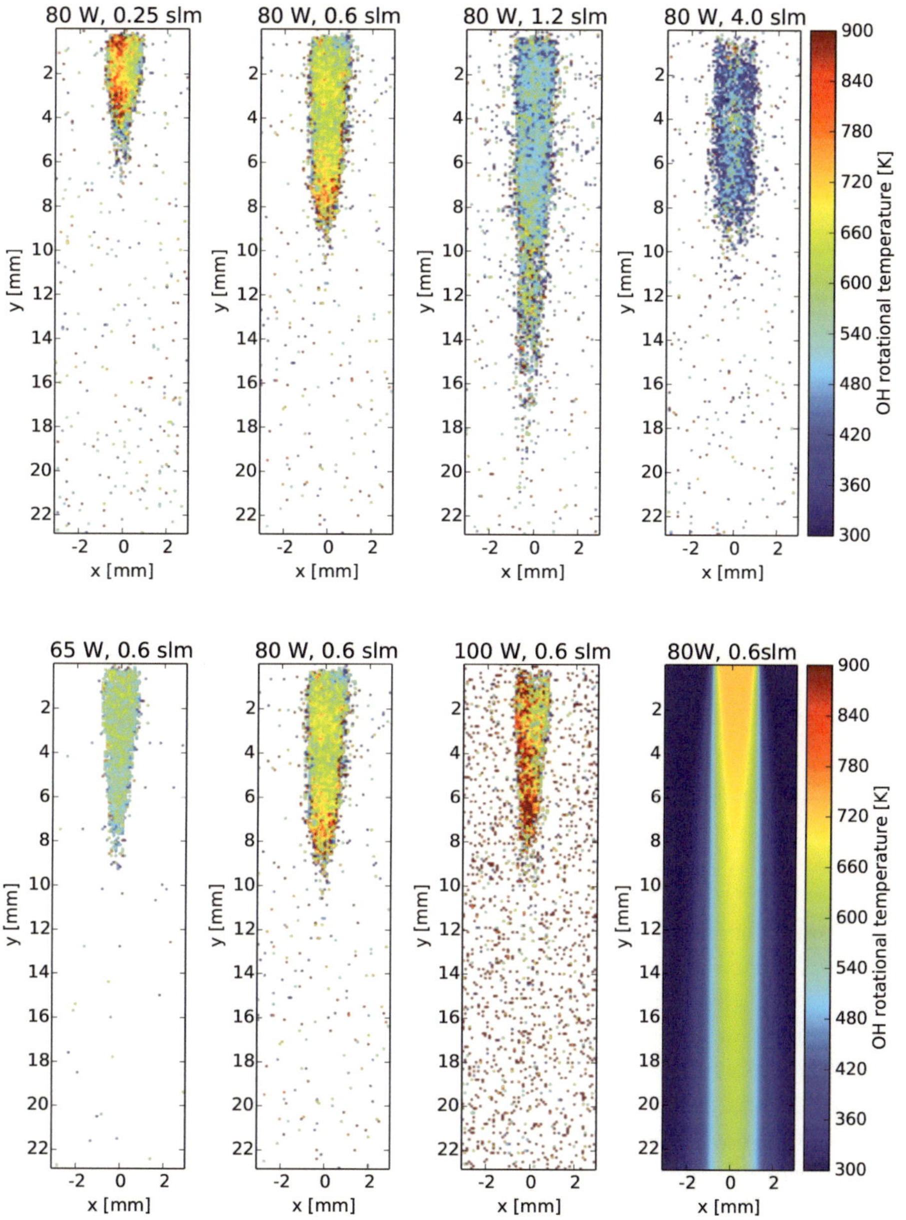

Figure 10. Measured discharge temperature for various discharge conditions. Bottom-right: Result of gas dynamics simulation (no plasma taken into account) for comparison. Reprinted from [14] © IOP publishing. Reproduced with permission. All rights reserved.

$$\frac{1}{\tau} = \sum_{J'J''\alpha'\alpha''} f_{v'=0J'\alpha'}(T) A^{v'=0J'\alpha'}_{v''=0J''\alpha''} + \sum_i n_i q_i(T) \tag{12}$$

where n_i is the concentration of collision partner i and $q_i(T)$ is its temperature-dependent quenching coefficient in $m^3\ s^{-1}$. To estimate the argon-air mixing ratio, a set of theoretical

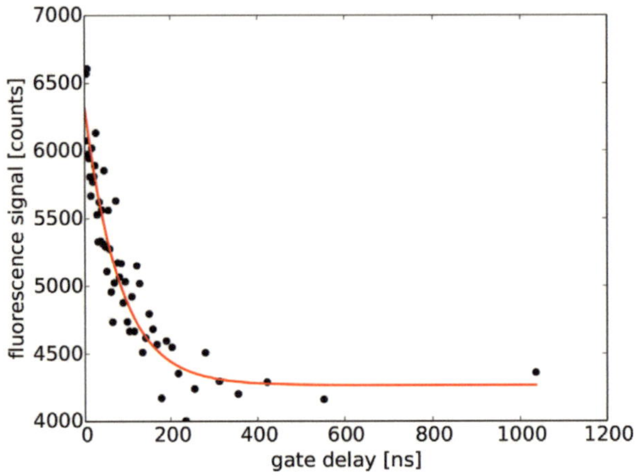

Figure 11. An example of the single exponential fit to the measured decay of the fluorescence signal at one pixel for RF driving power of 80 W and argon flow rate of 0.6 slm. The fitted lifetime is $\tau = (81 \pm 10)$ ns. Reprinted from [14] ©IOP publishing. Reproduced with permission.

fluorescence developments was simulated with LASKIN [56] for humid air and humid argon for the whole range of temperatures found in the preceding experiments in steps of 25 K. For each condition, a set of unique pairs temperature-lifetime was assembled. The amount of water in air was 1.3 volume percent (found by laboratory hygrometer), the amount of water in argon was dependent on the flow rate and was measured by the precise hygrometer coupled to the argon supply (ranging from 160 to 1030 volume ppm). The mixing ratio was then calculated by

$$\frac{n_{\mathrm{air}}}{n_{\mathrm{air}} + n_{\mathrm{Ar}}} = \frac{1/\tau_{\mathrm{meas}}(x,y) - 1/\tau_{\mathrm{Ar}}(T(x,y))}{1/\tau_{\mathrm{air}}(T(x,y)) - 1/\tau_{\mathrm{Ar}}(T(x,y))}, \tag{13}$$

where (x, y) are pixel coordinates, $\tau_{\mathrm{meas}}(x, y)$ is the measured lifetime as a result of the single exponential fit for each pixel, $\tau_{\mathrm{Ar}}(T(x, y))$ and $\tau_{\mathrm{air}}(T(x, y))$ are the theoretical lifetimes from the LASKIN simulations for humid argon and air, respectively. The temperature was taken from the rotational temperature measurement mentioned earlier. The resulting maps of mixing ratio are shown in **Figure 12**. The zone with low air admixture elongates with the increase of argon flow rate up to 2–3 slm, when transition to turbulent flow appears. A similar method was used by Yagi et al. for spatially resolved measurement of humidity based on quenching of NO radical [57]. For further investigations of gas dynamics in plasma jets, see also [58–60].

Finally, in **Figures 13** and **14**, the measured maps of OH concentration are shown with the images of the spontaneous plasma emission for comparison. It can be clearly seen that the regions with the highest OH concentration do not coincide with the regions with the strongest plasma emission. On the contrary, the highest OH concentration was found downstream from the tip of the visible discharge. Furthermore, we observe a hollow profile of OH concentration, which is the most pronounced for the highest argon flow rate of 4.0 slm. These conditions were also those with the lowest humidity of the feed gas, so this hollow profile is attributed to the amount of water molecules available for dissociation by collisions with the active particles of the plasma—mainly electrons and argon metastables.

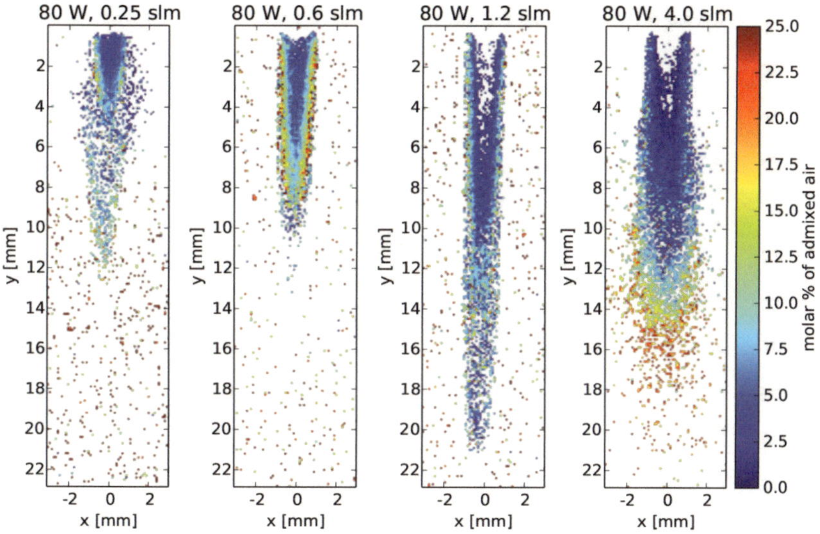

Figure 12. Maps of molar percent of ambient air admixed into the argon effluent of the plasma pencil for various conditions. The points with fluorescence lifetime uncertainty greater than 25% are displayed as white. Reprinted from [14] ©IOP publishing. Reproduced with permission. All rights reserved.

We have seen that the final result—map of OH concentration—consists of many steps, which means that this diagnostic method is not an easy one. On the other hand, proper evaluation of each of these steps has provided a very complex information about the active zone of the plasma jet effluent which can bring deep insight into the processes occurring there.

11. Surface DBD in water vapor

An example of a challenging LIF experiment is the OH measurement in a surface coplanar DBD ignited in water vapor at atmospheric pressure [33]. The coplanar DBD is made of a system of coplanar strip electrodes that are covered by a ceramic and plasma is created in the gas on the opposite side of the ceramic. The neighboring strip electrodes are charged with opposite polarity, which enables to cover a large area of the ceramic by macroscopically homogeneous plasma [61–63]. Since plasma layer is only 0.3-mm thick, the density of the power delivered to the discharge is high and the discharge is expected to generate high concentration of reactive radicals. Certainly, the small thickness complicates fluorescence investigation: all parasitic effects resulting from laser-surface interactions that are discussed in Section 5 can be expected and a good spatial resolution is required.

In order to generate a source of OH radicals that does not produce a number of other reactive species and that could be used for a specific surface treatment, the coplanar DBD was ignited in pure water vapor. In order to prevent water condensation at atmospheric pressure, a heated reactor was built, and both the electrode system and reactor walls were heated to 120°C. Again, water atmosphere complicates fluorescence measurements since water is a very

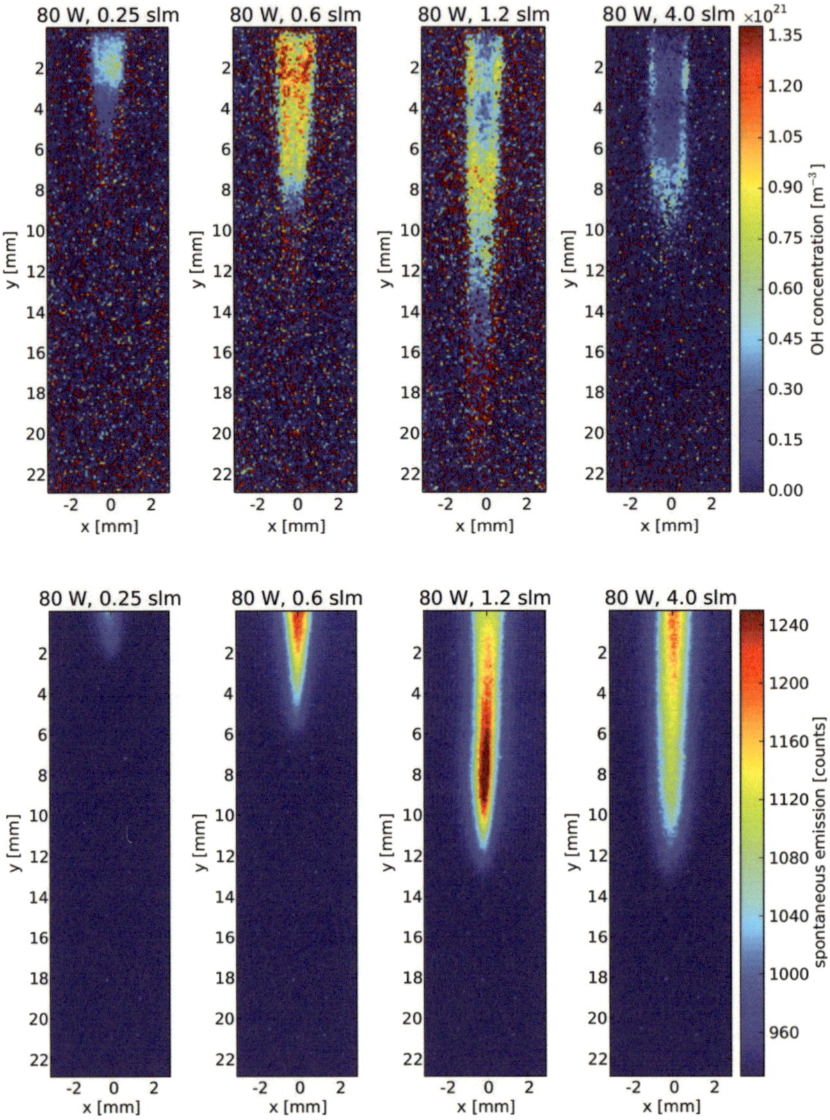

Figure 13. Maps of OH concentration for different argon flow rates (top) with images of spontaneous plasma emission (bottom). Reprinted from [14] ©IOP publishing. Reproduced with permission. All rights reserved.

effective quencher of excited OH radicals, which reduces their lifetime bellow 100 ps. In order to minimize reflections of laser beam, the windows for laser input and output were tilted at Brewster angle.

Due to the short lifetime of the excited OH state, it was necessary to excite OH radicals to the vibrationally excited state, and emission from both ($v' = 0, 1$) vibronic states was taken into account. The RET was not fast enough to establish rotational equilibrium during the short lifetime, which led to an overpopulation of the directly excited rotational level that had to be treated separately during the processing of measured data.

http://dx.doi.org/10.5772/intechopen.72274

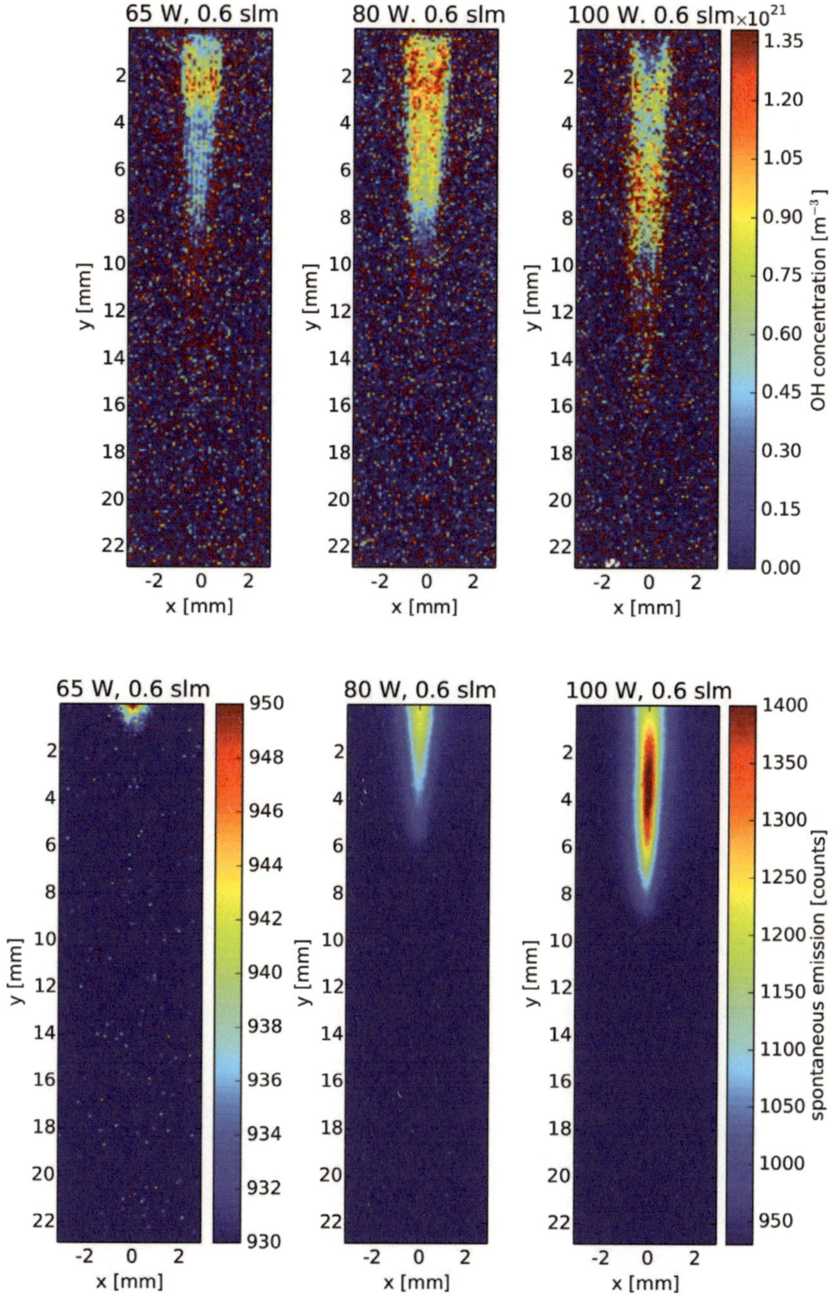

Figure 14. Maps of OH concentration for different values of driving power (top) with images of spontaneous plasma emission (bottom). Reprinted from [14] ©IOP publishing. Reproduced with permission. All rights reserved.

As discussed in Section 5, it was necessary to subtract the parasitic signal of scattered laser radiation and fluorescence of the dielectrics, which both can be quantified by means of the signal that was measured when laser wavelength was detuned from the absorption line of OH radicals. This measurement was further useful for detection of water droplets. Although they

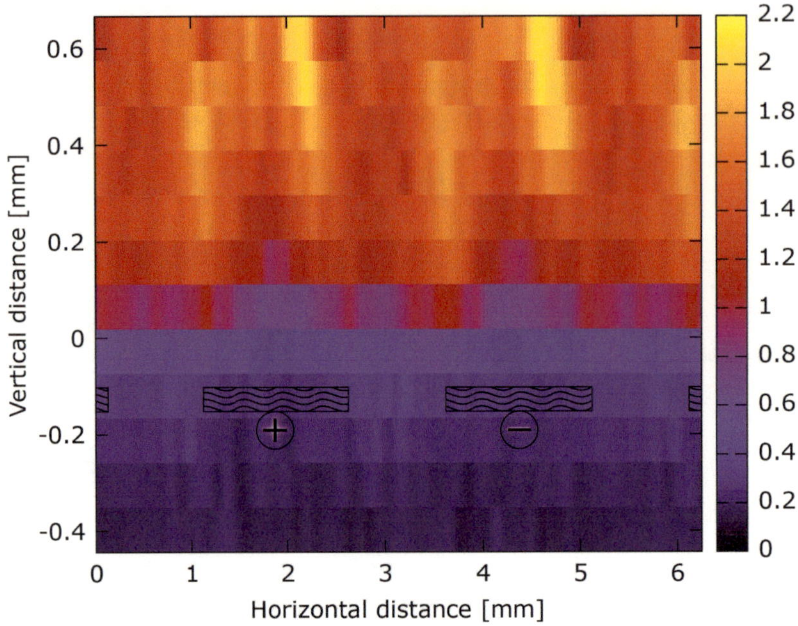

Figure 15. Laser radiation scattered on water droplets that were present in the coplanar DBD ignited in water vapor. Measured with laser wavelength detuned from the absorption line of OH radicals. Reprinted from [33] ©IOP publishing.

were not visible by naked eye, water droplets were really present in the gas phase as demonstrated in **Figure 15**. In order to eliminate spontaneous discharge emission, detection of droplets was realized in such a part of the period of the supplied voltage, when electric field was weak and it was not possible to ignite a discharge event. The shown distribution of droplets was not uniform—the scattering signal was weak in the discharge region, where gas was heated and droplets evaporated quickly. In addition, we can see an increased signal above electrode edges that demonstrates that besides direct gas heating there is another influence of discharge on the distribution of water droplets.

Another parasitic (from the fluorescence point of view) signal is the spontaneous discharge emission. Also, this signal is useful from another point of view and it enables, for example, to determine the spatial structure of discharge. In order to increase the signal-to-noise ratio in short measurements required for high temporal resolution, it is possible to synchronize the discharge event with the measurement by a laser shot with wavelength detuned from OH absorption line. In such a case, the signal of scattered laser radiation should be subtracted from the measured picture. An example of obtained spatial structure of the coplanar DBD is shown in **Figure 16**. It demonstrates that in water vapor the discharge is located namely above the area between electrodes.

After subtraction of parasitic signals, the LIF data can be finally used for the determination of OH concentration. As shown in **Figure 17**, the distribution of OH radicals followed well the shape of the visible discharge that bridged the interelectrode area. The OH concentration reached the value 5×10^{20} m^{-3}, which was by an order of magnitude higher than OH

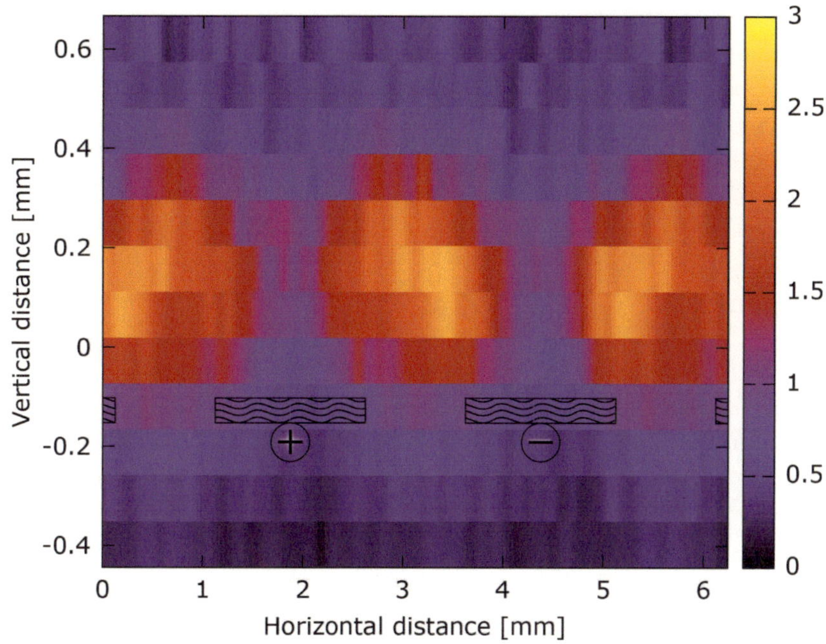

Figure 16. Emission of DBD ignited in water vapor at atmospheric pressure. Discharge ignition was synchronized with ICCD measurement by a laser shot. Reprinted from [33] ©IOP publishing. Reproduced with permission. All rights reserved.

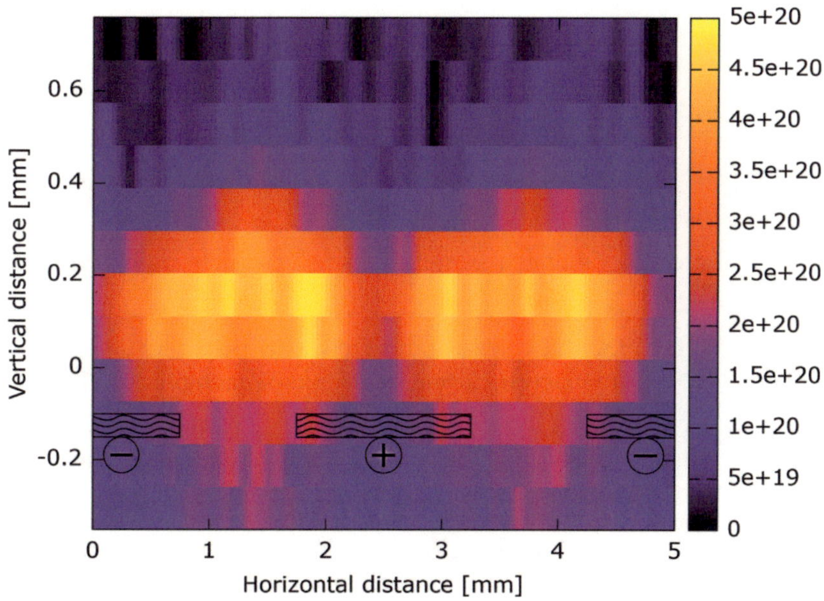

Figure 17. OH concentration determined by fluorescence measurement in the coplanar DBD ignited in water vapor at atmospheric pressure. Reprinted from [33] ©IOP publishing. Reproduced with permission. All rights reserved.

concentration in analogical discharge ignited in air (with relative humidity of 40%). By means of two-photon absorption LIF of atomic hydrogen and by optical emission spectrometry, it was verified that plasma in pure water vapor generated almost solely OH radicals, concentration of other reactive radicals was negligible. As a result, the LIF method proved that the investigated discharge can be used as a selective source of oxidizing OH radicals with relatively high concentration [33].

Acknowledgements

This research has been supported by the project LO1411 (NPU I) funded by the Ministry of Education, Youth and Sports of Czech Republic and by the Czech Science Foundation projects 16-09721Y and 17-04329S.

Author details

Jan Voráč*, Pavel Dvořák and Martina Mrkvičková

*Address all correspondence to: vorac@mail.muni.cz

Department of Physical Electronics, Faculty of Science, Masaryk University, Brno, Czech Republic

References

[1] Voráč J, Synek P, Procházka V, Hoder T. State-by-state emission spectra fitting for non-equilibrium plasmas: Oh spectra of surface barrier discharge at argon/water interface. Journal of Physics D: Applied Physics. 2017;**50**:294002

[2] Isaksen I, Dalsøren S. Getting a better estimate of an atmospheric radical. Science. 2011;**331**(6013):38-39

[3] Lee JH, Park JW, Lee HB. Cell adhesion and growth on polymer surfaces with hydroxyl groups prepared by water vapour plasma treatment. Biomaterials. 1991;**12**(5):443-448

[4] Sun B, Sato M, Clements J. Use of a pulsed high-voltage discharge for removal of organic compounds in aqueous solution. Journal of Physics D: Applied Physics. 1999;**32**(15):1908

[5] Repine J, Fox RB, Berger E. Hydrogen peroxide kills *Staphylococcus aureus* by reacting with staphylococcal iron to form hydroxyl radical. Journal of Biological Chemistry. 1981;**256**(14):7094-7096

[6] Tiedemann A. Evidence for a primary role of active oxygen species in induction of host cell death during infection of bean leaves withbotrytis cinerea. Physiological and Molecular Plant Pathology. 1997;**50**(3):151-166

[7] Visseren FL, Verkerk M, Van Der Bruggen T, Marx J, Van Asbeck B, Diepersloot R. Iron chelation and hydroxyl radical scavenging reduce the inflammatory response of endothelial cells after infection with chlamydia pneumoniae or inuenza a. European Journal of Clinical Investigation. 2002;**32**(s1):84-90

[8] Dieke G, Crosswhite H. The ultraviolet bands of oh fundamental data. Journal of Quantitative Spectroscopy and Radiative Transfer. 1962;**2**(2):97-199

[9] Luque J, Crosley DR. Transition probabilities in the $A^2\Sigma^+$-$X^2\pi i$ electronic system of OH. The Journal of Chemical Physics. 1998;**109**(2):439-448

[10] Luque J, Crosley D. Lifbase: Database and spectral simulation program (version 1.5). SRI international report MP. 1999;**99**(009)

[11] Herzberg G, Spinks JWT. Molecular Spectra and Molecular Structure: Diatomic molecules. Van Nostrand: Princeton, 1950

[12] Lefebvre-Brion H, Field RW. The Spectra and Dynamics of Diatomic Molecules. Elsevier Academic Press. 2004

[13] Luque J, Crosley D. Lifbase: Database and spectral simulation program (version 1.5), vol. 99, 1999

[14] Voráč J, Obrusník A, Procházka V, Dvořák P, Talába M. Spatially resolved measurement of hydroxyl radical (OH) concentration in an argon RF plasma jet by planar laser-induced fluorescence. Plasma Sources Science and Technology. 2014;**23**:025011

[15] Verreycken T, Mensink R, van der Horst R, Sadeghi N, Bruggeman PJ. Absolute OH density measurements in the effluent of a cold atmospheric-pressure Ar-H_2O RF plasma jet in air. Plasma Sources Science Technology. 2013;**22**:055014

[16] Li L, Nikiforov A, Xiong Q, Britun N, Snyders R, Lu X, Leys C. OH radicals distribution in an Ar-H2O atmospheric plasma jet. Physics of Plasmas. 2013;**20**:093502

[17] Lu X, Naidis GV, Laroussi M, Reuter S, Graves DB, Ostrikov K. Reactive species in non-equilibrium atmospheric-pressure plasmas: Generation, transport, and biological effects. Physics Reports. 2016;**630**:1

[18] Yonemori S, Ono R. Flux of OH and O radicals onto a surface by an atmospheric-pressure helium plasma jet measured by laser-induced fluorescence. Journal of Physics D: Applied Physics. 2014;**47**:125401

[19] K. Kohse-Höinghaus. Laser techniques for the quantitative detection of reactive intermediates in combustion systems. Progress in Energy and Combustion Science. 1994;**20**(3): 203-279

[20] Voráč J, Dvořák P, Prochazka V, Ehlbeck J, Reuter S. Measurement of hydroxyl radical (OH) concentration in an argon RF plasma jet by laser-induced fluorescence. Plasma Sources Science and Technology. 2013;**22**(2):025016

[21] Yonemori S, Nakagawa Y, Ono R, Oda T. Measurement of oh density and air-helium mixture ratio in an atmospheric-pressure helium plasma jet. Journal of Physics D: Applied Physics. 2012;**45**(22):225202

[22] Lee Y-H, Maus RG, Smith BW, Winefordner JD. Laser-induced fluorescence detection of a single molecule in a capillary. Analytical Chemistry. 1994;**66**(23):4142-4149

[23] Giamarchi P, Burel L, Stephan L, Lijour Y, Le Bihan A. Laser-induced fluorescence with an opo system. Part i. optimisation of the analytical system by use of experimental design methodology. application to the direct quantification of traces of benzo [a] pyrene. Analytical and Bioanalytical Chemistry. 2002;**374**(3):490-497

[24] Andresen P, Bath A, Gröger W, Lülf H, Meijer G, Ter Meulen J. Laser-induced fluorescence with tunable excimer lasers as a possible method for instantaneous temperature field measurements at high pressures: checks with an atmospheric ame. Applied Optics. 1988;**27**(2):365-378

[25] Vitelaru C, Aniculaesei C, De Poucques L, Minea T, Boisse-Laporte C, Bretagne J, Popa G. Tunable diode-laser induced fluorescence on al and ti atoms in low pressure magnetron discharges. Journal of Physics D: Applied Physics. 2010;**43**(12):124013

[26] Brockhinke A, Kreutner W, Rahmann U, Kohse-Höinghaus K, Settersten TB, Linne MA. Timewavelength-and polarization-resolved measurements of OH(A $^2\Sigma^+$) picosecond laser-induced fluorescence in atmospheric-pressure flames. Applied Physics B: Lasers and Optics. 1999;**69**(5):477-485

[27] Riès D, Dilecce G, Robert E, Ambrico P, Dozias S, Pouvesle JM. LIF and fast imaging plasma jet characterization relevant for NTP biomedical applications. Journal of Physics D: Applied Physics. 2014;**47**(27):275401

[28] Ambrico P, Ambrico M, Šimek M, Colaianni A, Dilecce G, De Benedictis S. Laser triggered single streamer in a pin-to-pin coplanar dielectric barrier discharge. Applied Physics Letters. 2009;**94**(23):231501

[29] Voráč J, Dvořák P, Procházka V, Morávek T, Ráhel J. Dependence of laser-induced fluorescence on exciting-laser power: Partial saturation and laser-plasma interaction. The European Physical Journal Applied Physics. 2015;**71**(2):20812

[30] Mrkvičková M, Ráhel J, Dvořák P, Trunec D, Morávek T. Plasma Sources Science and Technology. 2016;**25**:055015

[31] Teramoto Y, Kim H-H, Ogata A, Negishi N. Measurement of OH ($X^2\Sigma$) in immediate vicinity of dielectric surface under pulsed dielectric barrier discharge at atmospheric pressure using two geometries of laser-induced fluorescence. Journal of Applied Physics. 2014;**115**:133302

[32] Piepmeier E. Theory of laser saturated atomic resonance fluorescence. Spectrochimica Acta B: Atomic Spectroscopy. 1972;**27**(10):431-443

[33] V. Procházka, Tučekovà Z, Dvořàk P, Kovàčik D, Slavíček P, Zàhoranovà A, Voràč J. Coplanar surface barrier discharge ignited in water vapour - a selective source of OH radicals proven by (TA)LIF measurement. Plasma Sources Science and Technology. 2017; **27**:015001

[34] Vorac J, Prochazka V, Dvorak P. Vibrational and rotational energy transfer in a (2) sigma (+) state of oh radicals measured by laser induced fluorescence. CHEMICKE LISTY. 2012;**106**: S1504-S1507

[35] Dunn M, Masri A. A comprehensive model for the quantification of linear and nonlinear regime laser-induced fluorescence of oh under $a^2\sigma$+ $x^2\pi$ (1, 0) excitation. Applied Physics B. 2010;**101**(1-2):445-463

[36] Dilecce G, Ambrico PF, Simek M, Benedictis SD. LIF diagnostics of hydroxyl radical in atmospheric pressure He-H_2O dielectric barrier discharges. Chemical Physics. 2012;**398**: 142-147

[37] Copeland RA Crosley DR. Rotational level dependence of electronic quenching of OH (A $^2\Sigma^+$, v′ = 0). Chemical Physics Letters. 1984;**7**(3):295-300

[38] Copeland RA, Dyer MJ, Crosley DR. Rotational level dependence of electronic quenching of OH(A $^2\Sigma^+$,v′ = 0). Journal of Chemical Physics. 1985;**82**:4022-4032

[39] Copeland RA, Wise ML, Crosley DR. Vibrational energy transfer and quenching of hydroxyl (A $^2\Sigma^+$,v′ = 1). Journal of Chemical Physics. 1988;**92**:5710

[40] Wysong IJ, Jeffries JB, Crosley DR. Quenching of A $^2\Sigma^+$ OH at 300K by several colliders. Journal of Chemical Physics. 1990;**92**:5218-5222

[41] Bailey AE, Heard DE, Henderson DA, Paul PH. Collisional quenching of OH(A $^2\Sigma^+$, v′ = 0) by H2O between 211 and 294K and the development of a unified model for quenching. Chemical Physics Letter. 1999;**302**:132-138

[42] Williams LR, Crosley DR. Collisional vibrational energy transfer of OH(A $^2\Sigma^+$, v′ = 1). Journal of Chemical Physics. 1996;**104**:6507

[43] Hartlieb AT, Markus D, Kreutner W, Kohse-Hoeinghaus K. Measurement of vibrational energy transfer of OH(A $^2\Sigma^+$, v′ = 1→0) in low pressure. Applied Physics B. 1997;**65**:81-91

[44] Rahmann U, Kreutner W, Kohse-Hoeinghaus K. Rate-equation modeling of single and multiplequantum vibrational energy transfer of OH(A $^2\Sigma^+$, v′ = 0 to 3). Applied Physics B. 1999;**69**(1):61-70

[45] J. Burris, J. Butler, T. Mcgee, and W. Heaps, Collisional deactivation rates for (A $^2\Sigma^+$, v′ = 1) state of OH. Chemical Physics. 1988;**124**(2):251-258

[46] Burris J, Butler J, McGee T, Heaps W. Quenching and rotational transfer rates in the v′ = 0 manifold of OH(A $^2\Sigma^+$). Chemical Physics. 1991;**151**:233-238

[47] Joerg A, Meier U, Kienle R, Kohse-Hoeinghaus K. State-specific rotational energy transfer in OH(A $^2\Sigma^+$; v′ = 0) by some combustion-relevant collision partners. Applied Physics B. 1992;**55**:305

[48] Kienle R, Joerg A, Kohse-Hoeinghaus K. State-to-state rotational energy transfer in OH (A $^2\Sigma^+$; v′ = 1). Applied Physics B. 1993;**56**:249-258

[49] Miles RB, Lempert WR, Forkey JN. Laser rayleigh scattering. Measurement Science and Technology. 2001;**12**(5):R33

[50] Bucholtz A. Rayleigh-scattering calculations for the terrestrial atmosphere. Applied Optics. 1995;**34**(15):2765-2773

[51] Janca J, Klima M, Slavicek P, Zajickova L. HF plasma pencil-new source for plasma surface processing. Surface and Coatings Technology. 1999;**116**:547-551

[52] Slavíček P, Brablec A, Kapička V, Klíma M, Šíra M. Longitudinal emission diagnostics of plasma channel in rf barrier torch discharge. Acta Physica Slovaca. 2005;**55**:573-576

[53] Janča J, Zajíčková L, Klima M, Slavíček P. Diagnostics and application of the high frequency plasma pencil. Plasma Chemistry and Plasma Processing. 2001;**21**(4):565-579

[54] Novosád L, Hrdlička A, Slavíček P, Otruba V, Kanický V. Plasma pencil as an excitation source for atomic emission spectrometry. Journal of Analytical Atomic Spectrometry. 2012;**27**(2):305-309

[55] Winter J, Wende K, Masur K, Iseni S, Dünnbier M, Hammer M, Tresp H, Weltmann K, Reuter S. Feed gas humidity: A vital parameter affecting a cold atmospheric-pressure plasma jet and plasma-treated human skin cells. Journal of Physics D: Applied Physics. 2013;**46**(29):295401

[56] Bülter A, Lenhard U, Rahmann U, Kohse-Höinghaus K, Brockhinke A. Laskin: Efficient simulation of spectra affected by energy transfer. In: Laser Applications to Chemical and Environmental Analysis. p. TuE4: Optical Society of America; 2004

[57] Yagi I, Ono R, Oda T, Takaki K. Two-dimensional lif measurements of humidity and oh density resulting from evaporated water from a wet surface in plasma for medical use. Plasma Sources Science and Technology. 2014;**24**(1):015002

[58] Voráč J, Hnilica J, Kudrle V, Dvořák P. Spatially resolved measurement of hydroxyl (oh) radical concentration in a microwave plasma jet by planar laser-induced fluorescence. Open Chemistry. 2015;**13**(1)

[59] Voráč J, Potočňáková L, Synek P, Hnilica J, Kudrle V. Gas mixing enhanced by power modulations in atmospheric pressure microwave plasma jet. Plasma Sources Science and Technology. 2016;**25**(2):025018

[60] Voráč J, Synek P, Potočňáková L, Hnilica J, Kudrle V. Batch processing of overlapping molecular spectra as a tool for spatio-temporal diagnostics of power modulated microwave plasma jet. Plasma Sources Science and Technology. 2017;**26**(2):025010

[61] Černák M, Černáková L, Hudec I, Kováčik D, Zahoranová A. Diffuse coplanar surface barrier discharge and its applications for in-line processing of low-added-value materials. European Physical Journal Applied Physics. 2009;**47**:22806

[62] Černák M, Kovàik D, Ràhěl J, St'ahel P, Zahoranovà A, Kubincovà J, Tóth A, Černàkovà L. Generation of a high-density highly non-equilibrium air plasma for high-speed large-area at surface processing. Plasma Physics Control Fusion. 2011;**53**:124031

[63] Kelar J, Čech J, Slavíček P. Energy efficiency of planar discharge for industrial applications. Acta Polytechnica. 2015;**55**:109

Chapter 7

Study of Formation and Decay of Rare-Gas Excimers by Laser-Induced Fluorescence

Frédéric Marchal, Neermalsing Sewraj,
Jean-Pierre Gardou, Nofel Merbahi and
Mohammed Yousfi

Additional information is available at the end of the chapter

http://dx.doi.org/10.5772/intechopen.71942

Abstract

The aim of this chapter is to review the experimental and numerical techniques for the estimation of the laser-induced fluorescence (LIF) decay in rare gases using time-correlated single-photon counting. The advantages of single-photon counting technique are discussed by means of measurement uncertainty analysis. In addition, this chapter provides information concerning the application of this technique to filamentary dielectric barrier discharges (DBD) and radiation trapping of the resonant transitions.

Keywords: single-photon counting, VUV emission of rare gases, excimer formation and decay, laser-induced fluorescence, dielectric barrier discharges, radiation trapping

1. Introduction

Vacuum ultraviolet (VUV) radiation sources are of great interest for many applications such as photochemistry, surface treatment, sterilization, water treatment, photolithography and mercury-free lamps [1–4]. These applications often use dielectric barrier excimer lamps [5–11]. Optimization of these VUV sources requires numerical modeling associated with experimental characterization. The determination of the main mechanisms that lead to excimer formation or decay requires accurately determining the reaction rates.

The emission mechanisms of VUV radiations by rare gases are now reasonably understood [12–15]. Excimers correlated to the first metastable or resonant states are responsible for these emissions. Kinetic models of formation and decay of rare-gas excited states were developed and validated by laser-induced fluorescence (LIF) in the case of argon, krypton, xenon and krypton-xenon mixtures.

Indeed, kinetic studies of rare gases via VUV fluorescence decays, following a brief and selective multiphotonic excitation, recorded by using single-photon counting fluorescence are highly suitable for the determination of reliable kinetic models. Absorption and emission spectroscopies help to complete such studies based on time dependence of VUV fluorescence [12–16, 19].

In these experimental studies, VUV fluorescence was monitored by a VUV photomultiplier tube operating in the single-photon counting mode. The emission wavelength was selected by either a VUV monochromator for the recording of the emission spectra or a suitable VUV range passband interferential filter for the excitation spectra and fluorescence decay studies. A multichannel scaler operating with 16,384 channels and a time resolution of 5 ns were used to record the time variation of the VUV luminescence.

Due to the rather weak detected luminous flux, the histograms recorded are characterized by a large number of channels and a low counting rate per bin. The modeling of the fluorescence decay of excited species introduces a temporal function, representing the rate equation model of collisional and radiative decay in a pure gas or a gas mixture. As far as the number densities of excited species in the interaction volume remain low and the laser energy kept low enough to avoid multiphotonic ionization, this function can be fitted by a sum of exponential decays, each one characterized by its own decay frequency associated to a particular single excited state. The number of decay frequencies corresponds to the number of states involved in the formation-decay processes.

In order to determine the temporal evolution law, the experimental histograms must be processed by using the maximum likelihood method, applied to a Poisson distribution which is more suitable than the least squares method for histograms with a low counting rate per channel.

The aim of this chapter is to review the experimental and numerical techniques combined with spectroscopic studies for the estimation of the laser-induced fluorescence (LIF) decay in rare gases. The advantages of single-photon counting techniques are discussed by means of measurement uncertainty analysis. In addition, this chapter provides information concerning the application of these techniques to filamentary dielectric barrier discharges (DBD) and radiation trapping of the resonant transitions.

2. Mechanisms of VUV emissions by rare gases

2.1. Rare-gas excimers

The emission mechanisms of VUV radiation by rare gases are now reasonably understood. Excimers correlated to the first metastable or resonant states are responsible for these emissions. **Figure 1** gives an example of excited molecular states involved in VUV emission of the continua of krypton [17]. The first metastable state and the first resonant one are the precursors to the excimer states. Three-body collisional processes lead to the formation of the high vibrational levels of these excimers. In low-pressure conditions, the transition of these high vibrational levels correlated to the lowest atomic excited states toward the weakly bound ground-state dimers leads to the emission of the first continuum. The second continuum is observed at high pressures,

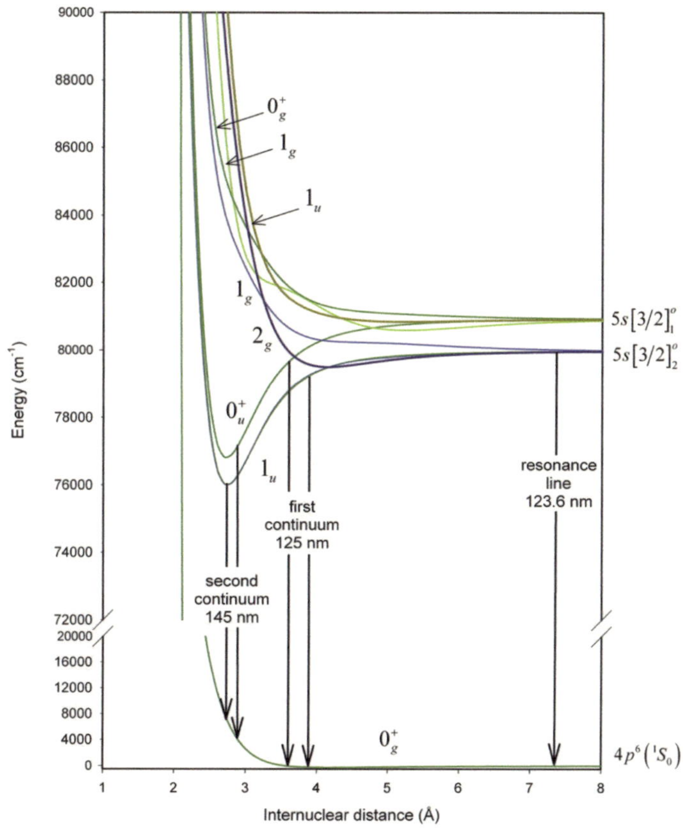

Figure 1. VUV emissions of the first and second continuum of krypton.

for which collisional relaxation of the high vibrational levels of these excimers is most likely. Thus, the second continuum, arising from radiative transitions of the lowest vibrational levels of the molecular excited states toward the ground state, dominates the spectrum.

2.2. Kinetic studies of VUV emissions of rare gases

Kinetic studies using laser-induced fluorescence (LIF) are highly suitable for the determination of kinetic models of formation and decay of rare-gas excimers.

The short-pulsed laser excitation populates only one atomic or molecular state (e.g. only the first resonant or only the first metastable states or the dissociative molecular states correlated to these atomic states). The initial time conditions are known at the beginning of the fluorescence decay, and the analysis of the time-correlated VUV fluorescence is greatly simplified. Non-selective excitation techniques such as electron beam, pulsed discharge or α-particles are not rigorous enough for the determination of kinetic models of the formation and decay of excimers because several processes compete (ionization, electron-ion recombination, electron-ion formation and destruction, second electron kinetics, etc.) [18, 21].

To illustrate the principle of kinetic studies of rare gases, we consider a simple atomic or molecular system of four states displayed in **Figure 2**. Just after the laser pulse, there is only

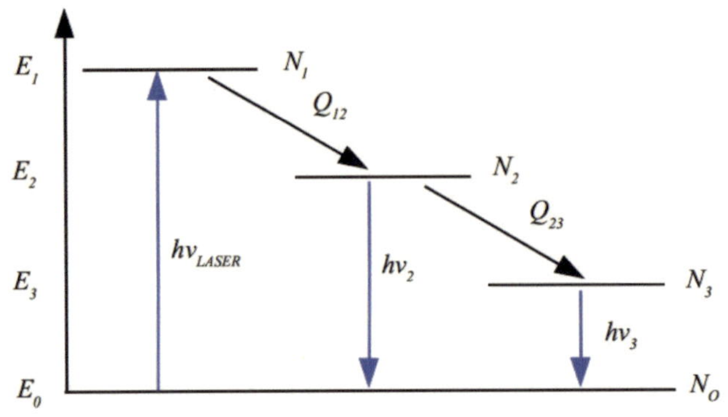

Figure 2. Simple model of laser-induced fluorescence for rare gases following the selective excitation of the first metastable state.

one exited state, for example, a metastable state noted N_1. The collisional decay of this state characterized by the quenching coefficient Q_{12} leads to populate a radiative state N_2. This state can also populate through a collisional reaction (characterized by the quenching coefficient Q_{23}), a lower level named N_3. This state can only decay by fluorescence emission. N_0 is the ground state.

This simple model can be used in the case of rare gases. The fluorescence coming from the N_2 molecular state is the first continuum, and the emission of the N_3 molecular state is called the second continuum. The exited state N_1 is the first metastable state.

The radiative decay frequencies α_1 and α_2 are relative to the two radiative states N_2 and N_3.

The number densities of the different states obey to the following ordinary differential equations:

$$\begin{cases} \dfrac{dN_1(t)}{dt} = -Q_{12}N_1(t) \\ \dfrac{dN_2(t)}{dt} = Q_{12}N_1(t) - (\alpha_1 + Q_{23})N_2(t) \\ \dfrac{dN_3(t)}{dt} = Q_{23}N_2(t) - \alpha_3 N_3(t) \end{cases} \tag{1}$$

provided that the laser pulse duration is short and the number densities of excited states can be comparatively negligible with respect to the number density of the ground state.

Thus, the initial conditions are

$$\begin{cases} N_1(t=0) = N_{10} \\ N_2(t=0) = 0 \\ N_3(t=0) = 0 \end{cases} \tag{2}$$

The number density N_0 of the ground state is supposed to remain constant. By solving the differential equations and considering the initial conditions, the time evolution of the number densities of the excited states are obtained:

$$\begin{cases} N_1(t) = N_{10}e^{-Q_1 t} \\ N_2(t) = N_{10}\dfrac{Q_{12}}{-Q_{12}+\alpha_1+Q_{23}}\left[e^{-Q_1 t} - e^{-(\alpha_2+Q_{23})t}\right] \\ N_3(t) = N_{10}Q_{12}Q_{23}\left[\dfrac{e^{-Q_1 t}}{(-Q_{12}+\alpha_1+Q_{23})(\alpha_3-Q_{12})} + \dfrac{e^{-(\alpha_2+Q_{23})t}}{(-Q_{12}+\alpha_1+Q_{23})^2} + \dfrac{e^{-\alpha_3 t}}{(-Q_{12}+\alpha_1+Q_{23})(-\alpha_3+Q_{12})}\right] \end{cases} \tag{3}$$

According to these expressions, the time decay of the VUV fluorescence intensities is given by

$$\begin{cases} I_{VUV2}(t') = \dfrac{\Omega}{4\pi}V_0 T\eta h\nu_2 \displaystyle\int_{t'}^{t'+\Delta t} \alpha_2 N_2(t)dt \\ I_{VUV3}(t') = \dfrac{\Omega}{4\pi}V_0 T\eta h\nu_3 \displaystyle\int_{t'}^{t'+\Delta t} \alpha_3 N_3(t)dt \end{cases} \tag{4}$$

where V_0 represents the observed volume, Ω is the solid angle of detection, T is the transmission of the optical system of detection, η is the efficiency of the photomultiplier at the fluorescence wavelength $\lambda_{VUV} = \frac{c}{\nu}$ and Δt is the time resolution of the detection system. The time acquisition of the fluorescence decay is supposed to be greater than the laser pulse width τ_L.

Overall, if the number densities of excited species in the interaction volume V_0 remains weak and the laser energy is low enough to avoid multiphotonic ionization, then collisions between excited species, ion recombination or electron collisions can be neglected. Thus, the simple model developed above can be generalized. The function *f*(t), which is the rate equation model of collisional and radiative decay in a neutral rare gas, is given by

$$f(t) = \sum_{p=1}^{p=D} a_p e^{-b_p t} + a_0 \tag{5}$$

where D is the number of excited states involved in the formation-decay processes, b_p is the decay frequency of each state and a_0 is the background noise of the detection system.

In pure rare gases, the decay frequency of the pth term with respect to the pressure p is

$$b_p = \alpha_p + Q_p = \alpha_p + k_{2p}N_{0+}k_{3p}N_0^2 \tag{6}$$

where α_p represents the radiative decay frequency, k_{2p} is the two-body collisional rate constant and k_{3p} is the three-body collisional rate constant. The sum $k_{2p}N_{0+}k_{3p}N_0^2$ is the quenching coefficient Q_p. In rare-gas mixtures, the expression of the quenching coefficient is more complex due to the three-body heteronuclear collisional rate constant.

Modeling the experimental decay of the fluorescent intensities at various pressures is the best way to determine the rate equation model of collisional and radiative decay in rare gases. So, a numerical method of computing the decay frequencies b_p from the experimental data fitted with a sum of exponential terms is required. The rate constants, α_p, k_{2p} and k_{3p}, can be determined by processing all the experimental data of the decay frequencies measured in the range of rare-gas pressures. Subsequently, kinetic schemes of formation and decay of rare-gas excimers can be proposed.

3. Single-photon counting method applied to the fluorescence decay recordings

3.1. Single-photon counting

As mentioned above, the time function of the fluorescence decay *f*(t) can be fitted by a sum of exponential terms if the number densities of excited space is very low compared to the number density of the ground state and if the laser excitation is short and selective enough. In order to fulfill these conditions, the laser intensity was considerably reduced to obtain a very low number density of the initially excited species. Thus, the intensity of the fluorescence becomes very weak, and the fluorescence photons are detected as separated pulses using a photomultiplier tube. In our experiments, only a few photoelectrons are detected per laser pulse. The average time intervals between signal pulses are wider than the time resolution of the detection system. In this case, the single-photon counting method using a photomultiplier is very effective and is superior to analog signal measurement in terms of stability and signal-to-noise ratio [22].

The single-photon counting method allows the number of photoelectrons detected to be in direct proportion of the fluorescence intensity. The signal pulses are counted by a multichannel scaler during the interval of two successive laser pulses, and the signal measured at each laser pulse period is accumulated at the laser pulse frequency to reproduce the fluorescence waveforms.

3.2. Photon counter system

The photon counter system is displayed in **Figure 3**. The fluorescence photons are collected by a photomultiplier tube (PMT). A non-inverting linear amplifier amplifies the negative output pulses of the PMT. The lower pulses are eliminated by the constant fraction discriminator, while the rest are reshaped. The output logic pulse corresponds to the point on the leading edge of the input pulse where the input pulse has risen to a fraction of its maximum amplitude. Thus, the time jitter is reduced, and the time resolution of the circuit system remains the same over a wide dynamic range of pulse amplitudes. The multichannel scaler counts events as a function of time. The laser trigger starts the counter, which segments photon count data into sequential time bins. The width of the bins can be set from a few nanoseconds to several milliseconds. The instrument records the number of photons that arrive in each bin. The multichannel scaler can be programmed to accumulate several records or set to free run. Each record is added to the current accumulator totals.

An example of time-resolved single-photon counting measurement of laser-induced florescence is shown in **Figure 4**. The experimental florescence is represented by a histogram with the bin number (proportional to the time) on x-axis and the counts on y-axis.

http://dx.doi.org/10.5772/intechopen.71942

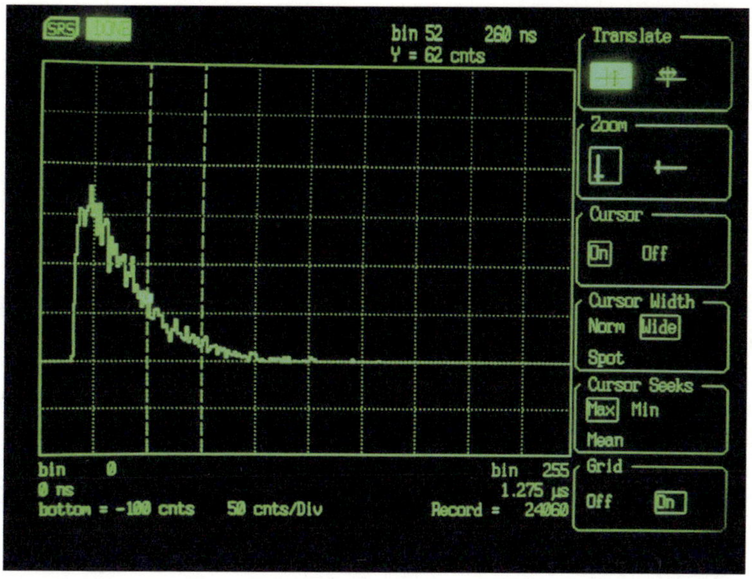

Figure 3. Photon counter system.

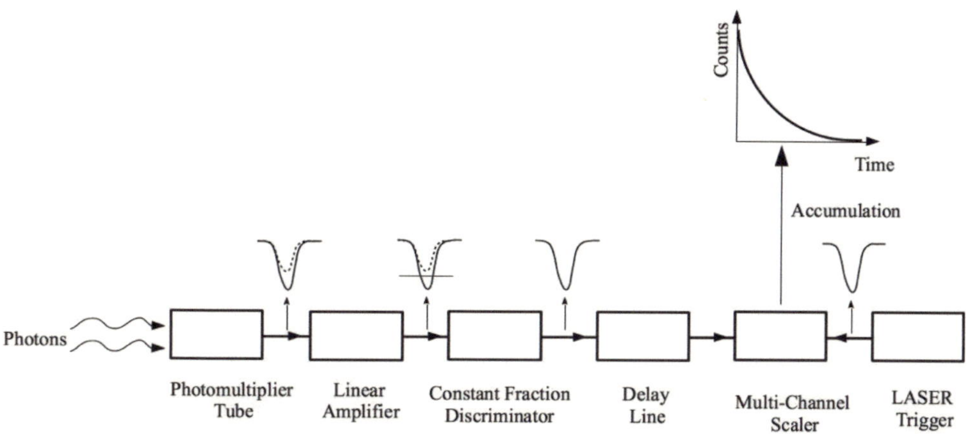

Figure 4. Laser-induced fluorescence decay recorded with a Stanford research (SR 430) multichannel Scaler (bin width, 5 ns; number of bins, 16,384; number of records, 4196).

In most cases, the luminous flux detected being rather weak, the histograms recorded are characterized by a large number of bins and a low counting rate per bin width.

The single-photon counting method is effective when the probability of detection of a single photon is greater than the probability of detection of a more than one photon. Therefore, each count corresponds to an individual incident photon, and these incident photons are detected as separate pulses. This is the case if the pulse-to-pulse interval is much greater than the pulse width.

3.3. Detection probability

Using a photomultiplier, when the average number of incident photons is $\overline{n}$, the probability of observing exactly y counts per time unit is given by the Poisson distribution:

$$P(y,\overline{n}) = \frac{(\eta\overline{n})^y}{y!} e^{-\eta\overline{n}} \quad (7)$$

where η is the quantum efficiency of the photomultiplier. η is the ratio of the average number of emitted photoelectrons from the photocathode per time unit to the average number of photon incident on the photocathode. For VUV fluorescence detection, a solar-blind photomultiplier was used with a Cs-I photocathode and MgF_2 window. The sensitivity of these photomultipliers is in the range of 110 to 200 nm, and the efficiency reaches typically 15% at maximum.

If the incident number of photons is sufficiently low, the probability of detection of photoelectrons is proportional to the number of incident photons:

$$P(y \geq 1, \overline{n}) \approx \eta\overline{n} \quad (8)$$

The detection of photoelectrons is unlikely, but when it occurs, a single photoelectron is detected because the probability of detection of more than a photoelectron vanishes.

4. Experimental methods

4.1. LIF setup

The updated experimental setup for LIF experiments shown in **Figure 5** is designed to the selective excitation by multiphotonic absorption of the first resonant or metastable states of rare gases (argon, krypton and xenon) and to the recording of the VUV fluorescence decay of rare-gas excimers or exciplexes.

The excitation wavelength range was obtained with the frequency-doubled light of a laser beam coming from a two-stage dye laser amplifier (Sirah Cobra Stretch) pumped by the second or the third harmonic beam of a neodymium YAG laser (Spectra-Physics). A spectral width of 0.002 nm was obtained with a double-grating oscillator. Typically, the dye laser operates in the visible spectral range, and the frequency of the dye laser beam is doubled by a BBO crystal followed by a compensator. The pulse duration of the frequency-doubled beam did not exceed $\tau_L = 7$ ns with a repetition rate of 30 Hz. The UV laser pulse energy is measured with an energy meter (Ophir Nova II) equipped with a pyroelectric energy sensor. The laser energy of the output beam is controlled by a motorized variable laser attenuator (Standard 10 MVAA) having an extremely low output beam deviation and a wide spectral range emission flux variation.

The vertically polarized VUV laser beam entered a cylindrical stainless steel chamber, and it provided excitation along the cell's axis. The luminescence was observed perpendicularly to the laser beam, through a magnesium fluoride porthole, by means of a VUV photomultiplier tube (PMT: EMR-510G 'extremely solar blind') working in a single-photon counting mode. The

http://dx.doi.org/10.5772/intechopen.71942

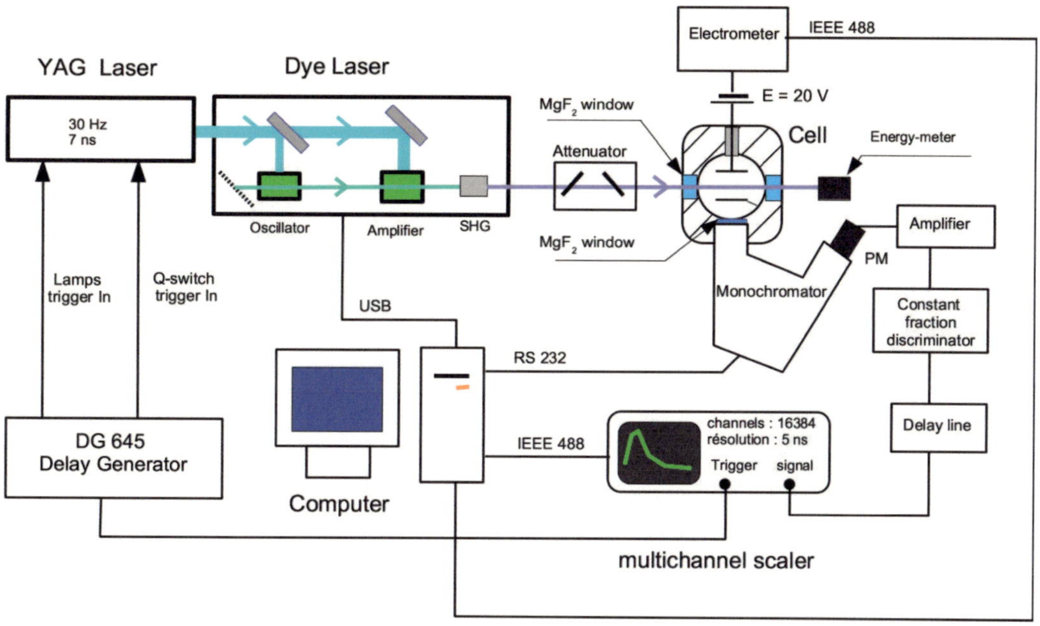

Figure 5. Schematic view of the LIF experimental setup.

detection wavelength was selected by either a VUV McPherson monochromator, for the recording of the emission spectra, or an interferential filter, for the excitation spectra and fluorescence decay recordings. The time dependence of the luminescence was monitored by using a multichannel scaler Stanford Research SR430 with 16,384 channels and a time resolution of 5 ns. The laser and the multichannel scaler were triggered by a digital delay-pulse generator (Stanford Research DG 645). An electrometer may also be added to record ionization spectra and to control multiphotonic ionization. The laser beam was on the axis of a plane capacitor where a dc electric field of 40 V cm^{-1} was applied. The charges were collected with a Keithley 617 electrometer, which could measure electric currents as low as 10 fA.

Table 1 displays some examples of configurations of the lasers used in our studies in accordance with the excited state in the case of argon, krypton and xenon (Racah notation is used).

4.2. Experimental histograms

Figure 6 shows an experimental histogram recorded at ≈128 nm representative of the fluorescence decay of the second continuum of argon following the excitation of the Ar4s[3/2]$_1$ state.

State	YAG wavelength	Dye	Output laser wavelength range	Multiphotonic process
Ar4s[3/2]$_1$	532 nm	DCM	300–330 nm	Three-photon absorption
Kr5s[3/2]$_2$	355 nm	Coumarin 307	240–270 nm	Two-photon absorption
Xe6s[3/2]$_2$	532 nm	Rhodamine B	284–307 nm	Two-photon absorption

Table 1. Laser dyes and multiphotonic process involved in kinetic studies of rare gases.

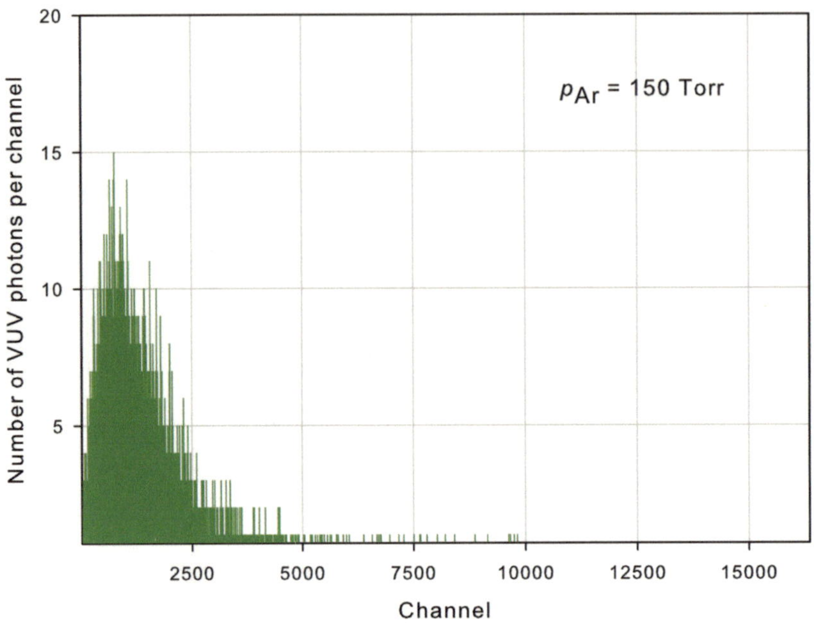

Figure 6. Experimental histogram of the VUV fluorescence following $Ar4s[3/2]_1$ selective excitation by three-photon absorption (LIF p_{Ar}, 150 Torr; λ_{em}, 128 nm; R, 5 ns) [15].

Typically, the recorded histograms are characterized by a large number of channels or bins (N_C = 16,384) and a low bin width (time resolution, R = 5 ns). For that emission wavelength, the quantum efficiency of the photomultiplier is η = 0.15. The photon counting system provides input pulses of width ≈2 ns. This histogram was recorded during 18,000 LASER pulses. This count is the accumulation number (N_A). The total VUV fluorescence acquisition time is $t_{ac} = N_C \times R \times N_A = 3.52$ s. The number of photoelectrons detected is about N_{PE} = 10,000, and the maximum of counts per channel is N_{PEmax} = 15. In average, during this experiment, the probability of counting one photoelectron during a bin width is less than 1.5×10^{-5} and reaches 3.5×10^{-4} at maximum. The pulse-to-pulse interval is more than 15 μs that is larger than the photoelectron pulse width. Thus, single-photon counting conditions are fully met.

To understand the mechanisms of formation and decay of rare-gas excimers, the histograms recorded at different pressures must be analyzed to determinate the frequency decay rates and also the lifetimes of the radiative states and the quenching rates.

5. Methods for fitting multiexponential histograms of the fluorescence decay

The luminescence decay curves follow a multiple exponential form after short laser excitation. Two methods for the curve fitting of multi-exponentials to experimental data have been compared. The chosen method needs to be accurate if the experimental signal contains exponential components with similar decay rates or both fast and slow decay rates.

5.1. Maximum likelihood method derived from a Poisson distribution

The luminescence decay curves $f(t)$ follow multiple exponential form:

$$f(t) = \sum_{p=1}^{p=D} a_p e^{-b_p t} + a_0. \tag{9}$$

The fitting method allows to obtain the amplitudes a_p and the decay frequencies b_p. The number D of exponential terms depends on the number of states involved in the process of decay. For the sake of concision and clarity, let $\theta = (a_0, a_1, \dots, a_D, b_1, \dots, b_D)$ be the $2D + 1$ dimension vector of unknown parameters of the function $f(t)$.

For fitting multi-exponential decay curves to experimental data, many methods of data analysis can be proposed to minimize the least squares deviation. The estimator to be minimized is

$$\Phi = \sum_{i=1}^{i=N_C} \frac{1}{\sigma_i^2} \left(y_i - f(t_i, \theta)\right)^2 \tag{10}$$

where N_C is the number of channels (bins) of the histogram, y_i is the number of counts for the ith channel observed at time t_i and σ_i is the standard deviation of the statistical distribution. Here, $\sigma_i^2 = y_i$ (Poisson distribution).

Taking into account the weak probability $p(y_i)$ of detection of y_i counts in the *ith* channel at time t_i given by Poisson's law

$$P(y_i) = \frac{(f(t_i, \theta))^{y_i}}{y_i!} f(t_i, \theta) \tag{11}$$

another estimator is more relevant: the likelihood function defined as the product of the probabilities:

$$L = \prod_{i=1}^{i=N_C} \left[\frac{(f(t_i, \theta))^{y_i}}{y_i!} f(t_i, \theta)\right] \tag{12}$$

The objective of fitting method is to determine the θ parameter that maximizes the likelihood function.

Gradient vanishing $\frac{\partial L}{\partial \theta} = 0$ leads to the non-linear systems of D equations:

$$\sum_{i=1}^{i=N_C} \left(\frac{y_i}{f(t_i, \theta)} - 1\right) \frac{\partial f(t_i, \theta)}{\partial \theta} = 0 \tag{13}$$

with

$$\frac{\partial L}{\partial \theta_p} = L\left[\sum_{i=1}^{i=N_C} \left(\frac{y_i}{f(t_i, \theta)} - 1\right) \frac{\partial f(t_i, \theta)}{\partial \theta_p}\right] \quad \textit{with } p = 1, \dots, 2D + 1 \tag{14}$$

The function $f(t)$ can be specified introducing the resolution R of the system of detection (e.g. the bin width of the histograms) and the channel number i:

$$f(i) = \sum_{p=1}^{p=D} a_p e^{-R(i-1)b_p} + a_0 \tag{15}$$

Thus, a new set of equations is obtained:

$$\begin{cases} \sum_{i=1}^{i=N_C} \left(\frac{y_i}{f(i)} - 1 \right) = 0 \\ \sum_{i=1}^{i=N_C} y_i \frac{e^{-R(i-1)b_p}}{f(i)} - \sum_{i=1}^{i=N_C} e^{-R(i-1)b_p} = 0 \qquad with\ p = 1, \ldots, D \\ \sum_{i=1}^{i=N_C} y_i (i-1) \frac{e^{-R(i-1)b_p}}{f(i)} - \sum_{i=1}^{i=N_C} (i-1) e^{-R(i-1)b_p} = 0 \end{cases} \tag{16}$$

These equations are solved numerically using the Newton-Raphson method. The iterative root-finding procedures of the Newton-Raphson method need an initial estimate θ_0 graphically obtained. Then, an improved estimate θ_u is produced after u iterations. The iterations go on until the relative variations of the estimate θ become negligible.

Linear equations are obtained with a first-order Taylor development of the non-linear equations about the estimate θ_u. Thus, the system of linear equations is solved with a classical Gauss algorithm.

5.2. Numerical simulations

For a simulated data derived from experimental fluorescence decays of the second continuum of krypton in krypton-xenon mixtures [17]. The exact value of the fluorescence decay is

$$f(t) = 10e^{-10^5 t} + 400e^{-2\times10^6 t} - 500e^{-1.5\times10^7 t} \tag{17}$$

Different data were simulated using a Monte Carlo method with a number of total accumulated counts varying from from 500 to 500 000. The simulated data were fluctuated by a Poisson distribution.

An example of simulated data is shown in **Figure 7**. Typically, the histograms recorded are characterized by a large number of N_C channels and a low counting rate per R = 5 ns width channel. This allows to measure with the same recording, the simultaneous measurement of both slow and fast decay rates with the same resolution and without time integration.

The decay frequency values (b_1, b_2, b_3) resulting from a least squares fit or a likelihood fit were compared to exact values. Estimation of uncertainties was performed through a statistical analysis (Type A evaluation) [25]. The standard deviation of the first convergent values was combined with the standard deviation of estimation of the decay frequencies obtained by processing the numerical method with a series of Monte Carlo-simulated histograms.

The mean values and standard deviations of the parameter estimations are summarized in **Figure 8**.

http://dx.doi.org/10.5772/intechopen.71942

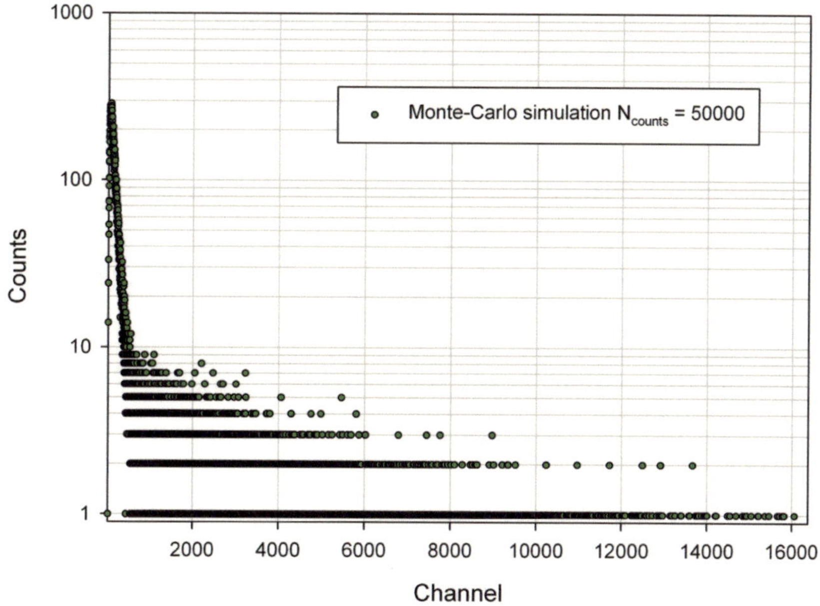

Figure 7. Simulated histogram of the VUV fluorescence (N_C, 16,384; R, 5 ns).

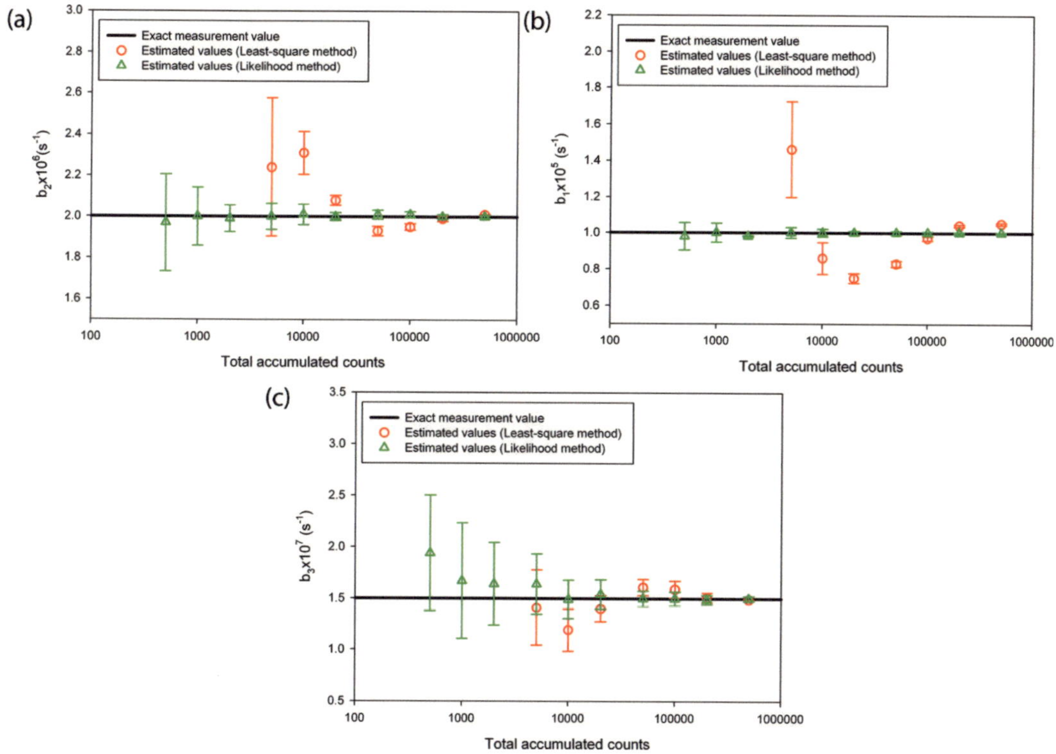

Figure 8. Estimated and exact values of the decay frequencies b_1 (figure 8-a), b_2 (figure 8-b), b_3 (figure 8-c).

Least squares method gives accurate results for high counting rates, but the method does not converge for low counting rates when the number count per channel is weak and many zeros are recorded per channel. As already demonstrated in the case of time-correlated single-photon counting (TCSPC) fluorescence decay analysis [24], the maximum likelihood method gives stable results over the whole count range, even for total counts less than 1000, where the least squares analysis delivers unreasonable values or does not converge.

For low counting rates, the maximum likelihood method gives excellent parameter estimations for multi-exponential fits of fluorescence decay curves. This method is ideally suitable for estimating the decay frequencies when the fluorescence decays are recorded with a single-photon counting system.

6. Application to the kinetic studies of rare gases

The kinetic studies following the selective excitation of the first metastable or resonant atomic states and of g or u molecular dissociative states correlated to these atomic states are highly relevant. Indeed, in operating conditions of applications, excimer production in discharges is obtained by populating relay states of higher configuration than the first metastable and resonant states, but the latter are quickly populated because the radiative and collisional decay frequencies of these higher states are fast. Finally, the $A1_u$ as well as the $B0^+_g$ molecular states are at the origin of VUV emissions. Molecular transitions can also intervene in these cascades, and the dissociative g states can serve as intermediates in populating the first metastable or resonant atomic states.

Laser-selective excitation of the first metastable state or the g parity molecular states can only be achieved with absorption of two photons, while selective excitation of the first resonant state or the u parity molecular states is done by absorption of three photons. The single-photon counting method adapted to laser-induced fluorescence kinetic studies of rare gases ensures to accurate determinations of the rate constants k_p for each decay frequency b_p involved in the VUV fluorescence decay. The values of the rate constants, k_{1p}, k_{2p} and k_{3p}, are determined using the least squares method by processing all the experimental data of the decay frequencies measured in a range of rare-gas pressures. The combined standard uncertainties of the constant rates were estimated according to the ISO Guide to the Expression of Uncertainty in Measurement [25]. Full details about the method used for the estimation of uncertainties can be found in Ref. [15].

6.1. Example of results from analyses of VUV emissions of xenon excimers in pure xenon

A comparative kinetic study of the first and second continua following the excitation of either the dissociative g molecular states correlated to $Xe6s[3/2]_1$ state or the $Xe6s[3/2]_2$ state by two-photon absorption has been performed by our team for the first time in a wide range of pressure up to 600 Torr.

Two or three frequency decay rates, depending on the initial state populated, characterize the time evolution of the second continuum of xenon. The pressure dependence (**Figure 9**) of these

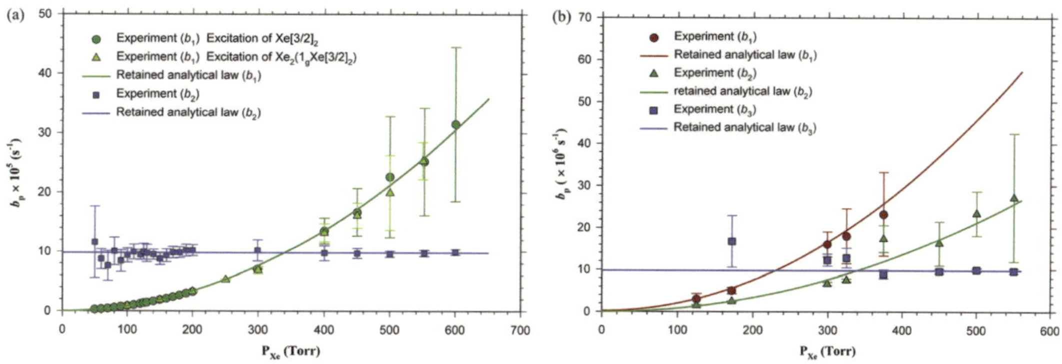

Figure 9. Decay frequencies b_1, b_2 and b_3 relative to the second continuum emission (λ_{em}, 168 nm; FWHM, 17.5 nm) versus the xenon pressure P_{Xe} following (a) an excitation of Xe[3/2]$_2$ and Xe$_2$[1 g [3/2]$_2$] and (b) excitation of Xe$_2$ [1 g [3/2]$_1$].

decay frequencies allows to determine the formation and decay mechanisms of the xenon excimers.

The first decay frequency b_1 is attributed to the decay of the metastable state by collisions with two ground-state xenon atoms:

$$\mathrm{Xe}[3/2]_2 + 2\mathrm{Xe}(^1S_0) \xrightarrow{k_{21}p^2_{Xe}} Xe_2(A1_u[3/2]_2)_{high\ v} + \mathrm{Xe}(^1S_0) \tag{18}$$

The second decay time b_2 is unambiguously identified as the lifetime of the lowest vibrational levels of the state $1_u[3/2]_2$ (or the $\mathrm{O}_u^-[3/2]_2$ state):

$$Xe_2(1u[3/2]_2)_{low\ v} \xrightarrow{b_2} Xe_2\left(BO_g^+(^1S_0)\right) + h\nu_{2nd\ continuum} \tag{19}$$

The third decay frequency, only observed when the resonant state is initially populated, corresponds to the decay of the resonant state by collisions with two ground-state xenon atoms:

$$\mathrm{Xe}[3/2]_1 \ + 2\mathrm{Xe}(^1S_0) \xrightarrow{k_{31}p^2_{Xe}} Xe_2(BO_u^+[3/2]_1)_{high\ v} \ + \mathrm{Xe}(^1S_0) \tag{20}$$

The kinetic study of formation and decay of xenon excimer combined with the spectroscopic studies allows establishing a kinetic scheme describing all the radiative and collisional mechanisms involved in the VUV emission of xenon [16] (**Figure 10**).

Three-body rate constants relative to the formation of excimers and radiative lifetime of $Xe_2(A1_u[3/2]_2)_{lowv}$ are consistent with those reported in the literature [26–28] and are more accurate.

Analysis of the amplitudes of the fluorescence decays allowed the determination of the binary collision rate constant for vibrational relaxation and energy transfer arising from the high vibrational levels of the xenon excimer correlated to the resonant state toward the metastable.

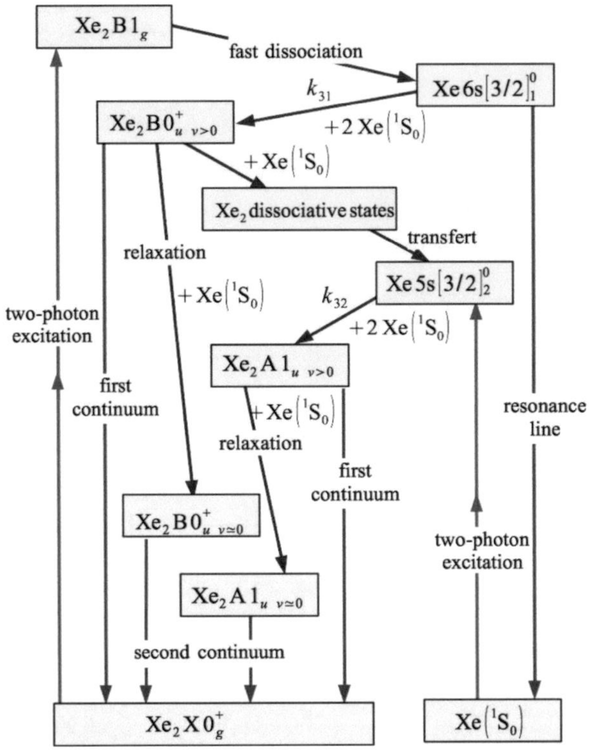

Figure 10. Kinetic scheme of formation and decay of xenon excimers when the first resonant or metastable states of xenon are initially populated.

6.2. Other rare gases and mixtures

Studies of VUV emission using two- or three-photon absorption laser-induced fluorescence (TALIF) technique were undertaken in argon and krypton. The decays of the first resonant and the first metastable state of argon Ar4s$[3/2]_1$ and Ar4s$[3/2]_2$ or Kr5s$[3/2]_1$ and Kr5s$[3/2]_2$ were explained with respect to the kinetic model already proposed for xenon [15, 20].

The decay of the first resonant state leads to very efficient population of the metastable via the molecular state $(B0_u^+)_{highv}$ and dissociative states correlated with the metastable state. Thus, at high pressure, emission of the second continuum comes from radiative de-excitation of the excimer states $(A1_u)_{lowv}$ toward the molecular ground state. The study of the temporal behavior of the second continuum of argon shows the formation of excimers only by three-body collisions. The three-body rate constant of formation was measured for the first time under selective excitation of the first resonant state of argon and the first resonant and metastable state of krypton (**Table 2**). All these studies allow the clarification of the mechanisms of VUV emission involved in the dielectric barrier discharge.

More complex kinetic schemes were proposed in Kr-Xe mixtures [17]. In mixtures, instead of the continuum of krypton, when a small amount of xenon was added, the first and second continua of xenon were observed, even though Kr5s$[3/2]_1$ was initially excited, proving the

Rare gas	k_{31} (10^{-32} cm^6 s^{-1})	k_{31} (10^{-32} cm^6 s^{-1})	τ_{1_u} (ns)
Argon		1.81 ± 0.18	3090 ± 50
Krypton	9.24 ± 0.80	3.81 ± 0.08	261.6 ± 5.7
Xenon	16.9 ± 1.4	7.87 ± 0.09	102.6 ± 1.3

Table 2. Three-body rate constants relative to the decay of the first resonant (k_{31}) and the first metastable state (k_{32}) and lifetime of the a[1u[3/2]$_1$] excimer.

efficiency of the energy transfer. From the temporal analysis in pure krypton, the three-body rate constants for Kr5s[3/2]$_1$ and Kr5s[3/2]$_2$ and the lifetime of the excimer Kr_2A1_u have been measured again. In the mixtures, we clearly showed the occurrence of an energy transfer from Kr5s[3/2]$_1$ to the Xe[5d(7/2)3] level, and the two-body collision rate constant has been estimated. A weak coupling could also intervene between these two states.

When the Xe5s[3/2]$_1$ or Xe6s[3/2]$_2$ were initially excited, the formation of the heteronuclear $KrXe^*$ exciplexes was clearly shown [18]. The decay frequency of heteronuclear excimers correlated to the xenon metastable state obeys different scaling laws depending on the xenon pressure. The roles of heteronuclear and homonuclear excimers in the formation of VUV emissions of the gas mixture are interpreted.

7. Rare-gas luminescence in dielectric barrier discharges

7.1. Monofilamentary dielectric barrier discharge

When performed in a neutral gas, kinetic analysis, following a brief and selective excitation of an identified state, is very reliable and proves to be a powerful tool. But, this analysis cannot be extrapolated with the same gas, subjected to discharge operating conditions, nor can it be considered as a mere juxtaposition of the kinetics of several distinct states. Nevertheless, if, on the one hand, the discharge excitation phase is quite short and, on the other one, time origin is perfectly defined, then an in situ kinetic analysis of the discharge can be achieved with the single-photon counting technique. A monofilamentary dielectric barrier discharge (MF-DBD) meets these requirements and can thus be a good candidate for such studies.

Figure 11 shows a schematic diagram of our experimental setup meant to produce spatially stable MF-DBDs. These micro-discharges were achieved in a glass cell, between two identical in-house-made electrodes in a classical double barrier discharge configuration. Each electrode was a 4 mm-diameter cylindrical aluminum rod covered with 0.5 mm-thick alumina, thus ensuring a monofilamentary micro-discharge (MF-DBD) [29]. The adjustable interelectrode distance d was set to 2 mm. A tuneable sinusoidal high voltage, $u$cell(t) ≈ $u$sup(t), was applied to the electrodes by means of a step-up ferrite transformer driven by a power amplifier. The current, $i$cell(t) ≈ $i$sup(t), flowing through the cell, consisted of a fast-rising pulse (pulsed current) superimposed on a displacement current, was obtained by measuring the voltage drop across a 50 Ω series-grounded resistance. Electric signals were recorded with a 10 GHz digital sampling 1 GHz oscilloscope (TDS7104). For each micro-discharge, the instant of origin

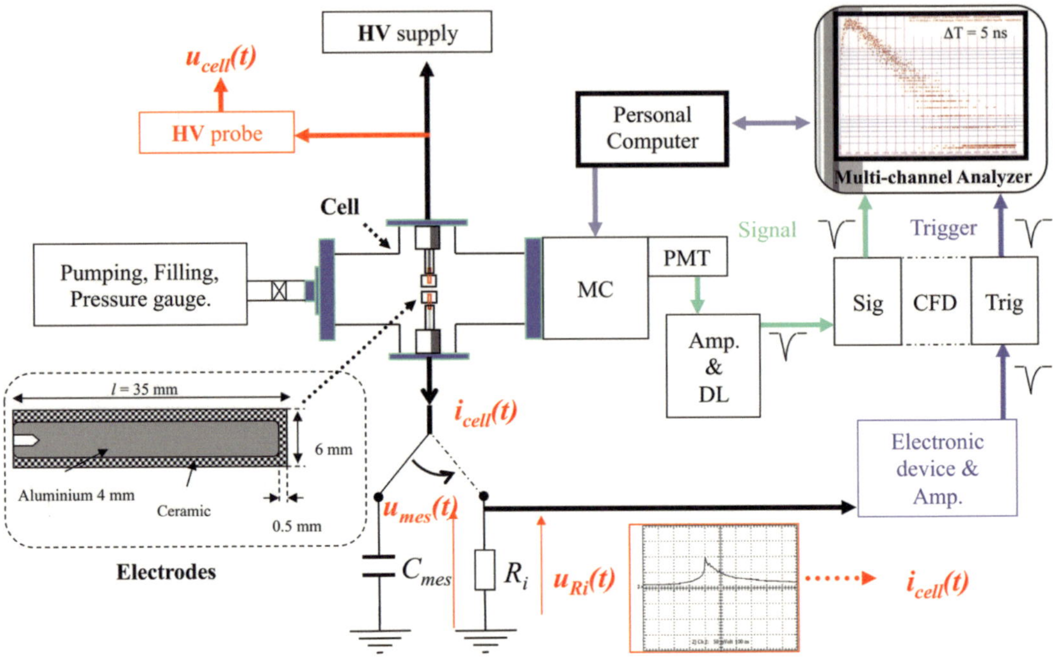

Figure 11. The experimental setup for electrical and emission spectroscopic analysis of MF-DBDs.

for photon counting was defined on the leading edge of the fast cell current pulse, using a constant fraction discriminator.

High-speed photography of the micro-discharge was performed with a 3 ns gated intensified charge-coupled device (ICCD) 512 × 512 array camera (PI-MAX Princeton Instruments) covering the visible to near-infrared spectrum (360–920 nm). Snapshots of the global visible luminescence were taken at regular intervals of 4 ns with an exposure time of 3 ns by accumulating 2000 micro-discharges on the positive half cycle. For each snapshot of the ICCD camera, triggering was performed on the leading edge of the preceding negative current pulse and with a constant delay of about half the signal period. During the high-speed photography of the discharge, the statistical fluctuations on the time lag between two successive current pulses for 2000 events corresponded to a pulse-to-pulse jitter (external jitter) of 3 ns. So, the time position of each snapshot was known with an uncertainty of about 3 ns. The cell was filled with laboratory-grade pure argon, xenon or krypton. Before filling up at the required pressure, the cell was evacuated down to at least 10^{-7} Torr. During filling up, the gas was drifted through an aluminum-zirconium getter pump for further purification.

The VUV emissions were recorded with the same device used for LIF experiments and processed with the same numerical techniques (**Figure 12**).

7.2. Electrical waveforms and discharge development

The development of a single micro-discharge is clearly shown by the successive snapshots given in **Figure 13a**, together with the cell current depicted in **Figure 13b** on an expanded time scale. The rhombuses correspond to the first six snapshots given in **Figure 13a**.

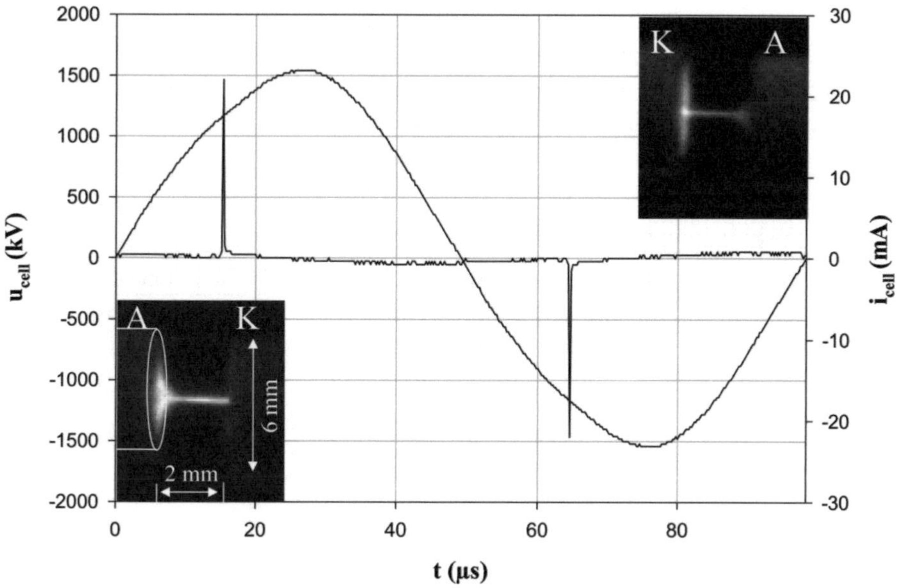

Figure 12. Supply voltage, cell current and 5 ns time-integrated snapshots of the discharge, in the visible range for P_{Xe}, 400 Torr; *f*, 10 kHz; *d*, 2 mm; and *U*max, 1.55 kV [36].

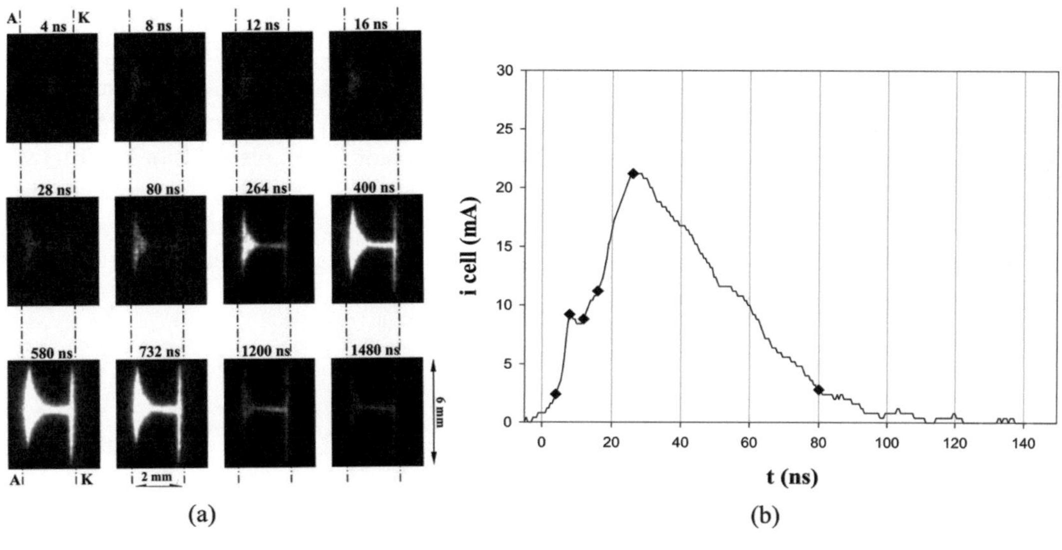

Figure 13. The micro-discharge for P_{Xe}, 200 Torr; f, 10 kHz; d, 2 mm; and Umax, 2.36 kV. (a) Evolution of the micro-discharge: 3 ns exposure time snapshots in the visible range (360 to 920 nm). The 264 ns snapshot is the reference for light intensity. A, anode; K, cathode. (b) Cell current with expanded time scale. Rhombuses correspond to the first six snapshots of (a) [36].

The cell current can be described as a main fast-rising pulse superimposed on a weak sinusoidal component. Above 200 Torr, its pulse duration is always less than 30 ns. Snapshots taken at t = 4 ns and at 8 ns show the propagation (its average speed is about 2.5×10^5 ms^{-1}) of the primary electronic avalanche reaching the anode at about t = 8 ns, for which icell shows the characteristic peak corresponding to the arrival of this avalanche at the anode. The positive streamer (cathode-directed streamer), which is, in fact, an ionization wave due to secondary avalanches, propagates toward and reaches the cathode at about t = 16 ns, forming a conducting channel between the anode and the cathode. The diameter of the luminous filament is 0.26 mm (the maximum current density is about 320 A cm^{-2}). This value is comparable with the current density proposed by Kogelschatz [30].

Operating at 10 kHz and at low pressures, between 25 and 150 Torr (**Figure 14a**), the xenon MF-DBD emitted, within 118–300 nm, both the first and second xenon continua centered, respectively, at 152 and 172 nm, the first continua being more significant up to 150 Torr. Above 300 Torr (**Figure 14b**), only the 172 nm emission was present. No significant emission was present above 180 nm in the VUV range.

These continua of xenon were also observed in other xenon discharges [23, 31–35].

7.3. Temporal decays of the observed VUV emissions

After a time $t = t_0$, always less than 80 ns, the luminescence decays of both the first and the second continua of xenon were fairly well described, over the whole pressure domain, by a sum of exponential terms. When we included the first 80 ns in the time range of our data processing, the histograms were not correctly represented by exponential terms: the maximum likelihood computation used for the estimation of the time constants did not converge. This feature is consistent with the current flowing through the cell, showing that the currently is nearly choked after 80 ns (**Figure 15**).

For all the working xenon pressures, the luminescence decays of the first continuum were well described, after t_0, by only two exponential terms. The first decay frequency, b_1, of this emission is attributed to the decay of the $Xe[3/2]_1$ resonant state through collisions with two ground-state xenon atoms, leading to the formation of $Xe_2[B0_u^+(3/2)_1]_{highv}$ excimers with $k_{31} = (16.7 \pm 2.7)10^{-32}$ cm^6 s^{-1} and $\alpha_{11} = (1.6 \pm 0.6) \times 10^6$ s^{-1}. α_{11} is the inverse of the apparent lifetime of the $Xe[3/2]_1$ due to radiation trapping of the resonant photons in the cell. Its value depends on the cell's configuration.

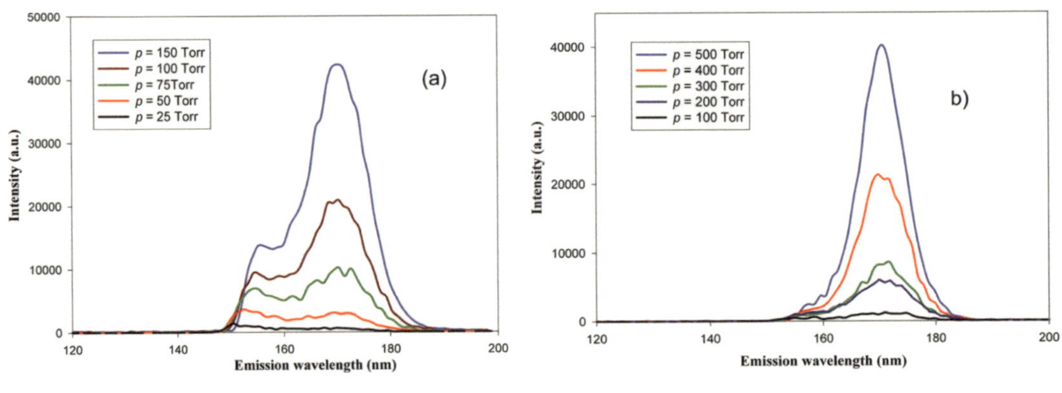

Figure 14. Emission spectra of the xenon MF-DBD at different pressures ($\Delta\lambda$, 1 nm): *f*, 10 kHz; *d*, 2 mm. (a) Low pressure and (b) high pressure [36].

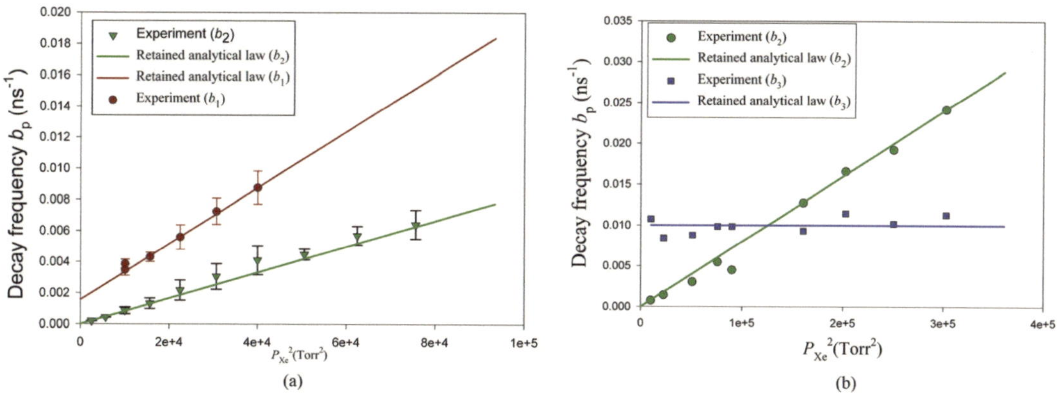

Figure 15. Decay frequencies *bp*, relative to (a) the first continuum (152 nm) and (b) the second continuum (172 nm) for *f* (10 kHz) and *d* (2 mm) versus the square of xenon pressure P_{Xe}^2 [36].

k_{31} is the three-body rate constant relative to the decay of Xe[3/2]$_1$ state, leading to the formation of $Xe_2[B0_u^+(3/2)_1]_{high\ v}$ excimer. Our measured value is consistent with other experimental [18, 37, 38] or theoretical works [39],where like us two-body collisions of the resonant state are not observed as well.

The second decay frequency, b_2, of the first continua is attributed to the decay of the Xe[3/2]$_2$ metastable state through collisions with two ground-state xenon atoms, leading to the formation of $Xe_2[A1_u(3/2)_2]_{high\ v}$ excimers [36], with $k_{32} = (7.7 \pm 1.1)\ 10^{-32}\ cm^6\ s^{-1}$.

After the initial active phase lasting at most 80 ns, the luminescence decays of the second continuum were fairly described, over the whole pressure domain (100 to 550 Torr), by only two exponential terms with decay frequencies b_2 and b_3. The decay frequency, b_3, of the second continua is attributed to the radiative decay of the $Xe_2[A1_u(3/2)_2]_{lowv}$ state itself, resulting from vibrational relaxation on the $Xe_2[A1_u(3/2)_2]_{highv}$ excimers, through very fast binary collisions with a ground-state xenon atom. This reaction is too fast to be observed by our photon-detecting device. The radiative lifetime of the $Xe_2[A1_u(3/2)_2]_{lowv}$ excimers is $\tau_{1u} = (96 \pm 6)$ ns. This value is concordant with literature values [16, 26, 27, 37, 38, 41, 42].

7.4. Kinetic scheme

The kinetic scheme given in **Figure 16** summarizes the main reactions occurring in the xenon MF-DBD during the post-discharge phase. Only the reactions (R-6), (R-7), (R-8) and (R-11) were directly identified in our kinetic study. Reactions (R-3), (R-4), (R-9) and (R-10) were indirectly identified but are too fast to be measured by our detecting device. Finally, reactions (R-1) and (R-2) were not observed in the kinetic study, but they were identified in our emission spectra at low pressures.

At early stages, electrons and ions, as well as higher excited states, are present in the discharge. Electrons likely excite the lowest resonant and metastable states, which are also indirectly populated via fast cascades issuing from these highly excited states. All these reactions occur

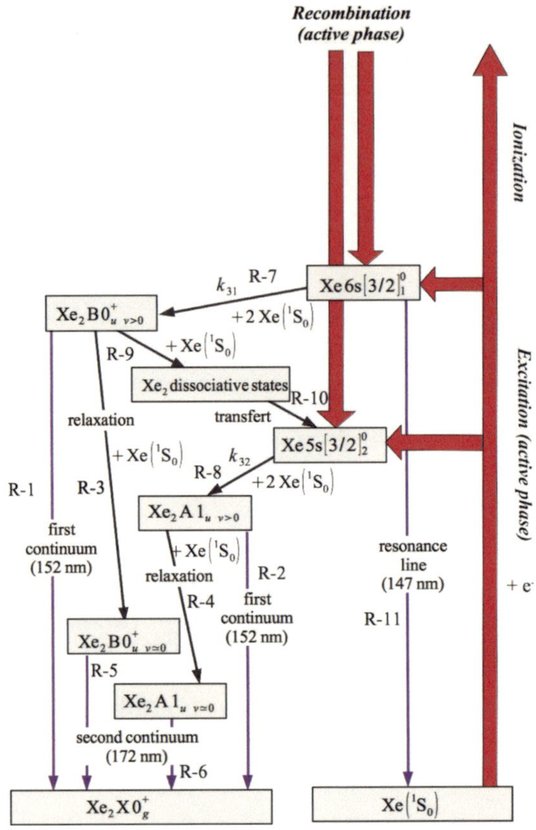

Figure 16. Kinetic scheme of the post-discharge in a high-pressure xenon MF-DBD.

during the first 80 ns, while the current is extinguished within 30 ns. Finally, these reactions lead to the formation of $Xe_2[A1_u(3/2)_2]_{high\nu}$ and then $Xe_2[A1u(3/2)_2]_{low\nu}$ which emits, nearly exclusively, the narrow energetic second continuum.

8. Radiation trapping in rare gases

Resonant states play an important role in excimer formations and are involved in emission of UV sources based on rare gases. Atoms excited in a resonant state will decay to the ground level by emission of resonant photons themselves having a good probability of being absorbed by surrounding ground-state atoms and then reemitted. Imprisonment of resonance radiation or radiative trapping is caused by multiple emission, absorption and re-emission of photons before they leave the cell [28, 40].

The escape factor g which characterizes this process is defined as the mean number of absorption-reemission processes. It depends on both the gas used and the geometrical characteristics of the experimental device. The measured apparent lifetime τ_a is expressed as a function of the natural lifetime τ_n: $\tau_a = g\,\tau_n$.

Each time a photon is reabsorbed, it can be lost by collisional quenching of the resonant atomic state. So, resonance radiation trapping allows collisional decay to prevail over radiative decay, permitting excimer formation from resonant states.

8.1. Experimental results in pure xenon

The selective excitation of $Xe[3/2]_1$ was achieved by three-photon excitation using a tunable laser. The luminescence of the first resonance line and the first continuum of xenon was observed through an interferential filter centered at 145 nm (FWMH = 17.5 nm). Luminescence was recorded by means of a VUV photomultiplier set in the photon counting mode.

When the pressure is less than 15 Torr, the temporal luminescence decay is well described by a single exponential term. The variation law of the decay frequency (b_1) is parabolic $(b_1 = \ \alpha_{11} + k_{31}N_{Xe}^2)$.

With $\alpha_{11} = 1/\tau a = 2{,}64.10^5\ s^{-1}$ and $k_{31} = (20.1 \pm 1.1)\ 10^{-32}\ cm^6\ s^{-1}$.

This time constant is attributed to the $Xe[3/2]_1$ state, and the apparent lifetime is 3.79 μs. The natural lifetime is equal to 3.98 ns [26]. The escape factor g = 952: a photon undergoes 952 absorption-re-emission processes, on average, until it leaves the cell in our experimental conditions. Constant k_{31} expresses excimer formation:

$$Xe(^3P_1\ ) + 2Xe(^1S_0\ ) \rightarrow \ Xe_2\ \ (O_u^+(^3P_1)high\,v\ ) + Xe(^1S_0\ ) \tag{21}$$

A second exponential term appears only for pressures greater than 15 Torr. The best fit of b_2 gives $(b_2 = k_{32}N_{Xe}^2)$.

The absence of a constant term suggests an excited state with a long lifetime. This time constant was attributed to the $Xe[3/2]_2$ metastable state. This state is created during the vibrational relaxation of the O_u^+ state.

8.2. Experimental results in xenon-krypton mixtures

This study was performed in the same excitation and detection conditions as for pure xenon. This study shows that the foreign gas (krypton) at high concentration modifies trapping phenomena and that van der Waals broadening by krypton should be taken into account. There exists no energy transfer between the two rare gases when the foreign gas (krypton) is lighter than the excited one (xenon).

On the contrary, efficient energy transfers occur from a lighter rare gas toward a heavier one, like in Kr-Xe mixtures following excitation of Kr.

9. Conclusion

Kinetic studies of rare-gas VUV fluorescence decays under short and selective multiphotonic excitation conditions using single-photon counting fluorescence are highly suitable for the determination of reliable kinetic models. The use of the maximum likelihood method offers the possibility to lower excitation energies and to determinate fast and slow decay frequencies in a same scan. This allows us to determine the main mechanisms involved in the formation and decay of rare-gas excimers.

For high pressures, emission of the second continuum comes from radiative de-excitation of the excimer states correlated to the first metastable state toward the molecular ground state. The study of the temporal behavior of the second continuum of pure rare gas shows the formation of excimers only by three-body collisions. The three-body rate constant of formation of excimers was measured for the first time under selective excitation. These studies allowed us to clarify the mechanisms of VUV emission involved in the dielectric barrier discharge. Resonant states do not seem to contribute to the production of excimers in a dielectric barrier microdischarge. The resonant state contributes to the formation of excimer through the transfer toward the metastable state at high pressures. The second continuum essentially comes from radiative transition of the low vibrational levels of the excimer correlated with the metastable state.

Author details

Frédéric Marchal*, Neermalsing Sewraj, Jean-Pierre Gardou, Nofel Merbahi and Mohammed Yousfi

*Address all correspondence to: frederic.marchal@laplace.univ-tlse.fr

LAPLACE, Université de Toulouse, CNRS, INPT, UPS, France

References

[1] ChanMay BP. Biomedical applications of photochemistry. Tissue Engineering Part B: Reviews. 2010;**16**(5):509-522. DOI: https://doi.org/10.1089/ten.teb.2009.0797

[2] Truica-Marasescu F, Guimond S, Wertheimer MR. VUV-induced nitriding of polymer surfaces: Comparison with plasma treatments in nitrogen. Nuclear Instruments and Methods in Physics Research Section B: Beam Interactions with Materials and Atoms. 2003;**208**:294-299. DOI: https://doi.org/10.1016/S0168-583X(03)00658-X

[3] Halfmann H, Denis B, Bibinov N, Wunderlich J, Awakowicz P. Identification of the most efficient VUV/UV radiation for plasma based inactivation of bacillus atrophaeus spores. Journal of Physics D: Applied Physics. 2007;**40**(19):5907. DOI: https://doi.org/10.1088/0022-3727/40/19/019

[4] Kristin Zoschke, Hilmar Börnick, Eckhard Worch. Vacuum-UV radiation at 185 nm in water treatment – A review. Water Research 2014;**52**:131-145. DOI: http://dx.doi.org/10.1016/j.watres.2013.12.034

[5] Eliasson B, Kogelschatz U. UV excimer radiation from dielectric-barrier discharges. Applied Physics B. 1988;**46**(4):299-303. DOI: https://doi.org/10.1007/BF00686452

[6] Kogelschatz U. Ultraviolet excimer radiation from nonequilibrium gas discharges and its application in photophysics, photochemistry and photobiology. Journal of Optical Technology. 2012;**79**(8):484-493. DOI: https://doi.org/10.1364/JOT.79.000484

[7] Oppenländer T. Mercury-free sources of VUV/UV radiation: Application of modern excimer lamps (excilamps) for water and air treatment. Journal of Environmental Engineering and Science. 2007;**6**:253-264. DOI: https://doi.org/10.1139/s06-059

[8] Wilkinson PG, Byram ET. Rare gas light sources for the vacuum ultraviolet. Applied Optics. 1965;**4**(5):581-588. DOI: https://doi.org/10.1364/AO.4.000581

[9] Jun-Ying Zhang, Ian W. Boyd. Efficient excimer ultraviolet sources from a dielectric barrier discharge in rare-gas/halogen mixtures. Journal of Applied Physics. 1996;**80**:633. DOI: http://dx.doi.org/10.1063/1.362871

[10] Xueji Xu. Dielectric barrier discharge - properties and applications. In Thin Solid Films. 2001;**390**(1-2):237-242. https://doi.org/10.1016/S0040-6090(01)00956-7

[11] Mildren RP, Carman RJ. Enhanced performance of a dielectric barrier discharge lamp using short-pulsed excitation. Journal of Physics D: Applied Physics. 2001;**34**(1):L1. DOI: https://doi.org/10.1088/0022-3727/34/1/101

[12] Wilkinson PG, Tanaka Y. New xenon-light source for the vacuum ultraviolet. Journal of the Optical Society of America. 1955;**45**(5):344-349. DOI: https://doi.org/10.1364/JOSA.45.000344

[13] Tanaka Y. Continuous emission spectra of rare gases in the vacuum ultraviolet region. Journal of the Optical Society of America. 1955;**45**(9):710-713. DOI: https://doi.org/10.1364/JOSA.45.000710

[14] Hoff PW, Rhodes CK. Introduction. In: Rhodes Charles K, editor. Excimer Lasers. Berlin Heidelberg, New York: Springer-Verlag; 1979. pp. 1-4. DOI: https://doi.org/10.1007/978-3-662-11716-3_1

[15] Marchal F, Merbahi N, Ledru G, Gardou JP, Sewraj N. The study of VUV emissions of Ar*2 excimers using three-photon absorption laser-induced fluorescence. Journal of Physics B: Atomic, Molecular and Optical Physics. 2008;**42**(1):015201. DOI: https://doi.org/10.1088/0953-4075/42/1/015201

[16] Ledru G, Marchal F, Sewraj N, Salamero Y, Millet P. Comparative study of the formation and decay of xenon excimers following selective excitation of the 5p56s states: Spectroscopic and kinetic analysis. Journal of Physics B: Atomic, Molecular and Optical Physics. 2006;**39**(8):2031. DOI: https://doi.org/10.1088/0953-4075/39/8/020

[17] Marchal F, Berejny P, Sewraj N, Salamero Y, Millet P. Energy transfers in Kr–Xe mixtures following selective multiphotonic excitation of Kr(3P1). Temporal analysis in Kr–Xe mixtures. Journal of Physics B: Atomic, Molecular and Optical Physics. 2004;**37**(6):1279. DOI: https://doi.org/10.1088/0953-4075/37/6/012

[18] Ledru G, Marchal F, Merbahi N, Gardou JP, Sewraj N. Study of the formation and decay of KrXe* excimers at room temperature following selective excitation of the xenon 6s

states. Journal of Physics B: Atomic, Molecular and Optical Physics. 2007;**40**(10):1651. DOI: https://doi.org/10.1088/0953-4075/40/10/002

[19] Marchal F, Sewraj N, Jabbour G, Rodriguez Akerreta P, Ledru G. Temperature dependence of xenon excimer formations using two-photon absorption laser-induced fluorescence. Journal of Physics B: Atomic, Molecular and Optical Physics. 2010;**43**(23):235210. DOI: https://doi.org/10.1088/0953-4075/43/23/235210

[20] Marchal F, Jabbour G, Yousfi M, Ledru G, Sewraj N, Rodriguez-Akerreta P. Study of VUV emission of krypton using two-photon absorption laser. In: Proceedings of the 29th International Conference on Phenomena in Ionized Gases (ICPIG); 12–17 July; Cancún, Mexico. 2009. p. A-1

[21] Moutard P, Laporte P, Subtil JL, Damany N, Damany H. Pressure effects on kinetics and decay processes in argon under selective photoexcitation. The Journal of Chemical Physics. 1987;**87**:4576. DOI: http://dx.doi.org/10.1063/1.452869

[22] Hamamatsu. https://www.hamamatsu.com/us/en/technology/innovation/photoncounting/index.html [Internet]. Photon counting. Available from: https://www.hamamatsu.com/us/en/technology/innovation/photoncounting/index.html [Accessed: 05–09-2017]

[23] Mildren RP, Carman RJ. Enhanced performance of a dielectric barrier discharge lamp using short-pulsed excitation. Journal of Physics D: Applied Physics. 2001;**34**(1):L1. DOI: https://doi.org/10.1088/0022-3727/34/1/101

[24] Michael Maus, Mircea Cotlet, Johan Hofkens, Thomas Gensch and Frans C. De Schryver. An experimental comparison of the maximum likelihood estimation and nonlinear least-squares fluorescence lifetime analysis of single molecules. Analytical Chemistry. 2001;**73**(9): 2078-2086. DOI: http://dx.doi.org/10.1021/ac000877g

[25] ISO/IEC Guide98:1995, editor. Guide to the Expression of Uncertainty in Measurement (GUM). Geneva, Switzerland: ISO; 1998

[26] Bonifield TD, Rambow FHK, Walters GK, McCusker MV, Lorents DC, Gutcheck RA. Time resolved spectroscopy of xenon excimers excited by synchrotron radiation. The Journal of Chemical Physics. 1980;**72**(5):2914-2924. DOI: http://dx.doi.org/10.1063/1.439490

[27] Wenck HD, Hasnain SS, Nikitin MM, Sommer K, Zimmerer G, Haaks D. Time and spectrally resolved fluorescence of Xe molecules excited with synchrotron radiation. Chemical Physics Letters. 1979;**66**(1):138-143. DOI: http://dx.doi.org/10.1016/0009-2614(79)80384-X

[28] Sewraj N, Gardou JP, Salamero Y, Millet P. Radiation trapping of the 3P1-1S0 resonant transitions of xenon and krypton in Xe-Kr, Xe-Ar, and Kr-Ar mixtures: Kinetic analysis and determination of the van der Waals broadening coefficients. Physical Review A. 2000;**62**(5):052721. DOI: https://doi.org/10.1103/PhysRevA.62.052721

[29] Merbahi N, Sewraj N, Marchal F, Salamero Y, Millet P. Luminescence of argon in a spatially stabilized mono-filamentary dielectric barrier micro-discharge: Spectroscopic and kinetic analysis. Journal of Physics D: Applied Physics. 2004;**37**(12):1664. DOI: https://doi.org/10.1088/0022-3727/37/12/011

[30] Kogelschatz U. 10th International Conference on Gas Discharge and Applications; 13–18 September ; Swansea. 1992. p. Vol. 972–980

[31] Stockwald K, Neiger M. Some properties of a novel far UV xenon excimer barrier discharge light source. Contributions to Plasma Physics. 1995;**35**(1):15-22. DOI: http://dx.doi.org/10.1002/ctpp.2150350103

[32] Adler F, Müller S. Formation and decay mechanisms of excimer molecules in dielectric barrier discharges. Journal of Physics D: Applied Physics. 2000;**35**(14):1705. DOI: https://doi.org/10.1088/0022-3727/33/14/310

[33] Gellert B, Kogelschatz U. Generation of excimer emission in dielectric barrier discharges. Applied Physics B: Lasers and Optics. 1991;**52**(1):14-21. DOI: https://doi.org/10.1007/BF00405680

[34] Kessler F, Bauer GH. VUV excimer light source for deposition of amorphous semiconductors. Applied Surface Science. 1992;**54**:430-434. DOI: http://dx.doi.org/10.1016/0169-4332(92)90082-9

[35] Wieme W, Lenaerts J. Excimer formation in argon, krypton, and xenon discharge afterglows between 200 and 400 K . The Journal of Chemical Physics. 1981;**74**(1):483-493. DOI: http://dx.doi.org/10.1063/1.440855

[36] Sewraj N, Merbahi N, Marchal F, Ledru G, Gardou JP. VUV spectroscopy and post-discharge kinetic analysis of a pure xenon mono-filamentary dielectric barrier discharge (MF-DBD). Journal of Physics D: Applied Physics. 2009;**42**(4):045206. DOI: https://doi.org/10.1088/0022-3727/42/4/045206

[37] Salamero Y, Birot A, Brunet H, Galy J, Millet P. Kinetic study of the VUV xenon emissions using selective multiphoton excitation. The Journal of Chemical Physics. 1984;**80**(10):4774-4780. DOI: http://dx.doi.org/10.1063/1.446550

[38] Dutuit O, Gutcheck R, Calvé JL. Spectral and kinetic studies of the second continuum fluorescence of xenon excited by synchrotron radiation. Chemical Physics Letters. 1978;**58**(1):66-72. DOI: http://dx.doi.org/10.1016/0009-2614(78)80318-2

[39] Janssens H, Vanmarcke M, Desoppere E, Lenaerts J, Bouciqué R, Wieme W. A general consistent model for formation and decay of rare gas excimers in the 10^{-2}–10^{+5} mbar pressure range, with application to krypton. The Journal of Chemical Physics. 1987;**86**(9):4925-4934. https://doi.org/10.1063/1.452662

[40] Molisch AF, Oehry BP, editors. Radiation Trapping in Atomic Vapours. Oxford University Press; 1998. 536 p

[41] Keto JW, Gleason RE, Bonifield TD, Walters GK, Soley FK. Collisional mixing of the lowest bound molecular states in xenon and argon., Chemical Physics Letters. 1976;**42**(1): 125-128. DOI:, http://dx.doi.org/10.1016/0009-2614(76)80566-0

[42] Barbet A, Sadeghi N, Pebay-Peyroula JC. Decay of metastable xenon atoms Xe*(3 P 2) in a xenon afterglow. Journal of Physics B: Atomic and Molecular Physics. 1975;**8**(10):1776. DOI: https://doi.org/10.1088/0022-3700/8/10/027

Section 3

Applications

Silicon Photomultiplier for the Plug & Imaging PET System: Physics, Technological Challenges and Application to Modern Nuclear Medicine

Nicola D'Ascenzo and Qingguo Xie

Additional information is available at the end of the chapter

http://dx.doi.org/10.5772/intechopen.73007

Abstract

We propose the design of a Silicon Photomultiplier at the 180 nm GLOBALFOUNDRIES BCDLITE CMOS technology node. We perform a characterization of the device, in comparison with other results obtained a CMOS technology node and we investigate the limits and strengths of this approach. Finally we show possible future applications of the SiPM in Nuclear Medicine, in particular to digital positron emission tomography systems.

Keywords: silicon photomultiplier, avalanche breakdown, positron emission tomography

1. Introduction

Low photon flux sensors are specific devices dedicated to the detection of a small number of photons. The classical example of such sensors is the photomultiplier tube (PMT), which was dominating the field during the last 70 years. PMT were broadly applied to high energy physics, nuclear medicine, space applications, among others, despite their complicated manufactory which includes vacuum technology, high bias voltage up to few thousand volts, large and not scalable size [1].

The advances in semiconductor technology during the last 20 years were beneficial for the development of modern light wave and telecommunication systems. The avalanche photodiode (APD) is currently one of the most spread semiconductor-based photo-detectors for low photon flux. APD allowed to study fundamental properties of the light in order to improve specific applications in photonics, including quantum coherence of multi-photon states and decoherence among others [2, 3].

PMT, APD and low photon flux sensors derived from them have a strong limitation. Their sensitivity to the detection of a single photon is affected by a large statistical fluctuation in the signal generation and do not allow de facto an efficient disentangling of single photon states.

In the late 1970s a new sensor, later known as silicon photomultiplier (SiPM), was invented [4]. It consisted of a space-distributed fine array of metal resistor semiconductor (MRS) micro-sensors with individual quenching and common output. Since then, a large number of groups contributed in the development of the modern SiPM, which is nowadays commercially available and started to substitute the PMT in the above-mentioned application fields [4–12].

Modern SiPMs are composed of an array of p/n junctions (microcell) operated in Geiger mode. A single photon interacts with a microcell through the photoelectric effect and generates an electron/hole pair, which initiates a self-sustaining avalanche. The quenching is realized passively with an integrated resistor in the microcell.

In parallel the similar technology of the single photon avalanche diode (SPAD) was developed [13]. The SPAD is a single photon detector operated in Geiger mode. Its optical sensing part is equivalent to a microcell of the SiPM. In addition the layout of a SPAD includes integrated active quenching electronics components [14]. In other words the SiPM is equivalent to an array of SPAD-like microcells with passive quenching.

The current advance in the development of SiPM and SPAD is represented by the complementary metal oxide semiconductor (CMOS) technology. The implementation of SiPM and SPAD sensitive cells in standard CMOS technology allows the combined integration of sensor and read-out electronics on the same device, with a significant reduction of power consumption and simplification of the operational conditions. In recent years a large number of advanced digital imagers and innovative concepts for light detection were proposed on this basis [15–18].

However some drawbacks of the application of the CMOS technology to photonics sensors operated in avalanche breakdown mode were identified in the last few years. CMOS processes introduce a higher level of noise and lower photon detection efficiency, with respect to custom-technologies [19–53].

Among the many fields of applications, photonics sensors realized in CMOS technology are expected to have a strong impact in Nuclear Medicine, and it is maybe better to introduce here this concept with a certain detail. Nuclear Medicine is dealing with the diagnosis and treatment of patients with ionizing radiation. From the diagnostic point of view computed tomography (CT) and positron emitted tomography (PET) machines are nowadays conventionally used in hospitals for the detection of cancer lesions among others [54, 55]. From the clinical point of view proton therapy is nowadays being used in various centers for a more effective cure of cancer with respect to chemotherapy.

In this chapter we provide a description of the technological challenges in the development of modern SiPM in CMOS technology. We base our discussion on two SiPM structures manufactured at the BCDLITE 180 nm GF CMOS technology node. We perform a characterization of the devices in dark and light illumination conditions, putting the results in the broader context of other structures realized in CMOS and in custom technology. Finally we provide an application of such optical sensors to Nuclear Medicine, showing how such small-sized novel photo-detectors have an impact in the technology for positron emission tomography (PET) machines in the new digital generation.

2. Physical principles of the SiPM

The basic component of a SiPM is a sensitive microcell. A simplified SiPM structure is shown on **Figure 1**. Within a n-doped region (nwell), a heavily p-doped implantation is formed (p+), generating a p+/nwell diode. The general schematics proposed here is not applicable to any working devices, due to many technological and operational constrains. However, this simplified layout of the microcell is used here in order to describe the physical background of the device. Several examples of real implementations of such device will be shown in the next section.

The SiPM is composed of an array of such microcells. The typical microcell pitch ranges between 20 and 100 μm. In analog SiPMs the microcells have common anode and cathode as reported in **Figure 2**. In digital SiPMs each microcell is read-out separately with a dedicated digital electronic processing as reported in **Figure 3**.

When the microcell is reverse-biased, a depletion region is formed between the p + and the nwell implantations. The typical depth of the depletion region is approximately 800 nm from the microcell surface and the typical width is approximately 1 μm. The size and the concentration of the depletion region defines also the capacitance of a single microcell, which reaches

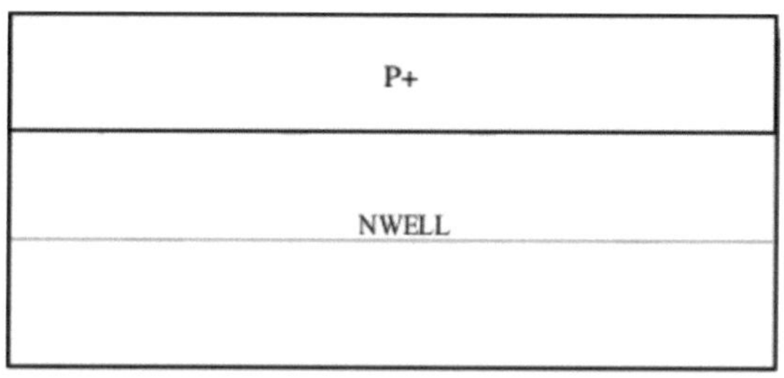

Figure 1. General simplified cross-section of a silicon photomultiplier microcell.

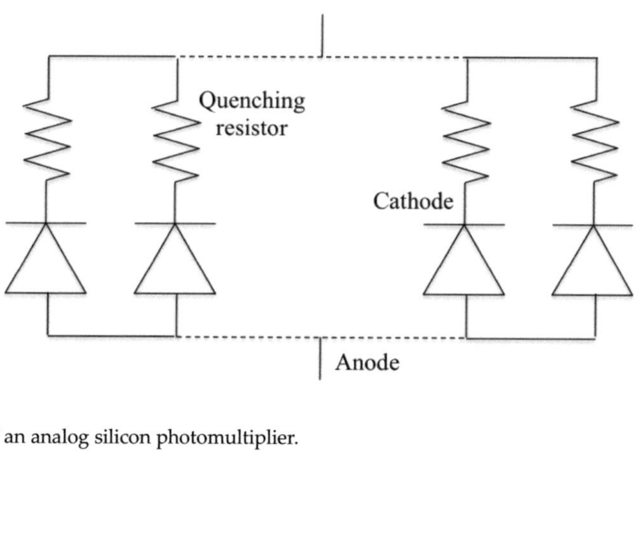

Figure 2. Schematics of an analog silicon photomultiplier.

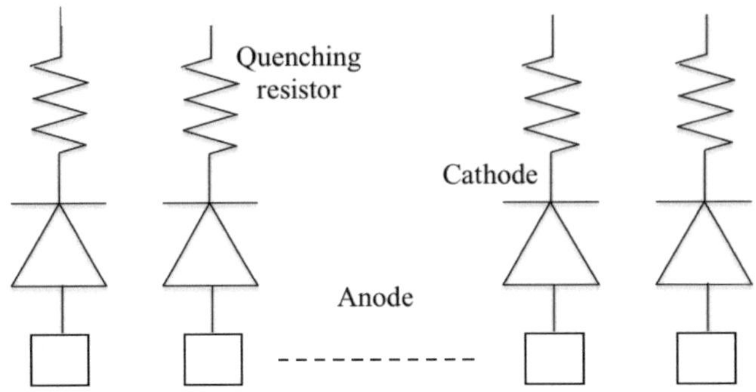

Figure 3. Schematics of a digital silicon photomultiplier with a dedicated read-out digital electronics for each microcell.

few hundreds fF. The reverse-biased microcell exhibits an electric field with maximal strength ranging between 10^5 and 10^6 V/cm. The strength of the electric field as a function of the total charge density is governed by the Poisson Eq.

A typical feature of the SiPM as quantum detector is its bi-stability, i.e. its operation allows two stability states with a continuous transition between them. The two states correspond to the physical processes of carrier recombination and generation occurring in the p+/nwell junction in the operational condition of the SiPM.

The state I of the SiPM microcell is determined by the electron-hole recombination mechanism. Electrons are dropping from the conduction to the valence band, transferring their excess energy to other electrons, phonons or photons. The involved physical processes can be radiative and

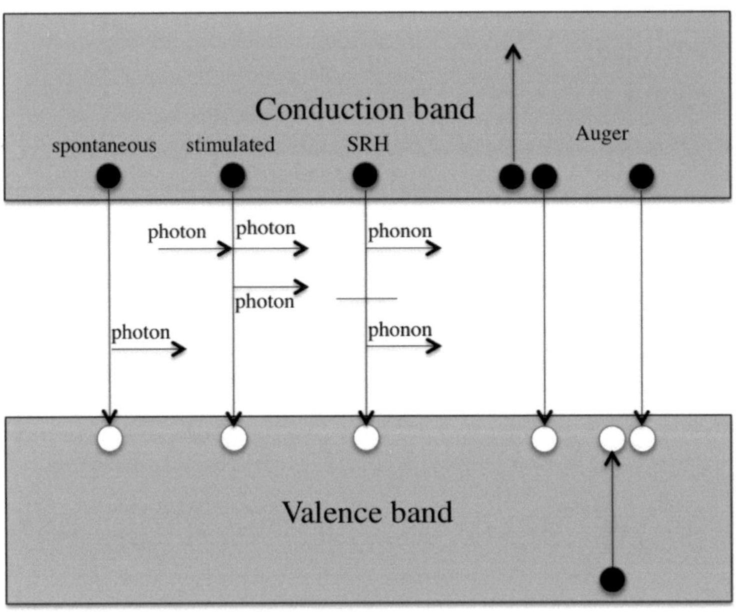

Figure 4. Radiative and non-radiative recombination processes in semiconductor devices.

non-radiative and are depicted in **Figure 4**. In the first case a photon is emitted, in the second case the energy is transferred only to phonons or electrons. The generation of photons by radiative recombination can be spontaneous or stimulated by an additional photon. The two main non-radiative processes are Shockley-Read Hall (SRH) recombination and Auger recombination [56]. SRH involves energy levels generated by crystal defects. Such energy levels capture electrons from the conduction band as well as holes from the valence band. The excess energy is released through phonons. In Auger recombination the excess energy is transferred to other electrons within the valence and the conduction band. A main result of the recombination process is that electrons, photons and phonons are released, which can either initiate further recombination process or generate electron/hole pairs. Because of the properties of the Fermi distribution function of carriers in silicon, the process of electron-holes generation through recombination is temperature-dependent and is often called thermal generation of electron/hole pairs.

The state II of the SiPM microcell is determined by the impact ionization process [57]. The impact ionization is the reverse Auger process as the energy of motion of an electron is absorbed in order to produce new electron/hole pairs. At the strong electric field inside the SiPM microcell, both electrons and holes participate to the charge multiplication process. In this condition an infinite number of charged carriers is produced by a single electron/hole pair. In other words the avalanche induced in the state II is self-sustaining.

In normal conditions the recombination process would bring the junction back to the equilibrium of the doping concentration, in the SiPM the produced electron/hole pairs are accelerated in the strong electric field and start a multiplication chain due to impact ionization. The transition between state I and state II is as fast as 100 ps. The transition back from state II to state I requires a feedback process. A passive quenching resistor is placed in series with a microcell. The current flow through the resistor causes a voltage drop on the junction, which reduces the actual junction bias to a value lower than the breakdown voltage. After the quenching occurs, the SiPM transits back to the state I within a recovery time determined by the size of the junction capacitance and of the quenching resistor. The typical value of the quenching resistor is between 100 kΩ and 1 MΩ.

Besides the thermal process, photon detection triggers the transition between the state I and the state II. When an optical photon is detected, an electron/hole pair is generated through photoelectric effect. Such e/h pair initiates the avalanche process through impact ionization.

A typical signal corresponding to the transition between the states I and II is shown on **Figure 5**. The current flowing through a single microcell is shown. At the beginning the microcell is in state I, with a low current of few picoamperes. A thermally generated pair causes the transition to the state II, which determines a fast rise of the signal with a peak at 30 μA, where the quenching time is reached. After that, within an exponential decay time governed by the quenching resistor and the intrinsic capacitance of the junction, the current reaches again the state I. The signal corresponding to a value of the quenching resistor 250 and 300 kOhm is reported. We observe that the only effect introduced by the change of quenching resistor consists of a larger decay time, being the total charge under the pulse constant. The total charge contained in the signal is corresponding to approximately 1.5×10^6 electrons. Thus from a single electron/hole pair produced in the sensitive area of the device, a signal corresponding to this large number of electrons is produced.

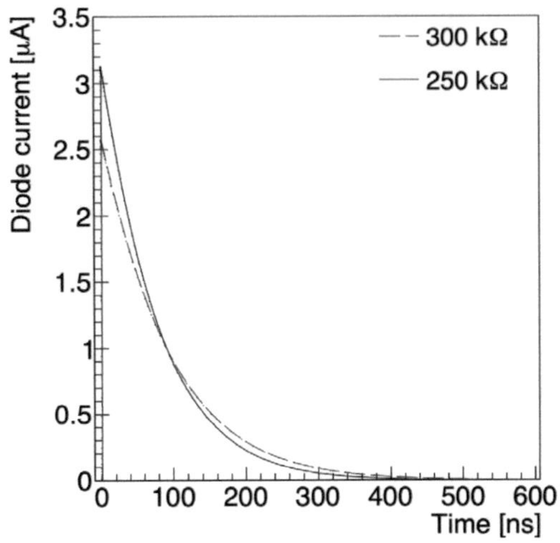

Figure 5. Typical diode current pulse corresponding to a thermally generated electron/hole pair. The transition between state I and state II is observed for two different quenching resistors.

The SiPM is hence a device which is sensitive to the energy deposited by a single visible photon and is able to produce a sizeable signal in correspondence to such energy. The signal is produced through a combination of quantum effects and a transition from the two stability states within a time of few tens picoseconds. Due to the ability detection of a single quantum of light and to the quantum working principle of the device itself, the SiPM is called sometimes "quantum detector".

3. SiPM in CMOS: challenges and limitations

The development of SiPM in the CMOS framework faces non-trivial issues. As we described in the previous section, SiPM is based on p+/n or n+/p junction structures working in two states defined by recombination and generation processes. Although the structure proposed in the previous section has a strong physical meaning, its realistic manufacturing in existing technology encounters several difficulties.

The most relevant problem in the integration of such structures within CMOS technology is that they can suffer of localized breakdown conditions on the locally concentrated high electric field at the junction edges. The use of guard ring structures around the sensitive area of each microcell is in this respect mandatory in order to obtain a uniform electric field across the whole sensitive area. Although CMOS compatible technology offers some possibility of implementing such guard rings [58–61], it is normally needed to add specific masks in order to allow the overlap between highly-doped and low- doped regions and to correct the doping profiles.

A second problem in the development of SiPM quantum devices within standard CMOS technology is the difficulty to control the thermal e/h pairs generating the dark noise of the

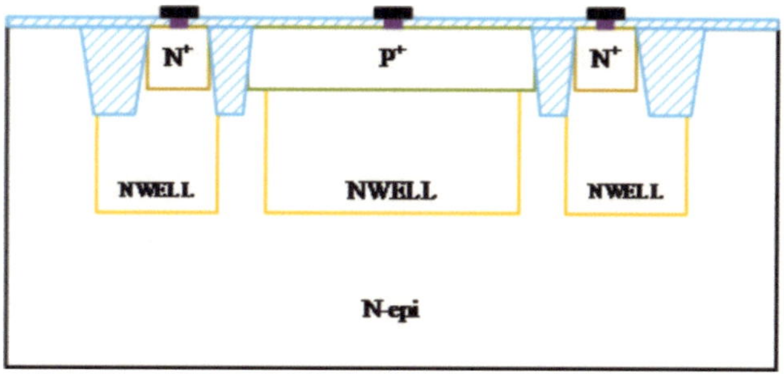

Figure 6. The concept of virtual guard rings.

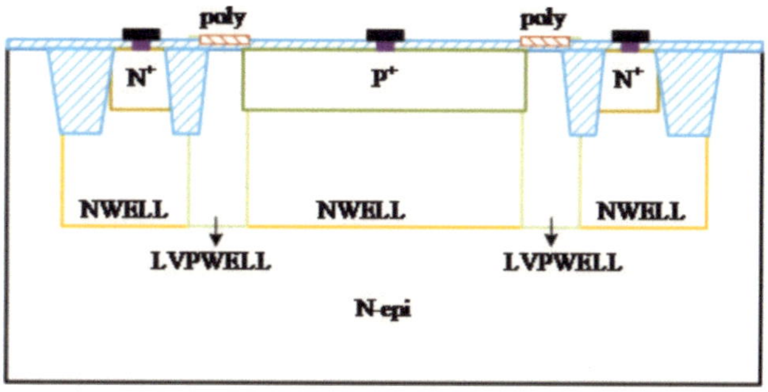

Figure 7. The concept of diffused lowly doped p-type guard rings.

1optical detector. The thermal e/h production rate depends on the concentration of the n- or p-well and on the impurities introduced in the production stage. There is a current trend to design APIX sensors within modern deep-sub-micron CMOS processes, below the 180 nm scale. In this case the increased concentration of the wells, needed to enable the reduced design rules, causes a lowering of the breakdown voltage down to approximately 10 V, with a significant increase of thermally generated e/h pairs due to band-to-band tunneling effect. Moreover, as we depicted in the previous section, defects, contributing to the thermally generated e/h pairs through SRH recombination, are of larger number in such modern CMOS technologies, due to lower annealing temperatures and to the presence of the shallow trench isolations (STI) close to the sensitive region of the optic detector.

A third challenge of the development of SiPM quantum detectors into standard CMOS is the correct design of the optical coupling, needed to optimize the photon detection efficiency at a visible wavelength. The production lines of standard commercial CMOS fabs are mainly oriented to large-scale market applications, as standard electronics components for automotive and energy industry among others. Although the optical communication industry plays a fundamental role in the actual panorama, it uses high power standard light sources, which are affected by low power losses and have different needs in comparison to weak visible light. Low photon flux sensors in the optical range occupy clearly a smaller market section and most of the standard CMOS technology fabs do not provide the necessary anti reflective coating (ARC).

The most severe limitation of CMOS processes for the design of the SiPM is that a series of design rules needs to be verified. In CMOS technology there is an enclosure hierarchy between active area, p + implantation and nwell. A minimal distance between p + implantation and pwell should be also preserved as well as between active area and polysilicon resistance elements. As an example at a scale λ = 180 nm CMOS the minimal distance between pwell implantations is approximately 2 λ and the enclosure of p + implantation and active area is approximately 1 λ. It is often happening that one is forced to violate some of these rules in order to produce a Silicon Photomultiplier device.

Several studies were performed at a CMOS scale ranging between 90 and 800 nm [19–55]. A possible implementation solution consists of designing diffusive guard rings, as shown on **Figure 6**. As an example such layout was developed using the Teledyne DALSA 800 nm HV and the BCDLITE GLOBALFOUNDRIES 180 nm CMOS processes [18, 55]. Such layout is based on an inverted enclosure of the p + and nwell implantations, so that the high electric field zones at the edges of the diode are corrected with the geometrical distribution of the carriers. In this case only the CMOS enclosure rule is violated. Another possibility is offered by the implementation of diffusive lowly doped p guard rings around the p + implantation, as depicted in **Figure 7**. Such layout was developed using the Teledyne DALSA 800 nm HV, 350 nm HV AMS and Chartered GLOBALFOUNDRIES 130 nm/Tezzaron CMOS technology [18, 37, 41]. In this case the lowly doped region is smoothing the electric field at the edges of the junction.

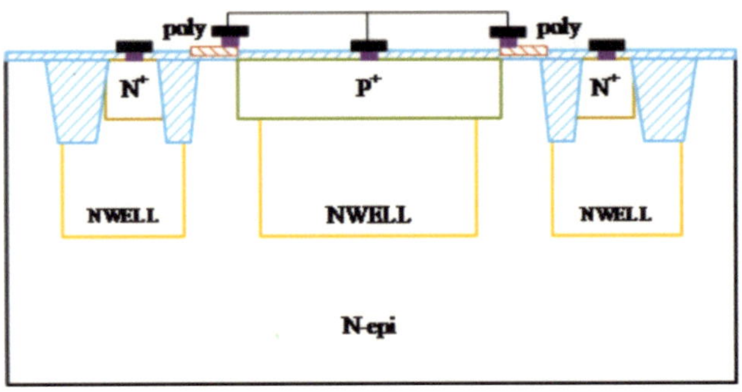

Figure 8. Structure I: a virtual guard ring with corrected STI.

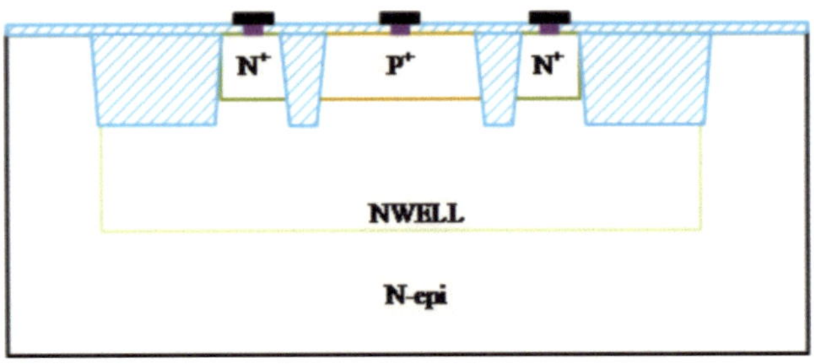

Figure 9. Structure II: a STI guard ring.

One way to prevent the violation of design rules is to include additional masks in the CMOS process, i.e. implementing new design rules accommodating the needs of the sensor, without a perturbation of the transistor development. As an example in [15, 17] an additional mask is implemented in order to produce the p-diffusive guard ring. Another way consists of finding innovative approaches for controlling the properties of the diode edge. In [28, 29] a layout is proposed with perimeter-gated guard rings. In this case a p+/nwell diode is surrounded by a gate. A gate bias ranging from 0 to 8 V is applied to the structure in order to shape the electric field at the edges of the diode and avoid localized breakdown.

Following these design ideas and suggestions, we present here two possible design solutions obtained at the BCDLITE LV GLOBALFOUNDRIES 180 nm CMOS node. Structure I on **Figure 8** is based on an n-epitaxial layer on which the SiPM sensor is formed. The microcell consists of a 50 × 50μm n+/pwell junction. GF does not share the exact doping of the wells with the customers. A virtual guard ring p+/nepi is formed on the periphery of the sensitive avalanche area. As a difference from [55], a polysilicon ring is placed around the p+ implantation in order to prevent the formation of STI close to the sensitive area. Structure II on **Figure 9** is based on an n-epitaxial layer on which the SiPM sensor is formed. The microcell consists of a 50 × 50 μm n+/pwell junction. STI is used around the sensitive area as guard rings. The aim of the study of these two structures is to verify up to which amount STI are deteriorating the performances of the SiPM.

Two series of measurements have been carried out for the SiPM sensor prototypes proposed in this chapter: a static and a dynamic characterization. In this section we show only the characterization in dark condition and we postpone the characterization of the light response to the next section. The benchmarking of the experimental results needs to follow a twofold approach. On the one side it is needed to benchmark against the mature available SiPMs, which are not always developed in standard CMOS technology but in dedicated lines with additional optimized masks. On the other side it is needed to benchmark against the above-mentioned experimental results, which are based on other CMOS processes. This second comparison allows us to understand how the chosen CMOS technology process performs with respect to other available ones.

The measurement of the current-voltage characteristic is performed on the single microcell test structure at wafer level. A Keithley 2636A source meter, connected to a computer, obtains measurements of current in reverse mode. This experiment is achieved by generating a sweep voltage between 0 and 25 V, then measuring the current and limiting it to 20 uA in order to avoid damaging the device.

The current-voltage curve of the first structure in reverse mode is shown on **Figure 10**. The structure exhibits a dark current below a few picoamperes before avalanche breakdown. At breakdown the current rises abruptly up to a few microamperes. After breakdown it gets limited by the quenching resistor and rises linearly. The breakdown voltage is approximately 13.2 V. On the plot few curves corresponding to measurements performed on chips obtained in different wafers are shown.

The current-voltage curve of the second structure in reverse mode is shown in **Figure 11**. The structure exhibits a dark current below a few picoamperes before avalanche breakdown.

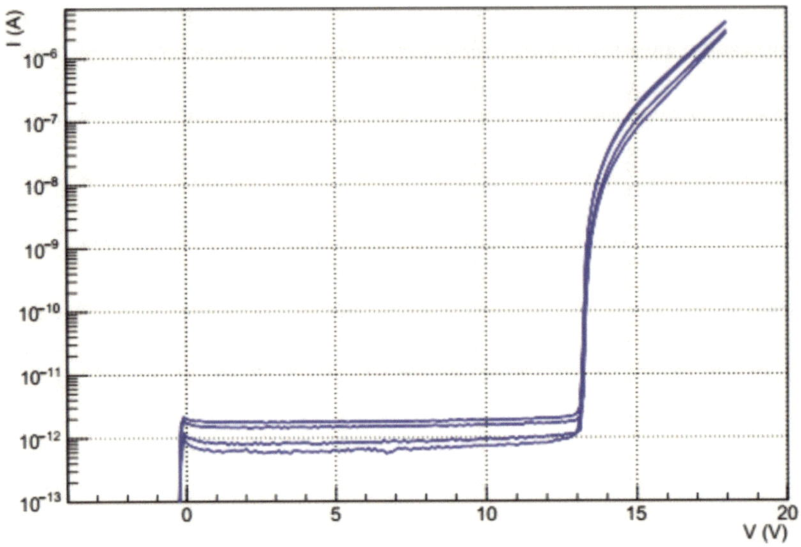

Figure 10. Current-voltage characteristics of structure I.

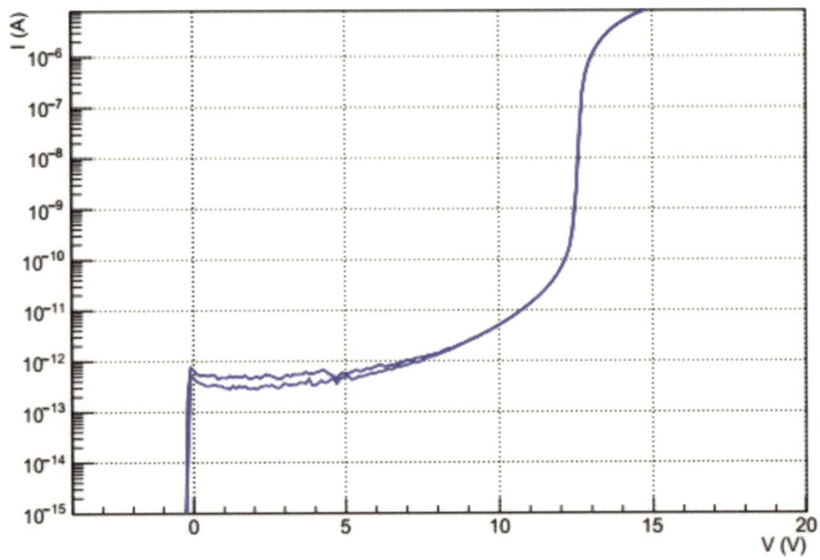

Figure 11. Current-voltage characteristics of structure II.

At breakdown the current rises up to a few microamperes. After breakdown it gets limited by the quenching resistor and rises linearly. The breakdown voltage is approximately 12.4 V. In comparison with the previous structure we notice here an additional noise source, which affects the result. Such noise source is due to the leakage at the sides of the STI. On the plot few curves corresponding to measurements performed on chips obtained in different wafers are shown.

The additional noise introduced by the STI has an impact also on the dynamic performance in dark conditions. The dynamic characterization consists of detecting the transition signal from the state I to the state II, which is normally called dark pulse. The verification of this signal and the measurement of its occurrence rate give an impression of the purity of the device. In fact, as the measurement is done in dark conditions and on a single microcell, the quality of the signal and its rate are directly proportional to the amount of impurities present in the device, which are triggering the thermal emission of electron/hole pairs trough SRH mechanism.. We measure the dark count rate by counting the number of avalanche signals produced in the SiPM sensor operated 2 V above breakdown. The dark rate is measured at room temperature and in dark conditions. The voltage amplitude of the signal of the SiPM is measured on a 50 Ohm load resistor. The output voltage is connected to a fast amplifier based on a two-stage voltage amplifier obtained with the Gali 5+ wide-band monolithic chip [62]. The total amplification gain is adjusted to 10 with a voltage divider between the two amplification stages. The signal is sent to a threshold discriminator (CAEN N844). The number of pulses above threshold are registered within a 1 s observation time window.

The signals observed at the oscilloscope within a 1 μs time window are shown on **Figures 12** and **13**. We observe that the structure I presents a clear signal corresponding to the production of a thermal electron/hole pair. The signal has a typical rise time of few hundred picoseconds and a decay time of 67 ns. The timing property is consistent with a quenching resistor of 250 kOhm and a capacitance of approximately 300 fF. Structure II shows a similar timing property. However dark pulses are occurring with higher frequency. This effect is due to the additional leakage coming from the STI walls. The dark rate measured at a overvoltage of 1.5 V at an amplitude of 0.5 dark pulse is respectively 20×10^6 and 160×10^6 and 80×10^6 kHz/mm^2 for the two structures.

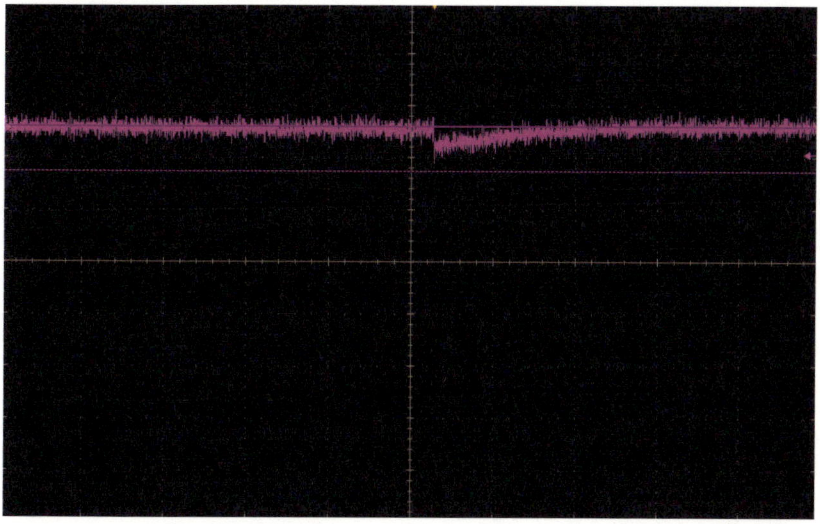

Figure 12. Structure I: signal corresponding to the thermal generation of electron/hole pairs within a 1 μs time window.

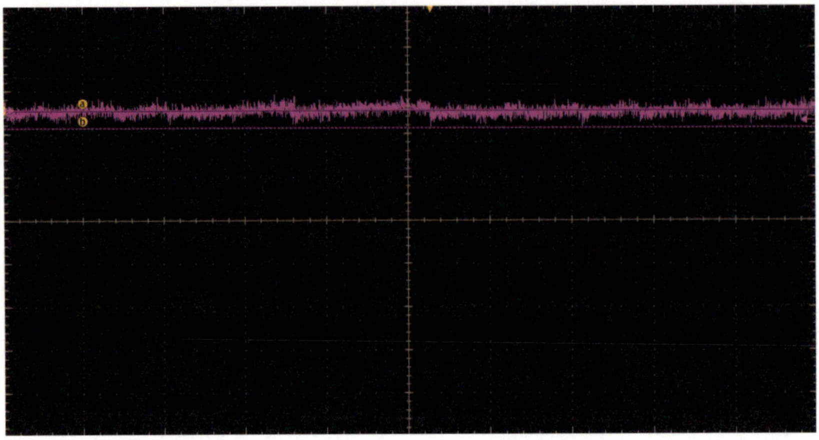

Figure 13. Structure II: signal corresponding to the thermal generation of electron/hole pairs within a 1 μs time window.

These measurements are well in agreement with the expectation from similar studies. Breakdown voltages ranging from 10 to 14 V are usually obtained at a scale lower than 180 nm. This value depends on the doping of the standard CMOS wells, which ranges from 2×10^{17} to 5×10^{17}. In comparison with custom technology the dark rate is few order of magnitudes higher. This is also a common feature of standard CMOS technology. Dark rate up to 100×10^5 kHz/mm^2 was observed at 800 nm CMOS node [20]. The typical dark rate of 20–30×10^3 kHz/mm^2 was observed at 180 nm CMOS node [55]. The results of the second structure show that the presence of the STI deteriorates the dark rate up to a factor 8 and confirms previous experiments [19–55]. These performances are worse that the state of the art SiPMs developed with custom technology, which obtain dark count rates as low as 30 kHz/mm^2 and show that current CMOS technology needs improvements in order to fully accommodate the needs of SiPM sensor development [15, 17].

4. The light response of the SiPM

The response of the SiPM to light was studied only for a small prototype based on Structure I. The prototype consists on a n array of 5×5 microcells. A low photon flux source is characterized by a Poisson-distributed number of photons with a certain average value depending on the absolute intensity of the source and on the photon detection efficiency of the SiPM. The binomial contribution due to the not perfect photon detection efficiency is not perturbate the Poissonian nature of the detected photon distribution.

A precise test set up was prepared for the experimental study. Its key-component is a fast LED with wavelength 550 nm. The light source is pulsed with a pulse width of 10 ns. The time duration is chosen to be approximately ten times less than the recovery time of the SiPM. In order to exclude the electromagnetic pick up noise, the light pulses are delivered to the operation position in the light protected area by an optic fiber. The value of the low photon flux is controlled both electronically and by tuning its relative position with respect to the sensor. The alignment can be obtained with a precision of 0.1 mm. Comparing the yeld of the developed SiPM with the yield of a commercially available SiPM with given characteristics we estimated that the photon detection efficiency of the developed SiPM is approximately 4%. Such small value is expected, due to the isolation layers present on the active area of the device and the absence of an optical coupling window.

The SiPM output signal is amplified with a high quality current-voltage amplifier with gain 15. Its charge is measured within an integration gate of 100 ns using the CAEN V1180 QDC in the VME Frame and stored in the control computer.

Figure 14 shows the experimental measured single photon spectrum. The black dots show clearly the specific form of the single photon spectrum. It consists of high resolved peaks correspondent to the number of photons distributed according to the Poisson statistics. Following the operation principle of the SiPM, every avalanche pixel detects one photon and gives as output the standard signal correspondent to a single photon. The common output of the SiPM is the analog sum of the signals from each avalanche pixel. In this condition, the first peak corresponds to 0 detected photons (electronic noise pedestal), the second one corresponds to 1 detected photon, the $(n + 1)^{th}$ one to n detected photons.

The number of peaks contains the information about the number of detected photons. The separation between two successive peaks has a constant magnitude and corresponds to the total number of electrons produced in the avalanches process. The area under each peak reflects the Poisson statistics of the photon detection in the SiPM structure.

The parameters of the expected Poisson distribution in the measured spectrum are estimated with a fit of the pedestal peak using the formula:

$$\mu = log\left(\frac{N}{N_0}\right) \tag{1}$$

where N is the total number of events and N_0 is the number of events under the pedestal peak. We find that $\mu = 4$ corresponds to the spectrum. As reported in the previous section, the SiPM

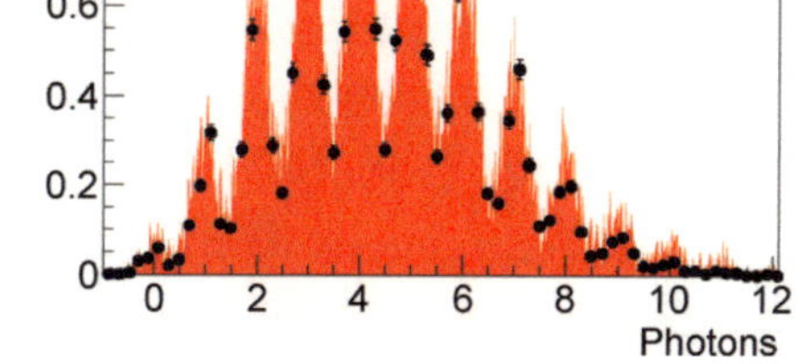

Figure 14. Measured spectrum corresponding to a low photon flux (black dots) and estimated Poisson distribution (filled area).

sensor considered here exhibits a sizeable dark rate. We observe a deviation between theory and experimental data. In particular we report an excess of events at values higher than 2 detected photons. This shift in the observed distribution is due to dark rate and cross talk, which are particularly high in the produced SiPM sensor.

5. Application of SiPM to digital positron emission tomography

One of the most promising area of the application of the SiPM is nuclear medicine. Its small size, low bias voltage, simple read-out circuitry and fast response of the state I-II transition are opening the way of the development of a new concept of nuclear medicine diagnostic systems. Positron Emission Tomography (PET) is a functional nuclear medicine technique. A tracer is prepared marking specific ligands with beta + emitting isotopes. When ligands interact through the target molecular process, they are locally metabolized. The positrons emitted by the ligands annihilate with electrons within a mean free path of approximately 0.5 mm and as a result two photons with 511 keV are emitted in opposite direction. Through the measurement of the photon directions it is possible to reconstruct the emission point and to have a detailed map of the activity of the ligands in the body.

A key-problem in PET is the technology used for the detection of the opposite photons and the reconstruction of their direction and energy. The basic block of modern PET is a detection system composed of a scintillator crystal read-out by a SiPM. When a 511 keV photon deposits its energy in the crystal though photoelectric effect, scintillation light is produced and is detected by a SiPM. Due to the small size of the SiPM it is possible to design detector models which are more and more compact.

The LYSO/SiPM unit developed by the PETLab at the Huazhong University of Science and Technology consists of a 3.9 × 3.9 × 20 mm LySO crystal read out by a 3.0 × 3.0 mm SiPM by SensL (FM30035). A PET block is composed of an array of 6 × 6 LySO/SiPM elements, with pitch 4.2 mm and with 0.3 mm crystal gaps filled in with barium sulfate to guarantee the crystal optical coupling. LySO crystal and SiPM are aligned and coupled using epoxy optical adhesive [63, 64].

On the basis of this pixelated block, a digital PET unit was developed. The key-component is here a digital electronic chain based on the Multi Voltage Threshold method. The basic idea of this new concept is that the digitalization of the signal produced when reading out the scintillation light from the crystal does not need a very high sampling. It is enough to sample 8 points corresponding to 4 thresholds. A double exponential fit to the signal shape through the interpolation of the sampled levels is enough to reconstruct the total integral of the signal, proportional to the number of detected scintillation photons and hence to the energy of the impinging gamma-ray, and the arrival time of the signal.

This novel digital PET system exhibited energy resolution between 10 and 15%, time resolution of less than 200 ps, space resolution of approximately 1–2 mm. In particular it enabled a new concept of Positron Emission Tomography: the Plug&Imaging (P&I) technique. FPGA-only implementation allows one to achieve a high channel density with greatly simplified electronics. A complete software package for data acquisition, coincidence detection and image reconstruction (allowing a flexible overall geometry of the PET imager) has also been developed to support the P&I feature and is shown on **Figures 15** and **16**. Such compact PET module can be in fact combined with other modules in different shapes and geometries, allowing a better match with the specific organs to be studied.

On the basis of this technology a full body digital PET scanner and a small animal PET scanner were successfully developed. However the modularity of the PET blocks allows a fast design of less standard approaches.

As an example of an easy combination of the P&I elements, Positron Emission Tomography may be used in the correct determination of proton range in Proton Therapy. Protons interact with human tissue mainly with ionization, releasing their energy in the typical Bragg peak mechanism. However, nuclear interactions of protons with O, C and N present in human tissues may have as results the production of neutrons associated with beta + emitters as ^{15}O and $^{11}C11$. PET systems will be dedicated to the reconstruction of the position of such nuclear interaction. A comparison with Monte Carlo allows reconstructing the position of the Bragg peak from the space distribution of the nuclear interactions in the patient's body.

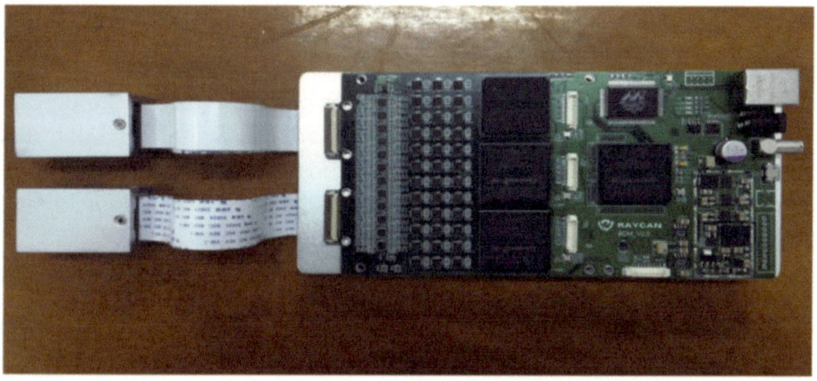

Figure 15. Plug & Imaging PET detector. Two PET heads and the digital readout electronic boards are visible.

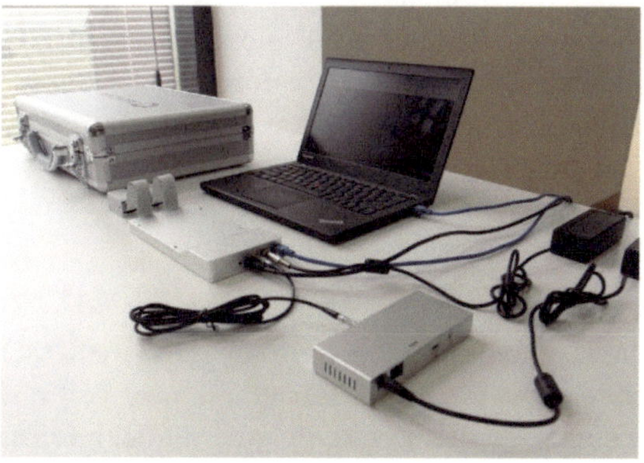

Figure 16. Plug & Imaging PET detector: the mounted system with software chain.

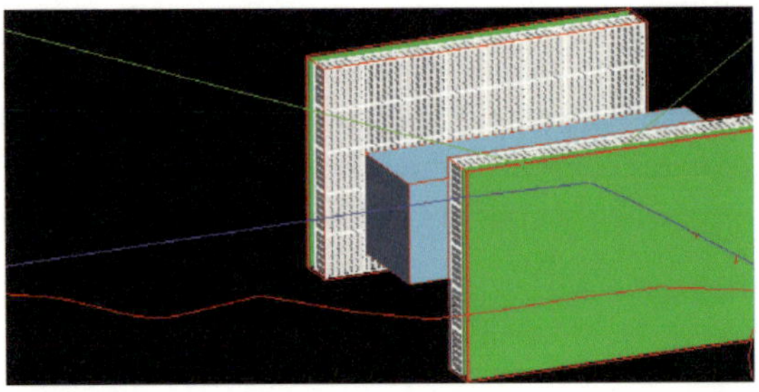

Figure 17. Simulation of a PET system for proton therapy monitoring and treatment.

Figure 17 shows the Monte Carlo of a PET system for Proton Therapy. It consists of two modules composed of 40 PET blocks, disposed on an array of 8 × 5. The total size of one head is 12.5 × 20 cm. Figure shows a typical event obtained with a proton beam energy of 160 MeV in a water phantom. The nuclear reaction $p + {}^{16}O \rightarrow {}^{15}O + p + n + gamma$ is followed by the β + decay of the ${}^{15}O \rightarrow {}^{15}N + e^{+} + \nu_{\varepsilon}^{-}$. The positron annihilates into two gamma rays, generates then the PET signal. The design of the proton therapy monitoring system is particularly suited to the cure of brain cancer.

However the P&I detection system based on the SiPM allows even more possibilities of dedicated PET scanners, as PET helmets for brain imaging. The technological challenges imposed by the development of this new kind of modular digital PET detection systems, together with the improvements in the medical diagnostic, such as the development of new ligands for the early detection of Alzheimer disease, will impose new benchmarks and new trends in modern Positron Emission Tomography and nuclear medicine.

One technological problem remains. The SiPM used in such new instrumentation for Nuclear Medicine are still obtained with custom technology. On the one hand, as we observed in the previous sections, the performances of mature custom technology is clearly better than the one achieved in standard CMOS. Nowadays also the price of mature technology is more attractive for PET producers. However, the possibility of the implementation of MVT digital electronics and SiPM sensor in a single chip offered by the CMOS implementation will significantly impact the compactness of the PET block realization, with a consequent miniaturization of the system and the possibility of even better performances of the P&I method in terms of scalability, modularity and system space resolution.

6. Conclusions

The development of Silicon Photomultipliers within standard CMOS technology reveals a clear multi-disciplinary interest and matches the requirement that nowadays are coming also from the clinician's community. Digital electronics and digital information is in fact driving a change in the predictivity of diseases and the development of key-digital-technologies, as advanced CMOS SiPMs, will be fundamental in order to develop accurate instrumentation for the combination of techniques in Nuclear Medicine. The development of the P&I platform is a clear example of such new paradigm for Positron Emission Tomography.

CMOS SiPMs still have space of improvement, which needs to be addressed. First the localized breakdown typical of SiPM structures could be corrected. However we found that only the SiPM structure with Shallow Trench Isolation placed far from the sensitive detector region exhibits a proper signal in correspondence to the production of an e/h pair, the other devices being dominated by additional noise sources due to impurities dislocation on the border of the edges of the STI.

Even in this case the minimal achieved dark rate is approximately 10 MHz/mm^2. This level is above the reached state of the art of current SiPM detectors produced in CMOS compatible processes with additional masks in the sensor region and suggests that the higher concentration of standard 180 nm CMOS processes is a limitation in the performances of the detector.

The photon detection efficiency obtainable in standard CMOS foundries is also very low. However, we observe that, although the standard 180 nm CMOS processes used to manufacture the APIX prototype does not allow us for an optimization of the optical window with a proper ARC technique, the manufactured structure I exhibits an excellent single photon spectrum, which is corresponding to a very good approximation to the Poisson statistics of the quantum state of detected photons.

These results shows that, although some improvements are needed with respect to concentration and STI technique, the standard CMOS technology is getting mature for a perspective, easy accessible, cost reductive SiPM optical detectors R&D. This topic is so sensitive, that standard CMOS facilities are currently introducing new specific processes for APIX-like optical detectors with improved performance, allowing us for further steps consisting of the optimization of the technological parameters as well as of the optical performances, including ARC processes for improved photon detection efficiency.

Acknowledgements

This work was supported in part by the Natural Science Foundation of China (NSFC) Grant #61425001, #61210003, #61604059, in part by the Natural Science Foundation of Hubei Province Grant #2016CFA005, in part by National Key Scientific Instrument and Equipment Development Project of China #2013YQ030923, in part by National Key Scientific Instrument and Equipment Development Project of Hubei Province #2013BEC050.

Author details

Nicola D'Ascenzo and Qingguo Xie*

*Address all correspondence to: qgxie@hust.edu.cn

Huazhong University of Science and Technology, Wuhan, China

References

[1] Dascenzo N, Saveliev V. The new photodetectors for high energy physics and nuclear medicine. In: Shi J-W, editor. Photodiodes. Croatia: Intech; 2011. pp. 261-284

[2] D'Auria V, Lee N, Amri T, Fabre C, Laurat J. Quantum decoherence of single photon counters. Physical Review Letters. 2011;**107**(5):050504. DOI: 10.1103/PhysRevLett.107.050504

[3] Achilles D, Silberhorn C, Silwa C, Banaszek K, Walmsley A. Fiberassisted detection with photon number resolution. Optics Letters. 2003;**28**(23):2387-2389. DOI: 10.1364/OL.28.002387

[4] Gasanov A, Golovin V, Sadigov ZY, Yusipov NY. Avalanche photodetector based on metal-resistive-layer semiconductor structures. Pisma V Zhurnal Tekhnicheskoi Fiziki. 1988;**14**:706-709

[5] Shushakov DA, Shubin VE. New solid state photomultiplier. In: Photonics West 95; International Society for Optics and Photonics; 1995. pp. 544-554

[6] Sadygov Z. Avalanche detector. Russian Patent. 1996; RU 2102820

[7] Golovin M, Akindinov A, Grigorev E, Martemyanov A, Polozov P. New results on MRS APD. Nuclear Instruments and Methods in Physics Research. 1997;**387**:231-234

[8] Saveliev V, Golovin V. Silicon avalanche photodiodes on the base of metal-resistor-semiconductor (MRS) structures. Nuclear Instruments and Methods in Physics Research. 2000;**442**:223-229

[9] Buzhan P, Dolgoshein B, Ilyin A, Kantser V, Kaplin V, Karakash A, Pleshko A, Popova E, Smirnov S, Volkov Y. An advanced study of silicon photomultiplier. ICFA Instrumentation Bulletin. 2001;**21**:28

[10] Piemonte C, Battiston R, Boscardin M, Dalla Betta G-F, Del Guerra A, Dinu N, Pozza A, Zorzi N. Characterization of the first prototypes of silicon photomultiplier fabricated at ITC IRST. IEEE Transactions on Nuclear Science. 2007;**54**:236-244

[11] Stewart A, Saveliev V, Bellis S, Herbert DJ, Huges P, Jackson J. Performance of a 1 mm^2 silicon photomultiplier. IEEE Journal of Quantum Electronics. 2008;**44**:157-164

[12] Ghassemi A, Sato K, Kobayashi K. Hamamatsu Technical Note 2017; KAPD9005E01

[13] Cova S, Lacaita A, Ghioni M, Ripamonti G, Louis T. 20 ps timing resolution with single-photon avalanche diodes. The Review of Scientific Instruments. 1989;**60**:1104-1110

[14] Villa F, Bronzi D, Zou Y, Scarcella C, Boso G, Tisa S, Tosi A, Zappa F, Durini D, Weyers S. CMOS SPADs with up to 500 μm diameter and 55% detection efficiency at 420 nm. Journal of Modern Optics. 2014;**61**:102-115

[15] Schwinger A, Brockherde W, Hosticka BJ, Vogt H. CMOSSiPM with integrated amplifier. Proceedings of SPIE. Feb. 2017;**10100**:101001A. DOI: 10.1117/12.2252516

[16] Della Rocca FM, Nedbal J, Tyndall D, Kristajic N, Day-Uei Li D, Ameer-Beg SM, Henderson RK. Real-time fluorescence lifetime actuation for cell sorting using a CMOS SPAD silicon photomultiplier. Optics Letters. Feb. 2016;**41**(4):673-676. DOI: 10.1364/OL.41.000673

[17] Zou Y, Villa F, Bronzi D, Tisa S, Tosi A, Zappa F. Planar CMOS analog SiPMs: Design, modeling, and characteriza- tion. Journal of Modern Optics. May 2015;**62**(20):1693-1702. DOI: 10.1080/09500340.2015.1049572

[18] Bérubé BL, Rhéaume VP, Parent S, Maurais L, Therrien AC, Charette PG, Charlebois SA, Fontaine R, Pratte JF. Implementation study of single photon avalanche diodes (spad) in 0.8 μm HV CMOS technology. IEEE Transactions on Nuclear Science. Jun. 2015;**62**(3):710-718. DOI: 10.1109/TNS.2015.2424852

[19] D'Ascenzo N, Zhang X, Xie Q. Application of CMOS technology to silicon photomultiplier sensors. Sensors. Sep. 2017;**17**:2204

[20] D'Ascenzo N, Brockherde W, Dreiner S, Schwinger A, Schmidt A, Xie Q. Design and characterization of a silicon photomultiplier in 0.35 um CMOS. IEEE Journal of the Electron Devices Society. Accepted Nov. 2017. DOI: 10.1109/JEDS.2017.2771145

[21] Niclass C, Rochas A, Besse P-A, Charbon E. Toward a 3-D camera based on single photon avalanche diodes. IEEE Journal of Selected Topics in Quantum Electronics. 2004;**10**:796-802

[22] Rochas A, Gosch M, Serov A, Besse P, Popovic R, Lasser T, Rigler R. First fully integrated 2-D array of single-photon detectors in standard CMPS technology. IEEE Photonics Technology Letters. 2003;**15**:963-965

[23] Tisa S, Zappa F, Labanca I. In On-chip detection and counting of single-photons, Electron Devices Meeting, 2005. IEDM Technical Digest. IEEE International: IEEE; 2005. pp. 815-818

[24] Stoppa D, Pancheri L, Scandiuzzo M, Gonzo L, Dalla Betta G-F, Simoni A. A cmos 3-d imager based on single photon avalanche diode. IEEE Transactions on Circuits and Systems. 2007;**54**:4-12

[25] Niclass C, Rochas A, Besse P-A, Charbon E. Design and characterization of a CMOS 3-D image sensor based on single photon avalanche diodes. IEEE Journal of Solid-State Circuits. 2005;**40**:1847-1854

[26] Stoppa D, Pancheri L, Scandiuzzo M, Malfatti M, Pedretti G, Gonzo L. In a single-photon-avalanche-diode 3D imager, Solid-State Circuits Conference, 2005. ESSCIRC 2005. Proceedings of the 31st European: IEEE; 2005. pp. 487-490

[27] Rochas A, Gani M, Furrer B, Besse P, Popovic R, Ribordy G, Gisin N. Single photon detector fabricated in a complementary metal-oxide-semiconductor high-voltage technology. Review of Scientific Instruments. 2003;**74**:3263-3270

[28] Bérubé B-L, Rhéaume V-P, Parent S, Maurais L, Therrien AC, Charette PG, Charlebois SA, Fontaine R, Pratte J-F. Implementation study of single photon avalanche diodes (spad) in 0.8 μm HV CMOS technology. IEEE Transactions on Nuclear Science. 2015;**62**:710-718

[29] Pancheri L, Stoppa D. In low-noise CMOS single-photon avalanche diodes with 32 ns dead time, Solid State Device Research Conference, 2007. ESSDERC 2007. 37th European; IEEE; 2007. pp. 362-365

[30] Habib MHU, McFarlane N. In a perimeter gated single photon avalanche diode based silicon photomultiplier as optical detector, Circuits and Systems (MWSCAS), 2015 IEEE 58th International Midwest Symposium on: IEEE: 2015. pp. 1-4

[31] Gu J, Habib MHU, McFarlane N. Perimeter gated single photon avalanche diodes: An information theoretic assessment. IEEE Photonics Technology Letters. 2016;**28**:701-704

[32] Niclass C, Sergio M, Charbon E. In a single photon avalanche diode array fabricated in 0.35 μm CMOS and based on an event-driven readout for TCSPC experiments. Proceedings Volume 6372, Advanced Photon Counting Techniques; 63720S, 2006

[33] Niclass C, Favi C, Kluter T, Monnier F, Charbon E. Single-photon synchronous detection. IEEE Journal of Solid-State Circuits. 2009;**44**:1977-1989

[34] Tisa S, Guerrieri F, Tosi A, Zappa F. In 100 kframe/s 8 bit monolithic single-photon imagers, Solid-State Device Research Conference. ESSDERC 2008. 38th European, 2008; IEEE; 2008. pp. 274-277

[35] Tisa S, Guerrieri F, Zappa F. Variable-load quenching circuit for single-photon avalanche diodes. Optics Express. 2008;**16**:2232-2244

[36] Stoppa D, Mosconi D, Pancheri L, Gonzo L. Single-photon avalanche diode cmos sensor for time-resolved fluorescence measurements. IEEE Sensors Journal. 2009;**9**:1084-1090

[37] Arbat A, Trenado J, Gascon D, Vilá A, Comerma A, Garrido L, Diéguez A. High voltage vs. high integration: A comparison between CMOS technologies for SPAD cameras. Proceedings of SPIE Optics and Photonics. 2010;**2010, 7780**:77801G

[38] Niclass C, Favi C, Kluter T, Gersbach M, Charbon E. A 128×128 single-photon image sensor with column-level 10-bit time-to-digital converter array. IEEE Journal of Solid-State Circuits. 2008;**43**:2977-2989

[39] Vilella E, Alonso O, Montiel A, Vilá A, Diéguez A. A low-noise time-gated single-photon detector in a HV-CMOS technology for triggered imaging. Sensors and Actuators A: Physical. 2013;**201**:342-351

[40] Nissinen I, Nissinen J, Lansman A, Hallman L, Kilpel AA, Kostamovaara J, Kogler, M, Aikio M, Tenhunen J. In a sub-ns time-gated CMOS single photon avalanche diode detector for Raman spectroscopy, Solid-State Device Research Conference (ESSDERC), 2011 Proceedings of the European: IEEE; 2011. pp. 375-378

[41] Schwinger A, Brockherdea W, Hostickaa BJ, Vogta H. In CMOS SiPM with integrated amplifier. Proceedings of SPIE. 2017:101001A-1101001

[42] Zou Y, Villa F, Bronzi D, Tisa S, Tosi A, Zappa F. Planar CMOS analog SiPMs: Design, modeling, and characterization. Journal of Modern Optics. 2015;**62**:1693-1702

[43] Vilella E, Diéguez A. A gated single-photon avalanche diode array fabricated in a conventional CMOS process for triggered systems. Sensors and Actuators A: Physical. 2012; **186**:163-168

[44] Jradi K, Pellion D, Ginhac D. Design, characterization and analysis of a 0.35 μm CMOS SPAD. Sensors. 2014;**14**:22773-22784

[45] Finkelstein H, Hsu MJ, Esener SC. Sti-bounded single-photon avalanche diode in a deep-submicrometer cmos technology. IEEE Electron Device Letters. 2006;**27**:887-889

[46] Faramarzpour N, Deen MJ, Shirani S, Fang Q. Fully integrated single photon avalanche diode detector in standard cmos 0.18-μ m technology. IEEE Transactions on Electron Devices. 2008;**55**:760-767

[47] Vornicu I, Bandi FN, Carmona-Galán R, Rodríguez-Vázquez Á. A CMOS digital SiPM with focal-plane light-spot statistics for DOI computation. IEEE Sensors Journal. 2017; **17**:632-643

[48] Pancheri L, Stoppa D. In low-noise single photon avalanche diodes in 0.15 μm CMOS technology, Solid-State Device Research Conference (ESSDERC). Proceedings of the European, 2011: IEEE; 2011. pp. 179-182

[49] Richardson JA, Grant LA, Henderson RK. Low dark count single-photon avalanche diode structure compatible with standard nanometer scale cmos technology. IEEE Photonics Technology Letters. 2009;**21**:1020-1022

[50] Gersbach M, Maruyama Y, Trimananda R, Fishburn MW, Stoppa D, Richardson JA, Walker R, Henderson R, Charbon E. A time-resolved, low-noise single-photon image sensor fabricated in deep-submicron CMOS technology. IEEE Journal of Solid-State Circuits. 2012;**47**:1394-1407

[51] Henderson RK, Richardson J, Grant L. Reduction of Band-to-Band Tunneling in Deep-Submicron cmos Single Photon Avalanche Photodiodes. Bergen, Norway: International Image Sensor Workshop (IISW 2009); 2009. pp. 26-28

[52] Karami MA, Gersbach M, Yoon H-J, Charbon E. A new single-photon avalanche diode in 90nm standard CMOS technology. Optics Express. 2010;**18**:22158-22166

[53] Webster EA, Richardson JA, Grant LA, Renshaw D, Henderson RK. A single-photon avalanche diode in 90-nm CMOS imaging technology with 44% photon detection efficiency at 690 nm. IEEE Electron Device Letters. 2012;**33**:694-696

[54] Herbert DJ, Moehrs S, D'Ascenzo N, Del Guerra A. The silicon photomultiplier for application to high-resolution positron emission tomography. Nuclear Instruments and Methods. 2007;**573**:84-87

[55] Colletti PM. SNMM: Highlights lecture: General nuclear medicine. JNM. 2017;**58**:11N-15N

[56] Shockley W, Read WT. Statistics of the recombination of holes and electrons. Physics Review. 1952;**87**:835

[57] Selberherr S. Analysis and Simulation of Semiconductor Devices. Wien, New York: Springer-Verlag; 1984

[58] Izhaky N, Morse MT, Koehl S, Cohen O, Rubin D, Barkai A, Sarid G, Cohen R, Paniccia MJ. Development of CMOS-compatible integrated silicon photonics devices. IEEE Journal of Selected Topics in Quantum Electronics. 2006;**12**:1688-1698

[59] Lee M-J, Rucker H, Choi W-Y. Effects of guard-ring structures on the performance of silicon avalanche photodetectors fabricated with standard CMOS technology. IEEE Electron Device Letters. 2012;**33**:80-82

[60] Lee M-J, Choi W-Y. Effects of parasitic resistance on the performance of silicon avalanche photodetectors in standard CMOS technology. IEEE Electron Device Letters. 2016;**37**:60-63

[61] Sul W-S, Oh J-H, Lee C-H, Cho G-S, Lee W-G, Kim S-D, Rhee J-K. Guard-ring structures for silicon photomultipliers. IEEE Electron Device Letters. 2010;**31**:41-43

[62] MINICIRCUITS. www.minicircuits.com

[63] Xie Q, Chen Y, Yuanbao Z. Implementation of LYSO/PSPMT block detector with all digital DAQ system. IEEE Transactions on Nuclear Science. 2013;**60**(3):1487-1494

[64] Xie Q, Kao C, Byrun K. Characterization of Silicon Photomultipliers for PET imaging, in 2016 IEEE Nuclear science conference records. Vol. 1–6; 2008. pp. 1199-1203

Chapter 9

Parallelized Integrated Time-Correlated Photon Counting System for High Photon Counting Rate Applications

Imane Malass, Wilfried Uhring,
Jean-Pierre Le Normand, Norbert Dumas and
Foudil Dadouche

Additional information is available at the end of the chapter

http://dx.doi.org/10.5772/intechopen.72273

Abstract

Time-correlated single-photon counting (TCSPC) applications usually deal with a high counting rate, which leads to a decrease in the system efficiency. This problem is further complicated due to the random nature of photon arrivals making it harder to avoid counting loss as the system is busy dealing with previous arrivals. In order to increase the rate of detected photons and improve the signal quality, many parallelized structures and imaging arrays have been reported, but this trend leads to an increased data bottleneck requiring complex readout circuitry and the use of very high output frequencies. In this paper, we present simple solutions that allow the improvement of signal-to-noise ratio (SNR) as well as the mitigation of counting loss through a parallelized TCSPC architecture and the use of an embedded memory block. These solutions are presented, and their impact is demonstrated by means of behavioral and mathematical modeling potentially allowing a maximum signal-to-noise ratio improvement of 20 dB and a system efficiency as high as 90% without the need for extremely high readout frequencies.

Keywords: SPAD, TCSPC, parallelized, FIFO, readout

1. Introduction

Time-correlated single-photon counting (TCSPC) is a mature and extremely accurate low light signal measurement technique that uses single quanta of light to provide information on the temporal structure of the light signal. The method was first conceived in nuclear physics [1] and was for a long time primarily used to analyze the light emitted as fluorescence during the relaxation of molecules from an optically excited state to a lower energy state [2]. Today,

TCSPC is widely used in many applications that require the analysis of fast weak periodic light events with a resolution of tens of picoseconds such as diffuse optical tomography (DOT) [3, 4], fluorescence lifetime imaging (FLIM) [5] and high-throughput screening (HTS) [6]. TCSPC is based on detecting single photons of a periodical light signal, measuring the detection times within the light period and reconstructing the light waveform from the individual time measurements after repeating the measurements for enough times. Traditionally, the TCSPC technique relied on vacuum tube technologies such as PMTs and MCPs. These mature technologies are capable of achieving very good performances, but they are expensive, cumbersome and fragile and require extremely high operating voltages, which make them unsuitable for the fabrication of miniaturized portable TCSPC imaging systems. In recent years, single-photon avalanche diodes (SPADs) have gained a wide popularity as a less expensive and more compact alternative for vacuum tube detectors. The integration of planar epitaxial SPADs in standard CMOS technology has significantly improved the level of miniaturization of SPADs and paved the way for SPAD arrays. These devices possess the typical advantages of microelectronics integrated circuits, such as small size, ruggedness, low operating voltages and low cost. Furthermore, they can be directly implemented with the necessary associated circuits on the same chip to realize an integrated, ultrasensitive, high-speed and low-cost TCSPC imaging system. Many SPAD-based TCSPC systems have been successfully demonstrated lately. Nowadays, state-of--the-art imaging sensors integrating thousands of single-photon detectors on the same chip have been demonstrated in standard CMOS technology [7, 8]. Most integrated TCSPC systems consist of 2D arrays or 1D arrays of SPADs with their associated electronics in the form of smart pixels resulting in a trade-off between high-photon detection efficiency and advanced electronic functionalities [9–11]. This approach allows for a better detection efficiency compared to a single commercial SPAD. However, such designs should be conceived such that the detection yield is optimized, i.e. ensure an optimal detection efficiency and a limited counting loss probability. In this chapter, we present these two issues and propose methods to quantify and limit their effects based on mathematical and behavioral modeling.

2. A parallelized macropixel structure for SNR optimization

Single-photon avalanche diodes (SPADs) operate in Geiger mode; in this mode, the p-n junction is biased beyond its breakdown voltage, as a result a high electric field exists in the charge space such that a charge carrier ideally created by photoelectric interaction is enough to generate a self-sustained avalanche. Indeed, unlike linear APDs, where stopping the light signal is enough to stop the avalanche, when an avalanche is triggered in an SPAD, the current will continue to increase until the destruction of the component as a result of overheating. Therefore, the avalanche must be swiftly quenched by an associated circuitry that senses the avalanche and stops it by reducing the reverse bias below the breakdown voltage, so that the avalanche cannot maintain itself, then returned it to its initial condition. The circuit used to accomplish these tasks is the quenching circuit, and the selection of such circuit is not a trivial task as it directly affects many of the SPAD performance metrics [12]. It is therefore important to choose a suitable quenching circuit for the desired application so it will not limit or deteriorate the SPAD characteristics.

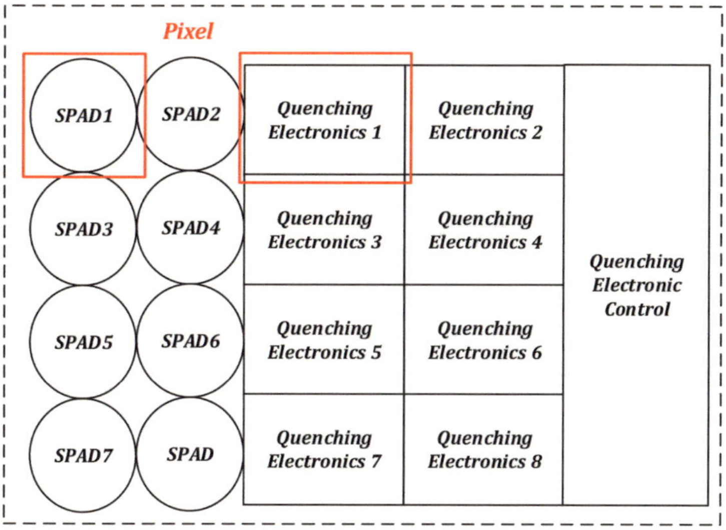

Figure 1. Simplified schematic of the parallelized macropixel is presented in [13].

Each SPAD with its associated electronics forms an independent pixel, and the quenching electronic is the main part of the SPAD-associated electronics; however, other smart functionalities could also be included in the pixel. In particular, it is possible to use a gating signal to activate or inactive the SPAD; this functionality is traditionally used to operate the SPAD in gated mode where it is enabled only during the gate-on window and disabled during the gate-off time interval such that absorbed photons do not trigger an avalanche. This functionality could also be used to deactivate SPAD showing an abnormal behavior that affects the system yield. In [13], a macropixel architecture that makes use of such approach was implemented, in this approach. The macropixel (**Figure 1**) is divided into eight pixels that could be activated or deactivated based on their activity levels. This option was added to ensure that the SNR is not affected by an undesirable effect that could decrease the detector's efficacy.

The signal delivered by a photon counting detector is affected by temporal fluctuations that are expressed as a Poisson distribution. If N is the average number of detected pulse, it includes a fluctuation expressed in the shot noise $n = \sqrt{N}$, while the other electronic noise can be ignored thanks to the infinity gain of the SPAD. The total signal N is given by $N=N_{ph} + N_d$ where N_{ph} is the total of detected photon and N_d is the number of counts caused by the dark count. The associated shot noises are $n_{ph} = \sqrt{N_{ph}}$ and $n_d = \sqrt{N_d}$. The number of photons is measured by subtracting the results of two measurements: one for the total number of counts (Nph + Nd) and the second for the dark ones (Nd). In this case, the total noise is given by

$$n_{tot} = \sqrt{N_{ph} + 2N_d} \tag{1}$$

If N_d is considered as a constant equal to the mean value $\overline{N_d}$ instead of being measured each time, the variance of the term comes to zero, and thus, the number of photons and its associated noise are given by

$$N_{ph} = N_{tot} - \overline{N_d} \tag{2}$$

$$n_{tot} = \sqrt{N_{ph} + \overline{N_d}} \tag{3}$$

Therefore, the signal-to-noise ratio is

$$SNR = \frac{N_{ph}}{n_{tot}} = \frac{N_{ph}}{\sqrt{N_{ph} + \overline{N_d}}} \tag{4}$$

In the case of a multi-SPAD macropixel, the SNR of the macropixel structure is the sum of each SPAD photon count divided by the total noise component:

$$SNR = \frac{\sum_i^8 N_{ph,i}}{\sqrt{\sum_i^8 \left(N_{ph,i} + N_{d,i}\right)}} \tag{5}$$

where Nph_i is the number of detected photons and Nd_i is the dark count rate of the ith SPAD ($SPAD_i$) in the macropixel. Consequently, the signal-to-noise ratio can be optimized by switching SPADs on/off such that pixels showing undesirable activity levels are deactivated. These undesirable pixels could be 'hot pixels' showing an above-average high dark count rate or 'dark pixels' showing a below-average low light sensibility.

2.1. Hot pixel elimination

These pixels could be identified through a calibration phase where the individual DCR of each SPAD Nd_i is measured in the dark and potentially eliminated based on a hot pixel elimination (HPE). To evaluate the benefit of such approach, we assume that the macropixels are uniformly lighted, i.e. all the Nph_i are equal to Nph, and all the SPAD's DCR are equal to Nd except for one $SPAD_j$ that presents a DCR m times higher than the rest of the SPADs. Thus, the signal-to-noise ratio is given by

$$SNR_{(7+m)} = \frac{8.\overline{Nph}}{\sqrt{8.\overline{Nph} + (7+m).\overline{Nd}}} \tag{6}$$

By turning off the noisy SPAD, the SNR becomes

$$SNR_{(7)} = \frac{7.\overline{Nph}}{\sqrt{7.\overline{Nph} + (7).\overline{Nd}}} \tag{7}$$

Consequently, disabling the noisy SPAD leads to a signal-to-noise ratio improvement of

$$\frac{SNR_{(7+m)}}{SNR_{(7)}} = \sqrt{\frac{7}{8}\left[1 + \frac{(m-1)}{8(\alpha+1)}\right]} \tag{8}$$

where $\alpha = \overline{N_{ph}}/\overline{N_d}$ is the mean photon count on the mean DCR ratio.

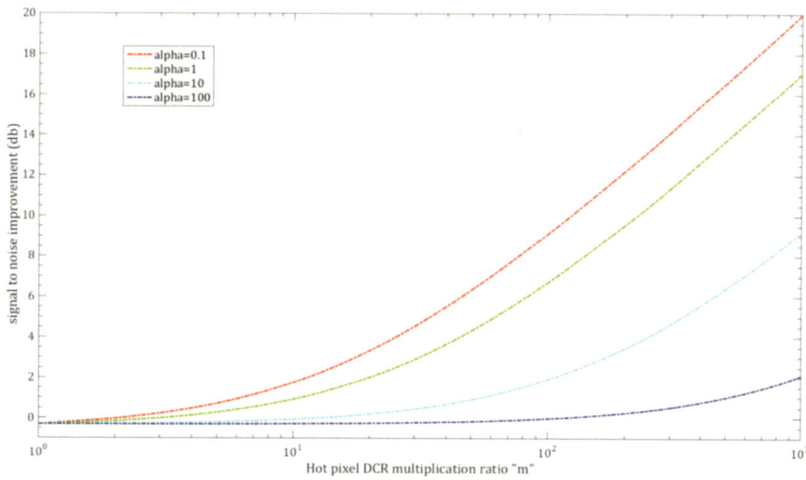

Figure 2. Signal-to-noise ratio improvement using the hot pixel elimination scheme.

Figure 2 shows the SNR gain versus the hot pixel DCR multiplication factor m for different α ratios. For a weak signal measurement ($\alpha = 0.1$), the gain can be as high as 20 dB. Nevertheless, this assessment clearly states that the SNR may be slightly lowered if the m coefficient is too low, and thus it is not advisable to remove SPADs with a DCR greater than the mean DCR. Based on these simulations, an efficient rule of thumb is to disable only SPADs with an m coefficient greater the α ratio, with obviously $m > 1$. Previous works have reported that about 20% of the SPADs integrated in an array have a dark count about 10 to 1000 times higher than the 80% other diodes [7, 14]. Consequently, there is a high probability of having a hot SPAD among the eight SPADs. Therefore, the proposed structure can lead to significant SNR improvement ranging from 0 to 20 dB.

2.2. Dark pixel elimination algorithm

The scenario that could lead to lower SNR is pixels with low light sensibility due to a manufacturing defect, a dust or as a result of the SPADs not being uniformly illuminated. To evaluate the SNR gain resulting from eliminating such pixels, we will consider the case where the eliminated SPADs are completely blind. This is the worst case of light sensibility and the elimination of these dark pixels results in the best SNR improvement (**Figure 3**). Assuming n as dark pixels, the corresponding SNR is

$$SNR_{(8-n_{blinded})} = \frac{(8-n).\overline{Nph}}{\sqrt{(8-n).\overline{Nph} + 8.\overline{Nd}}} \tag{9}$$

If all blinded SPADs are turned off, the SNR becomes

$$SNR_{(8-n_{blinded})} = \frac{(8-n).\overline{Nph}}{\sqrt{(8-n).\overline{Nph} + (8-n).\overline{Nd}}} \tag{10}$$

Consequently, for $n \neq 8$, the SNR gain is given by

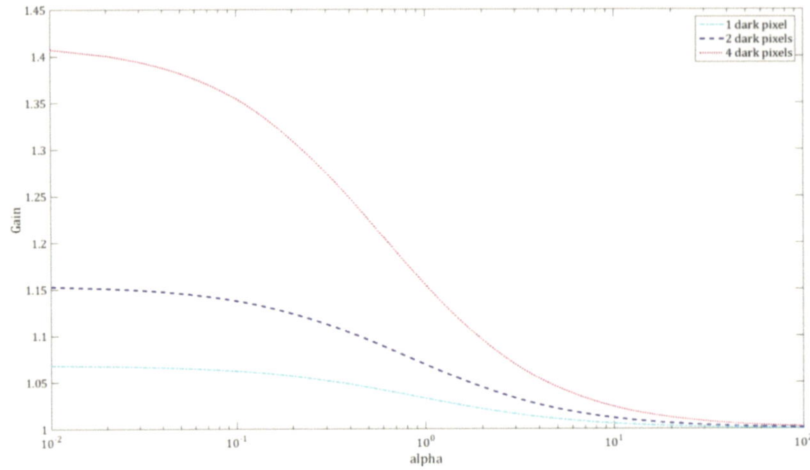

Figure 3. Signal-to-noise ratio improvement obtained through the dark pixel elimination scheme.

$$\frac{SNR_{(8-n_{blinded})}}{SNR_{(8-n_{off})}} = \sqrt{\frac{(8-n)\alpha+8}{(8-n)\alpha+(8-n)}} \tag{11}$$

2.3. SNR gain evaluation

A low SNR could be the result of a low signal levels or high noise levels; consequently, the SNR could be improved by elimination of pixels exhibiting high noise levels (hot pixel elimination) or pixels exhibiting low light sensibility (dark pixel elimination). Both schemes require a calibration phase. In the case of dark pixel elimination scheme, the counting rate of each pixel must be measured under illumination to detect SPADs with low sensibility levels, and these measurements should be repeated if the test conditions change. The hot pixel elimination scheme on the other hand requires a onetime calibration phase to measure the individual DCR for each SPAD and deactivate the too noisy SPADs based on their DCR levels. Both approaches resulted in an improved SNR; however, the dark pixel elimination efficiency was relatively low, whereas the hot pixel elimination was found to be useful in most cases.

3. Efficiency improvement of TCSPC systems

3.1. Counting loss in TCSPC systems

Typical TCSPC setup consists of a pulsed optical laser source, a photon detector such as a silicon photon multiplier (SiPM) or an SPAD, a time measurement block based on a time-to-digital converter (TDC) or time-to-amplitude converters (TAC) and an external CPU to process the measurement results. When a photodetection occurs, a certain time is required for data processing; such time interval is referred to as 'dead time' because the system is incapable of processing any additional photons collected by the SPAD resulting in counting loss and a reduction of the SNR caused by the decreased counting efficiency which is at best equal to

$$CV = \frac{1}{\sqrt{n}} \tag{12}$$

This issue is further complicated by the random nature of photon arrivals and the fact that TCSPC applications such as FLIM and HTS usually deal with high counting rates. In order to increase the rate of detected photons and improve the SNR, many parallelized imaging structures have been reported [5, 15], but this trend leads to an increased data bottleneck which requires the use of complex readout circuitry [7] as well as very high output frequencies to ensure a reasonable dead time [5]. Another solution for the high output rate is the use of an embedded FIFO to store the measurement results, while they have been processed; nevertheless, FIFOs are very demanding in terms of power and silicon area, and to our knowledge, there has been no study done to properly determine the exact FIFO length required to achieve optimum results. It is therefore important to evaluate the counting gain resulting from the use of an embedded FIFO as a function of its depth and the readout rate.

3.2. TCSPC system as a queuing model

TCSPC systems are based on measuring arrival times of single-photon events. Processing these measurements requires several additional operation steps such as quenching the photon detector, shaping the regenerated signal, converting the time to a digital value and sending it into a processing unit or memory. While these operations are being conducted, the system is unavailable to process another measurement for a certain time interval referred to as 'dead time'. To simplify the study of the TCSPC system, the readout period is considered equal to the system's dead time. The dead time as well as the random nature of the single-photon detection events leads to random counting losses as the system is busy processing a previous photon arrival, thus limiting the system efficiency. To evaluate the counting loss, the TCSPC can be modeled using a queuing model with an arrival rate λ representing the average number of photon arriving at the sensor's surface per second, a departure rate μ representing the readout data rate given in sample per second and a service rate ρ representing the rate at which the TCSPC system can process photon detections which is equal to (dead time)$^{-1}$. **Figure 4** illustrates this phenomenon; it is clear that even if the arrival rate λ is equal to or less than the departure rate μ, the random nature of the photon arrival leads to a quiet period followed by a peak of arrivals of photon, a well-known characteristic of a Poisson process. During this peak of activity, some photons will be lost as a result of the system's dead time. The simplest approach to limit such loss is the reduction of the dead time and the readout period, but reducing these times is limited by physical and electrical constraints to tens of nanoseconds. Another approach is the use of parallelized structures with the incoming light uniformly split (**Figure 5**); assuming an equal distribution of the photon arrivals, this is similar to the division of the arrival rate λ into M equal parts where M is the number of parallel modules. This approach leads to a reduction of the counting loss as well as the pile-up effect, but it also creates a data bottleneck at the end of the processing chain, thus requiring the use of high output frequencies to process the resultant high counting rate. Consequently, the loss problem is not resolved but only shifted towards the final output. This problem could be mitigated by integrating a FIFO in the TCSPC system which allows a better flexibility in processing the stochastic arrival events. Indeed, a TSCPC system without a FIFO can be modeled as one

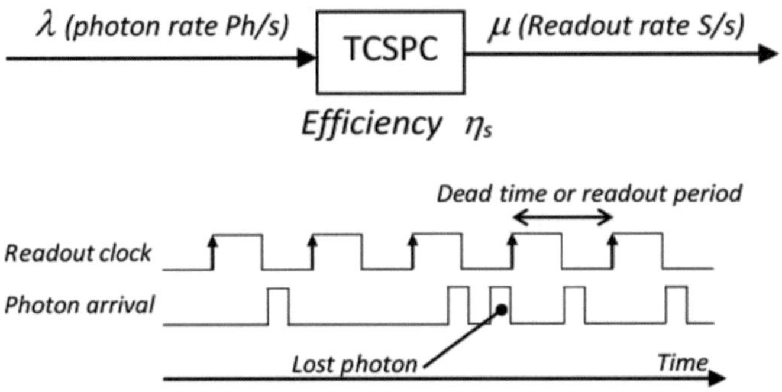

Figure 4. Counting loss in a TCSPC model.

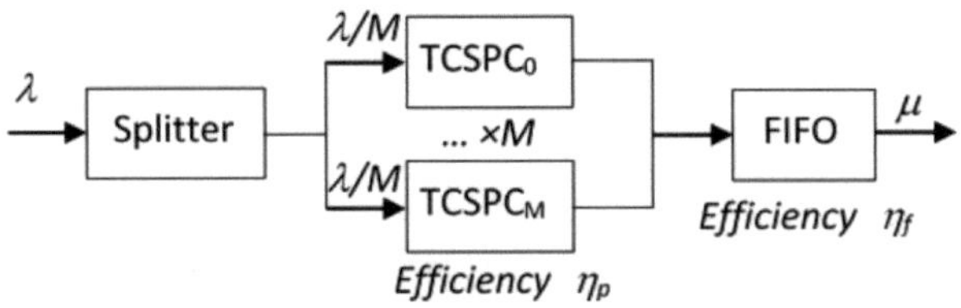

Figure 5. Parallelization principle of several TCSPC modules.

buffer queuing system; similarly, a TCSPC system integrating a FIFO with N rows can be modeled as an N cell queuing system. We will assume that the FIFO's input data follow a Poisson process, a reasonable assumption when the average photon arrival rate is significantly lower than the TCSPC's operating frequency. Giving the stochastic nature of the measured phenomena, i.e. the photon arrival Poisson process, the system's behavior must be studied in terms of the traffic intensity in and out of the FIFO to determine the impact of its limited capacity on the sensor's sensibility due to missed arrivals when the FIFO is full. The FIFO can be equated to a size N queuing system where the input is a Poisson arrival process with a mean arrival rate λ and the probability function of n arrivals occurring during the time interval $[t, t + \tau]$ given as

$$P[N(t+\tau) - \mathrm{N(t)}] = \frac{e^{-\lambda\tau}.(\lambda\tau)^n}{n!} n = 0, 1, 2, \ldots \tag{13}$$

The FIFO's output follows a periodic departure process with a departure rate μ and a readout period $T_d = \mu^{-1}$ which represents the time needed for one departure to be accomplished. The system can be modeled as a semi-Markov chain where $Q_n = Q(t = t_n)$ is the number of occupied cells in the FIFO immediately after departure moments $\{t_n, n = 0,1,2\ldots\}$ [16]. Giving that the FIFO's capacity is limited to N cells, the number of occupied cells in the system cannot exceed *N*-1, and the embedded Markov chain contains N states labeled according to the number of occupied cells left soon after a departure $S = \{n, n = 0,1,2\ldots N\text{-}1\}$. **Figure 6** shows the embedded Markov chain with all the possible transitions from a random state 'i'.

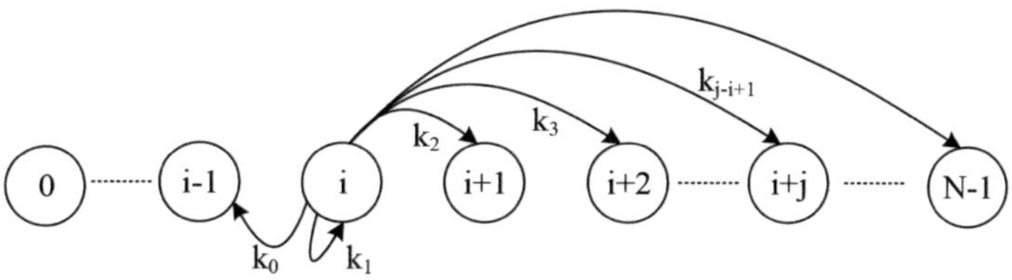

Figure 6. Markov chain states and possible state transitions from and into a state 'i'.

3.2.1. Steady-state probability evaluation

Let X_n be the number of arrivals during the readout period T_d giving the Poisson arrival property; the probability of j arrivals occurring during the readout period is

$$k_j = P[X_n = j] = \frac{e^{-\rho}.(\rho)^j}{j!} j = 0, 1, 2\ldots \quad (14)$$

where ρ defined as

$$\rho = \lambda.T_d = {}^{\lambda}/_{\mu} \quad (15)$$

is the photon rate to the readout rate ratio. The number of occupied cells after the *(n+1)th* period is increased by the number X_{n+1} of photon arrivals during this period and is reduced by one readout. If the number of photon arrivals overloads the FIFO, the number of occupied cells is clipped to *N-1* and a loss of measurement occurs. If the FIFO is empty, i.e. $Q_n = 0$, no readout occurs. Therefore, the relation between Q_n and Q_{n+1} is defined as

$$Q_{n+1} = \min(Q_n + X_{n+1} - 1,\ N - 1) \qquad if Q_n > 0 \quad (16)$$

$$Q_{n+1} = \min(X_{n+1},\ N - 1) \qquad if Q_n = 0 \quad (17)$$

And, the transition probability from the state i to the state j after m transitions is

$$P_{ij}^{(m)} = PP(Q_{n+m} = j/Q_n = i)\mathrm{i}, \mathrm{j} \in S \quad (18)$$

In particular the one-step transition probability is.

$$P_{ij}^{(1)} = P_{ij} = P(Q_{n+1} = j/Q_n = i) \quad (19)$$

$$P_{ij}^{(1)} = P(i + X_{n+1} - 1 = j) if\ i > 0 \quad (20)$$

$$= P(X_{n+1} = j) if\ i = 0 \quad (21)$$

which allows us to define the $K \times K$ transition probability Matrix *'P'* of the one-step transition probabilities $P_{i,j}$ [16]:

$$P = \begin{bmatrix} k_0 & k_1 & k_2 & \dots & k_{N-2} & \\ & & & & & 1-\sum_{i=0}^{N-2} k_i \\ k_0 & k_1 & k_2 & \dots & k_{N-2} & 1-\sum_{i=0}^{N-2} k_i \\ & & & & & 1-\sum_{i=0}^{N-3} k_i \\ 0 & k_0 & k_1 & \dots & k_{N-3} & \\ \vdots & \vdots & & & & \dots \\ 0 & 0 & 0 & \dots & k_0 & \dots 1-k_0 \end{bmatrix} \quad (22)$$

where the *i,j* of element $P_{i,j}$ of the matrix represents the probability of being in the state '*j*' giving that the system was in the state '*i*'. These probabilities describe the transient behavior of the system; however, as the system evolves, it will converge into a state of equilibrium known as the steady state with time-independent distribution [17] represented as a vector $\pi = (\pi_0, \pi_1, \pi_2 \dots \pi_{N-1},)$ where π_i is the probability to be in the state '*i*' once the system has reached its equilibrium.

The steady-state distributions satisfy the following equations:

$$\pi(\mathrm{j}) = \sum_{i=0}^{N-1} \pi(i).P(i,j) \quad (23)$$

and

$$\sum_{i=0}^{N-1} \pi(\mathrm{i}) = 1 \quad (24)$$

Furthermore, the vector π is the solution to the set of the linear Eq. [18]:

$$\pi.\mathrm{P} = \pi \quad (25)$$

resulting in a system of N equations with N variables π_i (26):

$$\begin{gathered} k_0.\pi_0 + k_0.\pi_1 = \pi_0 \\ k_1.\pi_0 + k_1.\pi_1 + k_0.\pi_2 = \pi_1 \\ k_2.\pi_0 + k_2.\pi_1 + k_1.\pi_2 + k_0.\pi_3 = \pi_2 \\ \dots \end{gathered} \quad (26)$$

3.2.2. Blocking probability

The main goal of using this queuing model is to evaluate the system efficiency based on the probability of an arrival finding the FIFO fully, and as a result of being lost, such probability represents the blocking probability P_B. In order to evaluate P_B, we need to have the state distribution at all moment and not only at departure moments. Let us define the following system probabilities:

P_k: Probability of the system containing k registered arrivals $(k = 0\ldots N)$.

π_k: State probabilities at departure instants $(k = 0\ldots N\text{-}1)$.

$\pi_{a,k}$: State probabilities at arrival instants regardless whether the arrival joins the queue or not $(k = 0\ldots N)$.

An important property of the Poisson arrival process is the Poisson Arrival See Time Averages [16] which implies that the distribution of occupied cells seen at arrival instants is the same as the distribution seen by a random observer:

$$P_k = \pi_{a,k} \tag{27}$$

On the other hand, the probability that an arrival finds $k < N$ occupied queue in the system is equal to the probability that a departure leaves k occupied cell giving that the new arrival is admitted:

$$P_k = \pi_k(1 - P_B) \tag{28}$$

In particular for $k = 0$, we have

$$P_0 = \pi_0(1 - P_B) \tag{29}$$

Furthermore, arrivals entering the system occur at a rate λ as long as they are admitted into the queue; hence, we define the effective arrival rate as

$$\lambda_e = \lambda(1 - P_B) \tag{30}$$

Simultaneously, departures out of the system continue to occur with a rate μ as long as the system is not empty which allows us to define the effective departure rate as

$$\mu_e = \mu(1 - P_0) \tag{31}$$

Given that in equilibrium the traffic entering the queue system is equal to the one leaving the queue [7], we have

$$\lambda(1 - P_B) = \mu[1 - \pi_0(1 - P_B)] \tag{32}$$

And, the blocking probability is

$$P_B = \frac{1}{\rho + \pi_0} \tag{33}$$

The described method was used to determine the blocking probability and the system efficiency η:

$$\eta = 1 - P_B = \frac{\pi_0}{\rho + \pi_0} \tag{34}$$

where π_0 is defined in (26).

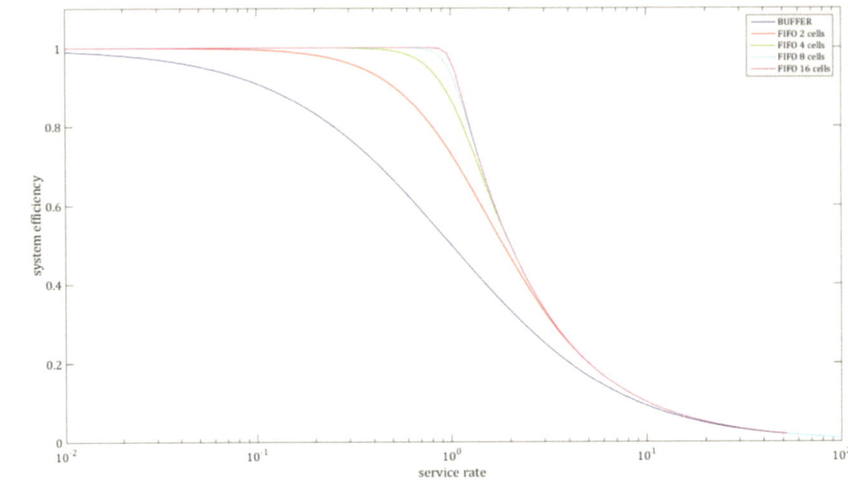

Figure 7. Simulation results of the system efficiency for a FIFO depth of 2, 4, 8 and 16 cells.

Figure 7 shows the system efficiency for the use of a buffer and a FIFO with *N* = *2, 4, 8, 16*. It is clear that the system's efficiency increases with the FIFO depth although the amount of the growth decreases. As a result, when taking in consideration the resources needed for an embedded FIFO, it is safe to say that a FIFO depth of 8 is enough to reduce the arrival input loss due to the blocking phenomenon.

3.3. Case study of a parallelized TCSPC system including an embedded FIFO

The TCSPC system illustrated in **Figure 8** was designed to be used for an HTS application that requires counting rates up to several MHz per channel. With a TDC dead time of 40 ns, the maximum data rate is equal to 25 MS/s. According to **Figure 7**, the use of a unique TCSPC module would lead to an efficiency η of, respectively, 98, 90 and 50% for a photon rate of 0.25, 2.5 and 25 MHz, i.e. a service rate of 0.01, 0.1 and 1. Obviously, for a service rate $\rho > 1$, the system's efficiency would tend to be $1/\rho$ regardless of the use of a FIFO. A photon rate of $\lambda = 25$ Mega photons/s is therefore not reasonable in the configuration of a single TCSPC module, but if the arrival rate is divided among the eight TCSPC (**Figure 8**) and assuming that the arrival process is equally distributed among the eight units, each $TCSPC_i$ receives an arrival rate:

$$\lambda_i = \frac{\lambda}{8} \tag{35}$$

resulting in a service rate $\rho_i = 0.0125$ and an efficiency $\eta_{ph} = 90\%$, i.e. an expected departure rate $\mu_{TCSPCi} = 2.8$ MHz out of each TCSPC unit which is similar to the value obtained in [19]. Giving the low service rate of each TCSPCi, the output of each TCSPC unit will have a distribution very similar to the Poisson process, and the resulting process is the sum of eight Poisson processes with their respective arrival rate $\lambda_i, i = 1, 2, \ldots 8$ and is therefore also a Poisson process with an arrival rate:

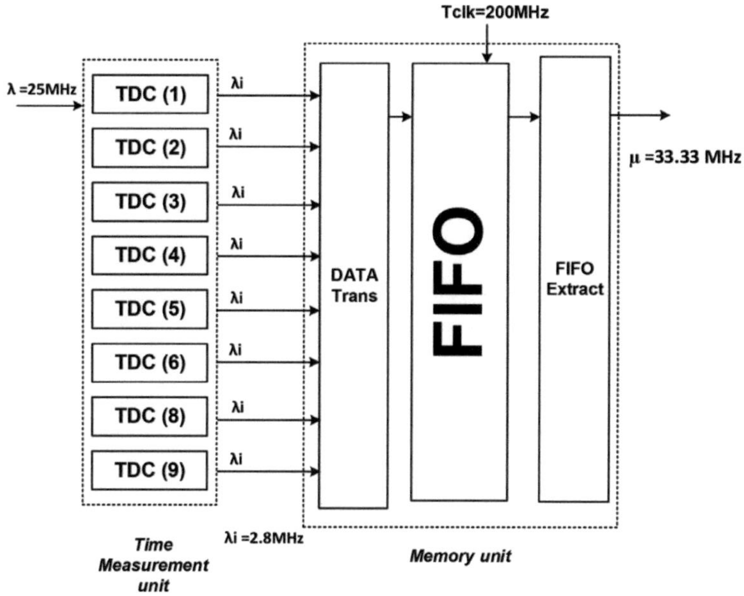

Figure 8. Parallelization scheme of the TCSPC system with the embedded FIFO as presented in [20].

$$\lambda = \sum_{i=1}^{8} \lambda_i = 8 \times 2.8\ MHz = 22.4\ MHz \tag{36}$$

Assuming an output frequency of only 33.33 MHz, the service rate will be $\rho_f = 0.67$. In the absence of the FIFO, the system can be assimilated to a buffer resulting in memory block efficiency $\eta_M = 0.6$ and a total efficiency:

$$\eta_{noFIFO} = \eta_{ph} \times \eta_M = 0.9 \times 0.6 = 54\% \tag{37}$$

The efficiency of the system is therefore not improved by the parallelization of the TCSPC even with the reduction of the pile-up effect. However, using the eight FIFO cells leads to a memory block efficiency of $\eta_M \cong 100\%$; the overall TCSPC system efficiency is maintained at about 90%. Such efficiency level can only be achieved with a 3 GHz output frequency without the use of the FIFO which proves the great impact including the FIFO in the TCSPC system.

4. Conclusion

The random nature of photon and applications involing a high counting rate require a specialized TCSPC system scheme to process the resulting data and improve the SNR. This requires the optimization of the photon detection process through the reduction of noise effects and low sensibility. It also requires the optimization of the system's architecture such that photon events are not lost due to the dead time following a previous photon arrival. In this chapter, we have discussed these two issues and presented solutions using mathematical models to assess the gain of such schemes. A low SNR could be the result of low signal levels or high noise levels. In the case of an SPAD, a low signal level is the result of low light sensibility, while a high noise level is the result of a high DCR. Thus, increasing the detector's SNR can be achieved by limiting the negative effect of these two cases. We presented a TCSPC macropixel architecture in which the SNR can be increased by deactivating dark pixels and/or hot pixels. A dark pixel is a pixel with an abnormally low sensibility level and a hot pixel is a pixel with high noise level in comparison to other pixel noises. The dark pixel elimination scheme requires a calibration phase to determine the activity level of each pixels and the low sensibility pixels that must be deactivated; this calibration phase should be conducted whenever the measurement conditions are changed and would lead to an SNR gain up to 1.5 times higher. The hot pixel elimination scheme on the other hand requires a onetime calibration scheme to determine the DCR of each pixel, and as a result, the pixels must be deactivated which allow an SNR improvement ranging up to 20 dB. The processing of detected photons can be optimized by means of a parallelized TCSPC architecture that make use of an embedded FIFO to limit counting loss due to photon detections' subsequent dead time. Using a queueing model, we demonstrated the impact of such approach and quantified the efficiency improvement as a function of the FIFO length, the counting rate and the readout rate. The proposed TCSPC architecture is capable of achieving a 90% efficiency rate with a counting rate of 25 MHz at a readout rate of 33 MHz. Without the use of the embedded FIFO; such efficiency would require the use of a 3 GHz readout frequency.

Author details

Imane Malass, Wilfried Uhring*, Jean-Pierre Le Normand, Norbert Dumas and Foudil Dadouche

*Address all correspondence to: wilfried.uhring@unistra.fr

ICube laboratory, University of Strasbourg and CNRS, Strasbourg, France

References

[1] Bollinger LM, Thomas GE. Measurement of the time dependence of scintillation intensity by a delayed-coincidence method. Review of Scientific Instruments. 1961;**32**:1044-1050

[2] O'Connor D, Phillips D. Time-Correlated Single Photon Counting. London: Academic Press; 1984

[3] Markovic B, Tisa S, Villa FA, Tosi A, Zappa F. A high-linearity, 17 ps precision time-to-digital converter based on a single-stage Vernier delay loop fine interpolation. Circuits and Systems I: Regular Papers, IEEE Transactions. 2013;**60**(3):557-569

[4] Pichette J, Lapointe E, Bérubé-Lauzière Y. Time-domain 3D localization of fluorescent inclusions in a thick scattering medium. Proceedings of SPIE. 2008;**7099**:709907

[5] Field RM, Realov S, Shepard KL. A 100-fps, time-correlated single-photon-counting-based fluorescence-lifetime imager in 130-nm CMOS. IEEE Journal of Solid-State Circuits. 2014;**49**(4):867-880

[6] Léonard J, Dumas N, Caussé J, Maillot S, Giannakopoulou N, Barre S, Uhring W. High-Throughput Time-Correlated Single Photon. Lab on a Chip, no. 14; 2015. pp. 4338-4343

[7] Veerappan C, Richardson J, Walker R, Li DU, Fishburn MW, Maruyama Y, Stoppa D, Borghetti F, Gersbach M, Henderson RK, Charbon E. A 160×128 single-photon image sensor with on-pixel 55ps 10b time-to-digital converter. In: Solid-State Circuits Conference Digest of Technical Papers (ISSCC). San Francisco, CA: 2011 IEEE International; 2011

[8] Villa F, Bronzi D, Bellisai S, Boso G, Shehata AB, Scarcella C, Tosi A, Zappa F, Tisa S, Durini D, Weyers S, Brockherde W. SPAD imagers for remote sensing at the single-photon level. Proceedings of SPIE. 2012;**8542**:85420G-1-85420G-6

[9] Niclass C, Charbon E. Single photon detector array with 64×64 resolution and millimetric depth accuracy for 3D imaging. Solid-State Circuits Conference, 2005. Digest of Technical Papers. ISSCC. 2005 IEEE International, San Francisco, CA, 1, pp. 364-604; 2005

[10] Villa F, Bellisai S, Bronzi D, Tosi A, Zappa F, Tisa S, Durini D, Weyers S, Paschen U, Brockherde W. SPAD smart pixel for time-of-flight and time-correlated single-photon counting measurements. IEEE Photonics Journal. 2012;**3**(4):795-804

[11] Perenzoni D, Gasparini L, Massari N, Stoppa D. Depth-Range Extension with Folding Technique for SPAD-Based TOF LIDAR Systems. in SENSORS, 2014 IEEEValencia; 2014

[12] Gallivanoni A, Rech I, Ghion M. Progress in quenching circuits for single photon avalanche diodes. IEEE Transactions on Nuclear Science. 2010;**57**(6):3815-3826

[13] Malass W, Uhring JP, Le Normand N. Zint DV, Dadouche F. SiPM based smart pixel for photon counting integrated streak camera. Design and Architectures for Signal and Image Processing (DASIP), 2013 Conference on, Cagliari, 2013, pp. 135-140

[14] Gersbach M, Maruyama Y, Trimananda R, Fishburn MW, Stoppa D, Richardson JA, Walker R, Henderson R, Charbon E. Time-resolved, low-noise single-photon image sensor fabricated deep-submicron CMOS technology. IEEE Journal of Solid-State Circuit. 2012;**47**:1394-1407

[15] Tamborini D, Villa F, Tosi A, 16-Channel module based on a monolithic array of single-photon detectors and 10-ps time-to-digital converters. IEEE Journal of Selected Topics in Quantum Electronics. 2014;**20**(6):1

[16] Stewart WJ. Markov chains, queues, and simulation. P. U. Press; 2009

[17] Abidi M, Koua Calliste K, Kanoun M, Panier S, Arpin L, Tétraul MA, Pratte JF, Fontaine R. A delay locked loop for fine time base generation in a positron emission tomography scanner. In: 5th International Conference on Design and Technology of Integrated Systems in Nanoscale Era (DTIS), Hammamet, 2010

[18] Sigman K. [Online]. Available: http://www.columbia.edu/~ks20/stochastic-I/stochastic-I-MCII.pdf.

[19] Léonard J, Dumas N, Caussé J, Maillot S, Giannakopoulou N, Barre S, Uhring W. High-throughput time-correlated single photon counting. Lab on a Chip. 2015;**14**:4338-4343

[20] Malass I, Uhring W, Le Normand JP, Dumas N, Dadouche F. Efficiency improvement of high rate integrated time correlated single photon counting systems by incorporating an embedded FIFO. New Circuits and Systems Conference (NEWCAS), 2015 IEEE 13th International, Grenoble, 2015. pp. 1-4

Chapter 10

Application of Fluorescence Spectroscopy for Microbial Detection to Enhance Clinical Investigations

Fardad Shakibaie, Laurent Lamard,
Halina Rubinsztein-Dunlop and Laurence J. Walsh

Additional information is available at the end of the chapter

http://dx.doi.org/10.5772/intechopen.73616

Abstract

Microbial biofilms are complex multi-layered communities of bacteria and fungi which cause a range of oral and other diseases. Efficient detection of biofilms is important for the clinical management of diseases they cause and for providing an endpoint to clinical treatments. For bacterial biofilms, bacterial metabolites such as porphyrins are important molecules for diagnostic purposes, since they fluoresce in the red and infrared regions of the spectrum. Fluorescence is a versatile and powerful diagnostic approach for detection of bacterial biofilms, particularly in dentistry. This chapter provides an overview of fluorescence spectroscopic methods for detection and analysis of biofilms and their derivatives such as deposits of dental calculus and how current technology can be extended using photon-counting detectors. Fluorescence can be used to help discriminate these from healthy tissues. The approaches described have broad applications to clinical and industrial situations where non-invasive detection of microbial biofilms is important.

Keywords: bacterial biofilms, clinical diagnosis, fluorescence spectroscopy, fluorophores, porphyrins

1. Introduction

The interactions of light with matter are heavily dependent on the wavelength of the light and the response of the target to that light. Major interactions include scattering, absorption and fluorescence. Various fluorescence spectroscopic methods have been used to analyse tissues and materials according to their fluorescence properties. In this chapter, the principles of fluorescence spectroscopy are discussed, with particular reference to the diagnostic values of porphyrin derivatives for the detection of infected biological tissues and applications of photon counting.

2. Fluorescence phenomena

Luminescence is a general term for the emission of radiation, which incorporates both fluorescence (a short-lived process) and phosphorescence (a long-lived process), as well as other phenomena such as bioluminescence in living organisms in which chemical reactions generate light. In fluorescence, the absorption of light of a particular wavelength results in the emission of light of a longer wavelength. This emission of light occurs as fluorophores get de-excited from a higher energy level to a lower energy level [1, 2]. When light is absorbed, the fluorophore becomes electronically excited, but the lifetime in the excited state is very short, and there is a rapid decay to a lower energy level. Fluorescence occurs if the transition is between states of the same electron spin and phosphorescence if the transition occurs between states of different spins. Fluorescence and phosphorescence phenomena are illustrated in the Jablonski energy diagram shown in **Figure 1**. Many naturally occurring substances fluoresce, including

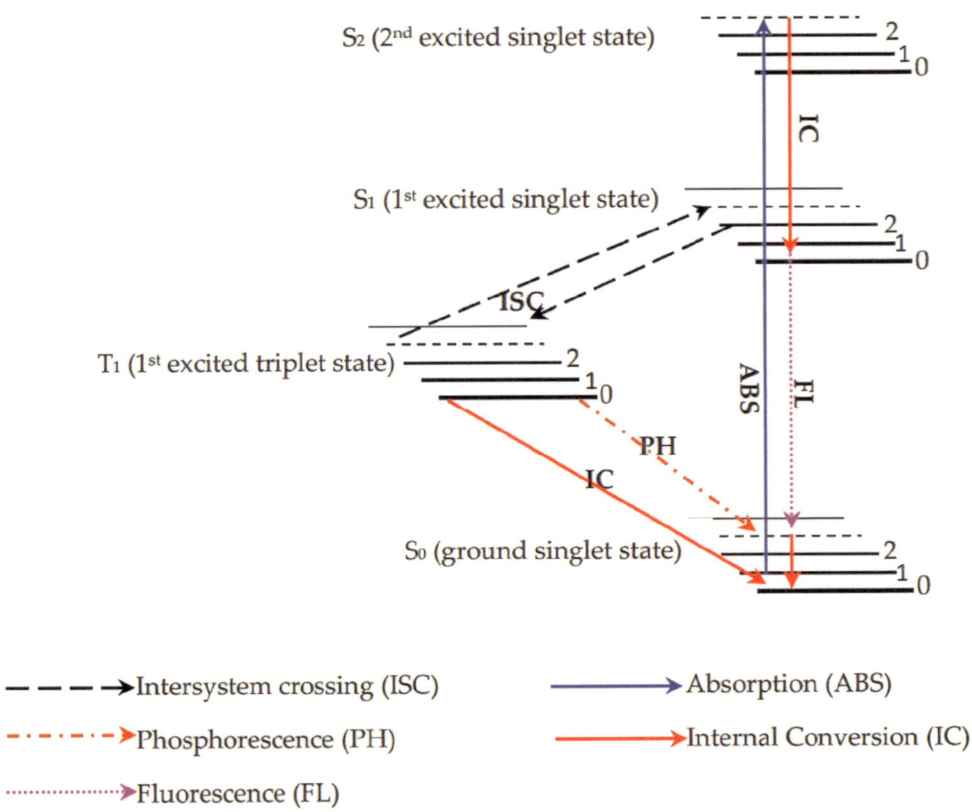

Figure 1. Jablonski energy diagram showing fluorescence and phosphorescence processes. Based on Ref. [3].

some minerals, fungi, bacteria, keratin, collagens and other components of body tissues. This is termed 'primary fluorescence' or 'autofluorescence'.

As the molecule absorbs energy, it transitions from the lower ground singlet state (S_0) to a vibrational level of an excited singlet state S_n (n = 1,2,...). The excited molecule loses energy partly through internal conversion without photon emission, and then it spontaneously releases a lower energy photon as it returns back to the singlet ground state [1, 4]. Light emission occurs within one microsecond of light exposure. Molecular fluorescence emissions persist only as long as the incoming stimulating radiation is continued, unlike phosphorescence, where light is emitted as a persisting 'afterglow' long after the incoming exciting light is no longer present.

The light that is emitted by fluorescence is readily distinguishable from the excitation light because it has a longer wavelength. This relationship is known as Stokes law and is named after Sir George Stokes, who published the first major paper on fluorescence [5]. For example, when a molecule absorbs short wavelength ultraviolet (UVA) light in the region of 315–400 nm, the emissions may be in the visible spectrum, such as visible red, in the case of porphyrins. Likewise, when excited by visible light wavelengths, porphyrins emit light in the near-infrared range.

3. Fluorescence spectroscopy

In fluorescence spectroscopy, also known as fluorometry or spectrofluorometry, fluorescence emissions from a sample are elicited using a range of wavelengths, and the emissions measured. The sample is typically in solution in a cuvette, and it is excited by near monochromatic light or by monochromatic light from a laser. Nearly monochromatic light can be produced using a monochromator, where a broad spectrum lamp such as halogen lamp is used (**Figure 2**), and

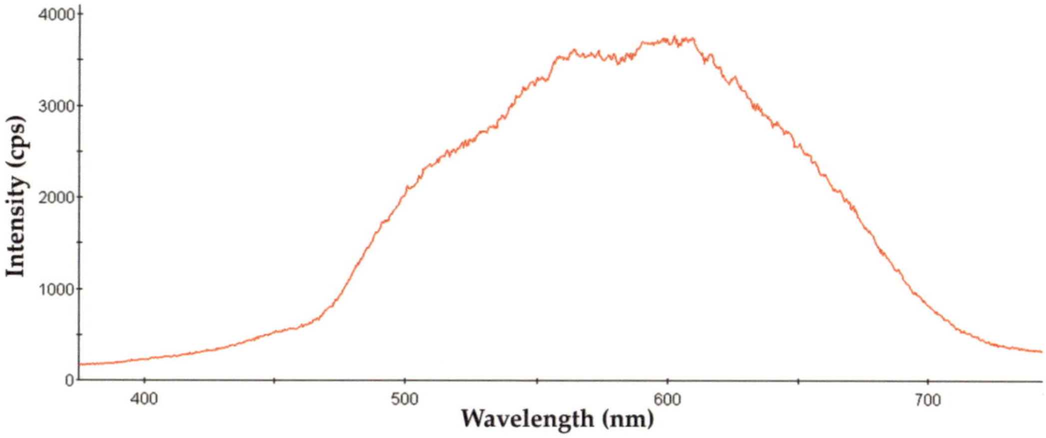

Figure 2. Emission spectrum from a halogen lamp.

the output is passed through a slit onto a diffraction grating, as shown in **Figure 3**. Even though fluorescence is emitted in all directions from the fluorophores within the sample, fluorescent emissions are detected normal to the incident beam path. This reduces the impact of stray light and the incident light wavelength. The detector (5) is usually a photomultiplier tube (PMT) or a photodiode array. These convert the intensity of the fluorescence emissions into an electrical output for subsequent analysis [1].

A spectrofluorimeter can show the range of fluorescence emissions at a particular constant excitation wavelength, or alternatively it can be used to record the excitation spectrum that gives rise to emission at a specific constant wavelength. The same principles as used in a spectrofluorimeter for sample analysis can be applied under field conditions or in clinical settings. The major challenges are in choosing the appropriate wavelength(s) of light to use, determining the most appropriate source of that light and ensuring that the detector system is sufficiently sensitive.

Fluorescence spectroscopy has become a useful analytical approach in many fields, including biochemistry, biophysics and biomaterial sciences. In recent years, molecular fluorescence analytical approaches have been developed to investigate fluorophores within biological samples [4].

A range of systems have been developed that give the user the choice of different excitation wavelengths, so that particular fluorophores of interest can be targeted. Using lasers as light sources, particularly semiconductor diode lasers, ensures that the emitted light is

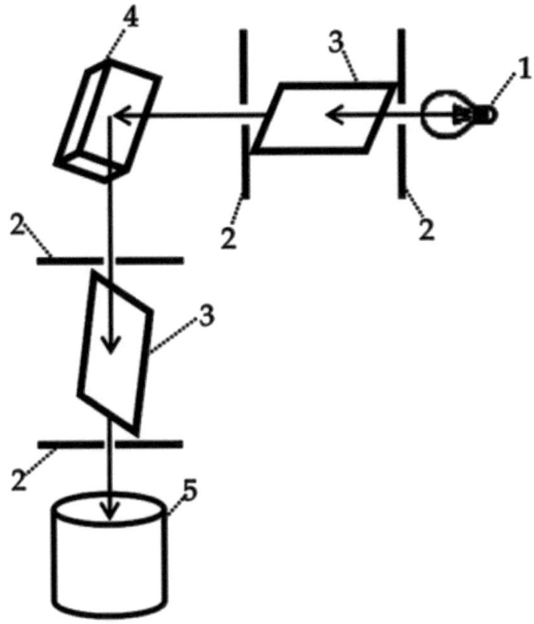

Figure 3. Schematic design of a spectrofluorimeter. 1 = excitation light source; 2 = slit; 3 = grating; 4 = sample cuvette; 5 = detector.

monochromatic, although there may be a drift in wavelength as the diode laser warms up from a cold start. The low power consumption and high electrical conversion efficiency of diode lasers make them well suited for use in portable systems. Diode lasers can be pulsed at a high frequency and can operate in chopped continuous waves as well as in superpulsed modes. By attenuating the intensity of the laser beam from a pulsed laser, the mean number of photons can be reduced dramatically and therefore it is possible to achieve single photon effects, if that is desired.

The detectors used in fluorescence diagnostic systems have included charge-coupled devices (CCDs), photodiodes and photon-counting detectors based on photomultipliers or avalanche diodes. Charge-coupled devices and photodiodes are well suited to fluorescence devices where the light source is running continuously or in long pulses, while photomultipliers and avalanche diodes are necessary to detect single photons.

4. Photomultiplier tube

A key component of a fluorescence system designed to work with faint laser light is a photomultiplier tube (PMT). This device amplifies the current generated by incident photons in the order of 100 million times, by using several dynode stages, enabling individual photons to be detected even when the incident light intensity is extremely low. Depending on the nature of the window through which the incident light enters the PMT, light in the ultraviolet, visible and near-infrared range may be detected, for photon-counting purposes or for high-sensitivity light detection techniques [6]. A simplified schematic design of a PMT is shown in **Figure 4** [7]. An electric field accelerates photoelectrons released from the photocathode, with increased numbers of secondary electrons released from successive dynodes as the incoming electrons collide with them, achieving amplification at each dynode through secondary emission. The electrons from the final dynode reach the anode, creating the signal current, which then gives final reading for light-induced fluorescence. A typical delay from an incoming photon striking the photocathode, causing electrons to be emitted through photoelectric actions, and a current pulse being measured from electrons reaching the anode, is in the order of 50

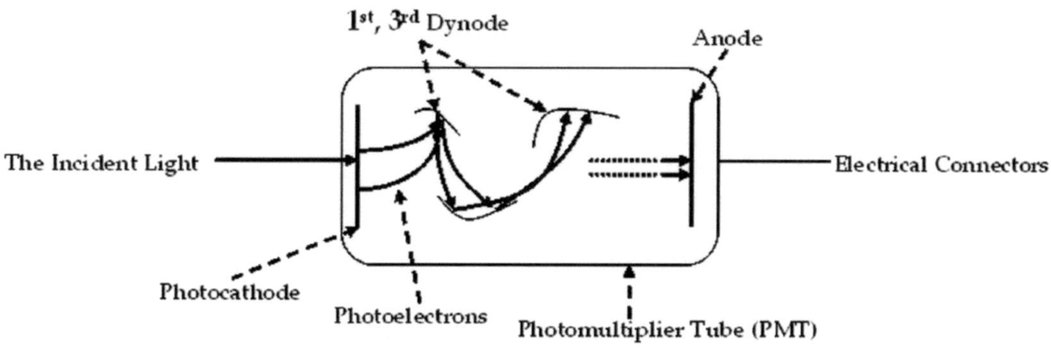

Figure 4. Schematic diagram of photon-counting photomultiplier tube. Based on Ref. [7].

nanoseconds. This brief time interval allows the distinction to be made between fluorescence and phosphorescence, since in the latter the emission of light continues for some time after the excitation pulse of light has ended.

Semiconductor devices such as avalanche photodiodes are now considered as alternatives to PMT for some applications. Their gain may however be less than that for PMTs. When very high gains are needed, single photon avalanche diodes can be used, applying voltages that are well above their breakdown voltage, for short intervals.

5. Fluorimeter equipment

The basis of many laboratory studies of fluorescence of biological materials is the spectrofluorimeter, which can determine the fluorescence characteristics of individual pure substances or mixtures of substances. A typical spectrofluorimeter is the FluoroMax-3 manufactured by JY-Horiba. This instrument is used widely to perform high-resolution fluorescence measurements. It has a standard configuration as shown in **Figure 3**, with a broad spectrum xenon arc lamp, excitation and emission monochromators before and after the sample and a photomultiplier tube for detection. Wavelength selection is achieved using the optical gratings of the monochromators. These diffract the incident beam, dispersing it into its constituent wavelengths. In addition, adjustable 'slits' are used at the entrance and exits of the grating, which can be used for resolving particular wavelengths. On the excitation monochromators, the slits control the bandpass (range) of light that is incident on the cuvette sample. On the other hand, the slits of the emission grating determine the intensity of the emitted fluorescence recorded by the PMT sensor. There is a reference photodiode which is used to correct for variations in the intensity of the emissions from the xenon lamp at different wavelengths.

The spectrofluorimeter interfaces with a computer, and a dedicated software package (DataMax) is used for data acquisition. A post-processing application manages the acquisition of emission and excitation profiles and allows the time course of fluorescence events to be followed. A constant wavelength analysis application allows multiple samples to be analysed at single wavelengths, similar to a microplate reader. A real-time display application allows individual hardware parameters such as slit width to be adjusted while immediately viewing the consequential changes in emission intensity.

6. Fluorescence analysis software

A typical example of software used for fluorescence spectroscopic analysis is MicroCal™ Origin. This software from OriginLab Corporation operates on a Windows® platform. The front end of the software has a spreadsheet design that is column oriented. The user has access to various templates to simplify workflow. The software can generate a range of 2D and 3D graphs. Data analysis in Origin includes curve fitting and peak analysis. Curve fitting is achieved through a non-linear least squares approach. There are several platform-independent open-source programmes with similar functions, such as QtiPlot and SciDAVis.

7. Bacterial biofilms

Bacterial biofilms have a complex 3D architecture. They may be composed of a single bacterial species or of multiple species living within distinct microenvironments. Biofilms are ubiquitous in moist environments [8, 9], where they allow bacteria to resist environmental physical stresses such as shear stresses from fluid flow, as well as chemical stresses from adverse environmental pH or eH. The biofilm structure provides a physical barrier to the diffusion of most biocides [10]. Organisms located deep within in biofilms exist in a dormant or quiescent metabolic state. The low rate of proliferation makes them resistant to antibiotics that target bacterial replication [11] or the synthesis of new bacterial cell membranes [12].

In the industry, biofilms are a constant problem in the food processing industry [13–15], as well as in ventilation systems [16] and in water treatment [17]. Biofilms pose major problems in healthcare because they adhere to surgical implants of various types, as well as to most body surfaces, causing chronic infections when the opportunity arises [18]. Examples of diseases related to biofilms include periodontitis which causes loss of bone and connective tissue attachment of teeth [19–21] and chronic lung infections in individuals with cystic fibrosis [22]. Medical devices that suffer problems from biofilms include central venous catheters [23], endocardial pacemaker leads [24], prosthetic heart valves [25], orthopaedic devices [26] and urinary catheters [27].

In the polyvinyl chloride (PVC) tubing used in much industrial and biomedical equipment, biofilms form readily when these are exposed to reticulated water. Surface colonisation is enhanced by calcium compounds (such as calcium carbonate) and adherent organic molecules present as contaminants in water [28]. Small diameter tubing gives laminar flow characteristics, with a high central flow rate and slow peripheral flow rates, with bacteria segregated near the walls of the tubing. The surface-to-volume ratio is high in small diameter tubing, since less than 100 mL of water may be spread over more than 1500 cm^2 of available surface. Tubing which has a diameter of 2 mm or less is particularly problematic in terms of the rapid formation of dense biofilms when connected to reticulated water or fluids which are not first rendered sterile.

Current management strategies that are used to control biofilms include flushing, purging with air or other gases, treatment with nitric or other acids and application of biocides including glutaraldehyde, sodium hypochlorite, hydrogen peroxide, ozone, silver ions or iodine [28, 29]. Such treatments are designed to reduce the problem of blockage of narrow tubing from biofilms. In addition to equipment failure, biofilms create health risks for patients through renal dialysis equipment [30]. They also form in the tubing in dental chairs [31], from where water may be aerosolized with pathogenic legionella or mycobacteria.

Because bacterial biofilms in pipes and tubing resist many chemical agents, treatments which have a physical removal action (such as scraping or agitation) are commonplace [12, 14]. Laser-generated cavitation can have powerful cleaning actions and can detach biofilms [32]. The same approach using lasers can be used to ablate biofilms, provided an appropriate laser wavelength and exposure parameters are used [33].

With regard to the detection of bacterial biofilms within pipes and tubing, levels of loosely attached bacteria can be quantified indirectly by using as a surrogate measurement the levels of bacteria in the fluid that exit the pipe in question. It is generally not practical to sample the tubing itself for the presence of bacteria, as this could be destructive. While the sampling of exit fluids for viable bacteria is used widely [34], this method is time-consuming since such samples require at least several days of incubation in the laboratory.

Real-time assessment of biofilm levels would be of great advantage in allowing precise control over the dosing of biocides and the timing of purging and other biofilm control measures. Ideally, such an assessment would be undertaken externally (i.e. through the tubing) [35, 36] in real time, without having to shut down or interrupt the system for testing and maintenance. A useful approach for determining the presence of bacterial biofilms in tubing made of PVC and similar transparent polymers may be light-induced fluorescence, applied externally (i.e. passing through the tubing walls) or applied internally using an optical fibre. This diagnostic approach has been used successfully within the narrow confines of the root canals of teeth [37]. By applying coherent (laser) or near-coherent light, fluorophores within the bacterial biofilm or the overlying fluid become excited. Not only could the levels of bacteria be assessed in a quantitative manner, but it should be possible to apply another laser to create cavitation and shockwaves inside the tubing to fragment and disrupt the biofilm. This concept of laser-generated internal shock waves has been applied successfully to debriding the root canals of teeth [38].

By selecting appropriate excitation sources and filters, fluorescence-based analysis systems can identify and quantify the target of interest in a tissue or on a surface. This selective fluorophore approach has been used for kidney stones, tumours, dental filling materials [39, 40], dental caries [41], dental plaque biofilms [2] and dental calculus [42]. The latter four sample types have been identified in diseased and healthy sites, employing optimal excitation wavelengths for fluorescence detection and then coupling this to a feedback-controlled second laser system for ablation. For biofilm detection within tubing, issues such as fluorescence from the liquid carried in the tubing and from the tubing itself need to be addressed. This is why it is essential to determine the excitation-emission ranges for various target materials using fluorescence spectroscopy under defined conditions in the laboratory using the type of spectrofluorimeter equipment and software described earlier.

A key objective is to disrupt and inactivate bacterial deposits without damaging the internal structure of the pipe or tubing. For effective disruption or ablation of biofilms, laser energy can be absorbed in both solid and fluid components. Key elements of bacterial biofilms from this perspective include water, calcific deposits and bacterial porphyrins. One can assume that at least 65% of bacterial biofilm volume is water because this is the typical water content of individual bacteria. The amount of energy absorbed by bacterial biofilms will vary according to the laser wavelength used, the concentration of the absorbing fluorophores and their absorption coefficient [43]. The absorption of fluorophores in biofilms can be assessed across a range of possible excitation wavelengths via spectroscopy [44]. The absorption of light by the tubing is also assessed using the same approach.

A challenge in the use of fluorescence in very small confined environments where there are low levels of microorganisms is the detection limit of the system used to detect fluorescence emissions. This is where changing from a conventional photodiode to an avalanche photodiode can be considered. Moreover, the choice of semiconductor used will be affected by the wavelength range of interest. For detecting levels of planktonic bacteria of around 5 colony-forming units per mL in volumes of 20 μL, such as in the case of the root canal system of a molar tooth root, a system using conventional photodiodes to measure fluorescence from pulsed laser light is working at its limits. Moving to an alternative approach using faint laser emissions and thus single photon counting should increase the overall sensitivity of the system dramatically. Using such an approach, it should be possible to achieve detection of a single organism, provided it is in the range of the optical detection system. Specific improvements such as micro-patterned optical fibre tips have been developed to allow wide-angle detection of microorganisms in confined narrow canals.

An elegant example of a widely deployed fluorescence device is the DIAGNOdent [45, 46]. This was developed for detecting dental caries (tooth decay) [47] and uses a pulse 655-nm diode laser as the light source and a photodiode detector to collect near-infrared light that is filtered through a long pass filter. The detection system is gated so that only fluorescence emissions that correspond with laser pulses are assessed, to thus remove the effects of ambient light and background noise. Work in our laboratory showed that it can also be used to detect infections within the root canals of teeth [48]. The fluorescent yield of a healthy surface decreases much more than the infected region, as excitation wavelength increases in the red spectral region [49, 50]. For removing biofilms *from the outside* of teeth, a pulsed Er:YAG laser has been combined with the DIAGNOdent system and linked to a feedback control system [51]. More recently, this approach has also been used for addressing biofilms and planktonic bacteria *inside* teeth [37], which is technically much more challenging because of issues of access.

The current level of technology deployed in dental practice for fluorescence diagnostics includes systems with LED illumination and charge-coupled device (CCD) or complementary metal oxide semiconductors (CMOS) sensors, such as in intra-oral cameras as well as diode laser-based systems such as the DIAGNOdent Classic, DIAGNdent Pen and KEY-3 laser (all from KaVo, Biberach, Germany) [52]. The intra-oral cameras use continuous wave emissions from multiple LEDs as the light source. A challenge for using CMOS sensors with such devices is the so-called 'rolling shutter' effect seen when the handpiece is being moved, due to the refresh rate used. CCD image sensors are considered to have better sensitivity for light detection than CMOS but are more expensive. They can operate well for detecting light emissions in the near infrared, which is useful for detection of bacteria [41, 52, 53], provided that a long-pass filter is used to remove reflected excitation light as well as ambient daylight and work-place lighting. In the case of the DIAGNOdent, only light wavelengths above 680 nm are measured [45–47].

8. Porphyrins and biofilm fluorescence

The literature identifies porphyrin derivatives as a potential fluorophore in bacterial by-products, based on the peak fluorescence of bacterial biofilms compared to the known emissions of porphyrins at a particular wavelength range [54, 55]. Porphyrins are derivatives of haemoglobin-related molecules known as tetrapyrrole porphyrins. These are involved in the biosynthesis of metalloporphyrin heme [ferroprotoporphyrin (Fe^{2+})], hemin [ferriprotoporphyrin (Fe^{3+})] and chlorophyll [54]. Aerobic cells can synthesise hemeproteins (**Figure 2**). The main porphyrins in biosynthesis are protoporphyrin IX (PP IX), coproporphyrin III (CP III), uroporphyrin III (UP III) and hematoporphyrin IX (HP IX). These are linked through heme biosynthesis, as shown in **Figure 5**.

Solutions of porphyrin derivatives show fluorescence upon red excitation, particularly PP IX, which has strong near-infrared fluorescence around 825 nm when irradiated at 655 nm. The fluorescence yield increases linearly with the concentration of PP IX [56]. Testing various fractions derived from high-performance liquid chromatography (HPLC) of carious dentine

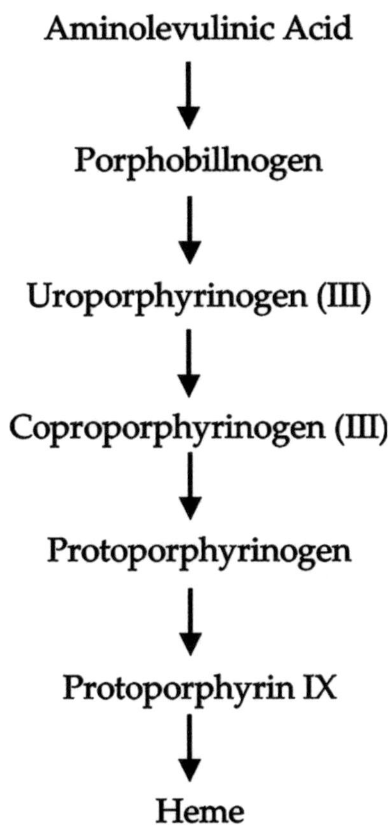

Figure 5. Schematic pathway of heme biosynthesis. Based on Ref. [54].

using 406 nm excitation has identified porphyrin, protoporphyrin, coproporphyrin and uroporphyrin as the main fluorescing compounds [54].

König had previously explored a similar concept while studying carious tooth tissue and had found emissions mainly in the visible red spectral region when samples were excited by a 407 nm (UVA) krypton ion laser [57]. Most carious teeth in König's experiments displayed a fluorescence maximum at 635 nm, and fewer than 10% of carious teeth examined showed additional maxima around 590 and 620 nm. These emission maxima correspond to the known emission peaks of protoporphyrin (633 nm), coproporphyrin (623 nm) and Zn protoporphyrin (593 nm) [57], as shown in **Figure 6**.

The fluorescence decay time (fluorescence lifetime) is the mean time during which the fluorophore remains in the excitation level before returning to the ground state. As shown in **Table 1**, the fluorescent decay kinetics of both protoporphyrin and carious tooth sample region are quite comparable, with a similar proportion of molecules having lifetimes of 3 and 17 ns [57].

In dental caries, the endogenous porphyrins are derivatives from bacteria [52]. As shown in **Table 2**, their presence is not a unique property of bacteria associated with dental caries. Bacterial strains such as *Bacteroides intermedius* and *Pseudomonas aeruginosa*, not associated with dental caries, when grown on agar plates and excited by 407 nm light, also display emission maxima at 635 and 700 nm [57]. Likewise, *Corynebacterium* species emit fluorescence around 620 nm, which corresponds to coproporphyrin fluorescence. Importantly, some key Gram-positive bacteria involved with dental caries, such as *Streptococcus mutans* and various *Lactobacilli* species, do not show strong porphyrin fluorescence in the red spectral region.

Table 3 presents summary details of the major and minor peak wavelengths for dental plaque biofilms growing on contaminated tooth surfaces. At 400–500 nm excitation wavelengths, the major and minor fluorescent peaks are mostly within the range of 610–614 nm, whereas

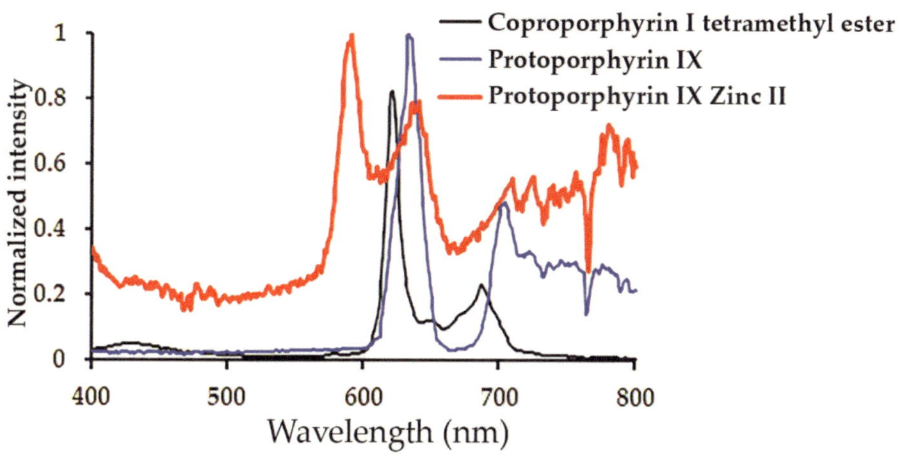

Figure 6. The normalized fluorescence spectra of various porphyrin derivative solutions at 350 nm excitation in the solvent dimethyl sulphoxide.

Sample type	Fluorescence lifetime (ns)	Fluorescent decay time (%)
Coproporphyrin	20	100
Protoporphyrin	3	11
	17	89
Zn-protoporphyrin	13	8
	2	92
Non-carious region	0.5	15
	9.8	39
	3.2	46
Carious region	0.31	7
	2.3	11
	17.3	62

Based on Ref. [57].

Table 1. Fluorescence decay kinetics of different porphyrins and samples from carious and non-carious regions of teeth, and their percentage of occurrence.

Bacteria	Peak fluorescence
Actinomyces odontolyticus	635 nm
Bacteroides intermedius	636, 708 nm
Pseudomonas aeruginosa	636, 618, 703 nm
Streptococcus mutans	Non-fluorescent
Streptococcus faecalis	Non-fluorescent
Lactobacterium casei	Non-fluorescent
Lactobacterium acidophilus	Non-fluorescent
Candida albicans	620 nm
Corynebacterium	620 nm

Based on Ref. [57].

Table 2. The peak fluorescence of different oral microorganisms at 407 nm excitation.

with excitation at wavelengths above 500 nm, the major and minor peaks are now mainly in the near-infrared spectrum, particularly around 825 nm. The fluorescence profiles for dental plaque or biofilms on the tooth surfaces are less intense in the visible red spectrum than those for dental caries and dental calculus [53].

Excitation wavelengths	Major peaks	Minor peaks
400 nm	—	—
425 nm	—	610 nm.
450 nm	614 nm	625, 704 nm
475 nm	610 nm	708 nm
500 nm	610 nm	625, 740, 800, 822 nm
525 nm	—	706, 733, 768, 798, 800, 822 nm
550 nm	—	708, 748, 759, 783, 824 nm
575 nm	—	708, 742, 765, 779, 822, 849 nm
600 nm	—	757, 766, 794, 828 nm
625 nm	825 nm	762 nm
650 nm	825 nm	761, 794 nm

Table 3. Major and minor fluorescence emission peaks for dental plaque biofilm on teeth.

9. Conclusions

Fluorescence spectroscopy has a significant value for laboratory assessment of complex materials and mixtures, including biofilms. The principles can be applied directly to clinical devices that use fluorescence principles for improved diagnosis and clinical care [47, 58] in detection and diagnosis of bacterial biofilms from target biological samples. Fluorescence has particular applications for detecting bacteria because of their porphyrin derivatives, both within planktonic bacteria and within bacterial biofilms, and there already is good support for the presence of porphyrins within target tissue samples [59–61]. These porphyrin derivatives generate visible red emissions from bacterially contaminated sites, whereas healthy tissue sites that are free from bacteria lack such fluorescence. A key direction for further work is to move towards more sensitive methods for analysis, such as using faint laser emissions as an excitation source and either avalanche diodes or PMTs as detectors. This should allow detection thresholds to move down to the level of single bacteria.

Fluorescence spectroscopy can be used to extend the use and application of optical methods [62] and particularly light-induced fluorescence devices [63] in clinical practice. Fluorescence can be used to identify infected target surfaces and to guide clinicians by providing feedback during ablation. This allows infected sites to be detected and ablated using an autopilot approach with maximum accuracy [51, 64].

Past studies have shown that visible red (655 nm) laser-induced fluorescence has clinical applications for guiding bacterial removal, on the basis that near-infrared emissions are likely to be from porphyrins of bacterial origin [21, 65]. Overall, visible light has applicability for eliciting fluorescence from porphyrins for detecting bacteria and their products present in infected tissues, with emissions in the visible red region. This highlights the value of fluorescence as a non-invasive adjunct to conventional clinical examination in detection and diagnosis of infected surfaces.

Acknowledgements

This work was supported by a National Health and Medical Research (NHMRC) Dental Postgraduate Scholarship and by research grants from the Australian Dental Research Foundation (ADRF) and the Australian Society of Periodontology Research Foundation. The authors acknowledge with gratitude the input of Dr. Roy George in relation to biofilm fluorescence detection systems using fibre optic systems. Author LJW holds patents on particular optical fibre tips for dental diagnosis. The authors report no other conflicts of interest related to this work.

Author details

Fardad Shakibaie[1], Laurent Lamard[2], Halina Rubinsztein-Dunlop[3] and Laurence J. Walsh[1]*

*Address all correspondence to: l.walsh@uq.edu.au

1 School of Dentistry, The University of Queensland, UQ Oral Health Centre, Australia

2 Division of Oral Laser Applications, Department of Dental Science, Faculty of Medicine, University of Liege, Belgium

3 School of Mathematics and Physics, The University of Queensland, Brisbane, Australia

References

[1] Anil K, Anand A. Fundamentals and Applications of Biophotonics in Dentistry. Hackensack, NJ: London Imperial College Press, World Scientific; 2007. pp. 105-117

[2] Walsh LJ, Shakibaie F. Ultraviolet-induced fluorescence: Shedding new light on dental biofilms and dental caries. Australasian Dental Practice. 2007;**18**:56-60

[3] Shakibaie F, George R, Walsh LJ. Applications of laser induced fluorescence in dentistry. International Journal of Dental Clinics. 2011;**3**:26-29

[4] Hof M, Hutterer V, Fidler V. Fluorescence Spectroscopy in Biology: Advanced Methods and their Applications to Membranes, Proteins, DNA, and Cells. Berlin: Springer; 2005. pp. 3-25

[5] Stoke GG. Über die Änderung der Brechbarkeit des Lichtes. Philosophical Transactions. 1852;**107**:11

[6] Flyckt SO, Marmonier C. Photomultiplier Tubes: Principles and Applications. Brive, France: Philips Photonics; 2002

[7] Engstrom RW. Photomultiplier Handbook. RCA/Burle; 1980

[8] Allison DG, Gilbert P, Lappin-Scott HM, Wilson M. Community Structure and Co-Operation in Biofilms. Cambridge: Cambridge University Press; 2000. pp. 1-23

[9] Costerton JW, Lewandowski Z, Caldwell DE, Korber DR, Lappin-Scott HM. Microbial biofilms. Annual Review of Microbiology. 1995;**49**:711-745. DOI: 10.1146/annurev.mi.49.100195.003431

[10] Cos P, Toté K, Horemans T, Maes L. Biofilms: An extra hurdle for effective antimicrobial therapy. Current Pharmaceutical Design. 2010;**16**:2279-2295. DOI: 10.2174/138161210791792868

[11] Lewis K. Persister cells and the riddle of biofilm survival. Biochemistry (Moscow). 2005;**70**:267-274. DOI: 10.1007/s10541-005-0111-6

[12] Stewart PS, Costerton JW. Antibiotic resistance of bacteria in biofilms. Lancet. 2001;**358**:135-138. DOI: 10.1016/S0140-6736(01)05321-1

[13] Wong ACL. Biofilms in food processing environments. Journal of Dairy Science. 1998;**81**:2765-2770. DOI: 10.3168/jds.S0022-0302(98)75834-5

[14] Kumar CG, Anand SK. Significance of microbial biofilms in food industry: A review. International Journal of Food Microbiology 1998;**42**:9-27. DOI: 10.1016/S0168-1605(98)00060-9

[15] Austin JW, Bergeron G. Development of bacterial biofilms in dairy processing lines. The Journal of Dairy Research. 1995;**62**:509-519. DOI: 10.1017/S0022029900031204

[16] Costerton B. Biofilms: A growing problem. In: Seminar Transcript. Kingston, Ontario, Canada: Maunco seminar; 21 September 2000. p. 1-20

[17] Lin SM, Svoboda KK, Giletto A, Seibert J, Puttaiah R. Effects of hydrogen peroxide on dental unit biofilms and treatment water contamination. European Journal of Dental Education. 2011;**5**:47-59

[18] Costerton JW, Stewart PS, Greenberg EP. Bacterial biofilms: A common cause of persistent infections. Science. 1999;**284**:1318-1322. DOI: 10.1126/science.284.5418.1318

[19] Shakibaie F, Gemmell E, Bird PSA. Mouse model to study pathogenicity of *Bacteroides forsythus*. Periodontology. 2001;**22**:5-8

[20] Darveau RP, Tanner A, Page RC. The microbial challenge in periodontitis. Periodontology 2000. 1997;**2000, 14**:12-32. DOI: 10.1111/j.1600-0757.1997.tb00190.x

[21] Bird PS, Shakibaie F, Gemmell E, Polak B, Seymour GJ. Immune response to *Bacteroides forsythus* in a murine model. Oral Microbiology and Immunology. 2001;**16**:311-315

[22] Singh PK, Schaefer AL, Parsek MR, Moninger TO, Welsh MJ, Greenberg EP. Quorum-sensing signals indicate that cystic fibrosis lungs are infected with bacterial biofilms. Nature. 2000;**407**:762-764. DOI: 10.1038/35037627

[23] Passerini L, Lam K, Costerton JW, King EG. Biofilms on indwelling vascular catheters. Critical Care Medicine. 1992;**20**:665-673

[24] Marrie TJ, Nelligan J, Costerton JW. A scanning and transmission electron-microscopic study of an infected endocardial pacemaker lead. Circulation. 1982;**66**:1339-1341

[25] Hyde JA, Darouiche RO, Costerton JW. Strategies for prophylaxis against prosthetic valve endocarditis: A review article. The Journal of Heart Valve Disease. 1998;**7**:316-326

[26] Gristina AG, Shibata Y, Giridhar G, Kreger A, Myrvik QN. The glycocalyx, biofilm, microbes, and resistant infection. Seminars in Arthroplasty. 1994;**5**:160-170

[27] Morris NS, Stickler DJ, McLean RJ. The development of bacterial biofilms on indwelling urethral catheters. World Journal of Urology. 1999;**17**:345-350

[28] Marion-Ferey K, Pasmore M, Stoodley P, Wilson S, Husson GP, Costerton JW. Biofilm removal from silicone tubing: An assessment of the efficacy of dialysis machine decontamination procedures using an in vitro model. The Journal of Hospital Infection. 2003;**53**:64-71

[29] Mun S, Jeong JS, Kim J, Lee YW, Yoon J. Inactivation of *Pseudomonas aeruginosa* biofilm by dense phase carbon dioxide. Biofouling. 2009;**25**:473-479. DOI: 10.1080/08927010902874876

[30] Man NK, Degremont A, Darbord JC, Collet M, Vaillant P. Evidence of bacterial biofilm in tubing from hydraulic pathway of hemodialysis system. Artificial Organs. 1998;**22**:596-600. DOI: 10.1046/j.1525-1594.1998.06195.x

[31] Coleman DC, O'Donnell MJ, Shore AC, Swan J, Russell RJ. The role of manufacturers in reducing biofilms in dental chair waterlines. Journal of Dentistry. 2007;**35**:701-711. DOI: 10.1016/j.jdent.2007.05.003

[32] George R, Meyers IA, Walsh LJ. Laser activation of endodontic irrigants with improved conical laser fiber tips for removing smear layer in the apical third of the root canal. Journal of Endodontia. 2008;**34**:1524-1527. DOI: 10.1016/j.joen.2008.08.029

[33] Shakibaie F, Diklic S, Walsh LJ. An assessment of changes in dentine permeability following irradiation with a pulsed erbium:YAG laser. Periodontology. 2002;**23**:4-7

[34] Boe-Hansen R, Martiny AC, Arvin E, Albrechtsen HJ. Monitoring biofilm formation and activity in drinking water distribution networks under oligotrophic conditions. Water Science and Technology. 2003;**47**:91-97

[35] Jarrett WA, Ribes J, Manaligod JM. Biofilm formation on tracheostomy tubes. Ear, Nose, & Throat Journal. 2002;**81**:659-661

[36] Marion-Ferey K, Enkiri F, Pasmore M, Husson GP, Vilagines R. Methods for biofilm analysis on silicone tubing of dialysis machines. Artificial Organs. 2003;**27**:658-664. DOI: 10.1046/j.1525-1594.2003.07148.x

[37] Ho QV, George R, Sainsbury AL, Kahler WA, Walsh LJ. Laser fluorescence assessment of the root canal using plain and conical optical fibers. Journal of Endodontia. 2010;**36**: 119-122. DOI: 10.1016/j.joen.2009.09.024

[38] Walsh LJ. Laser analgesia with pulsed infrared lasers: Theory and practice. Journal of Oral Laser Applications. 2008;**8**:7-16

[39] Shakibaie F, Walsh LJ. Fluorescence imaging of dental restorations using the VistaCam intra-oral camera. Australian Journal of Forensic Sciences. DOI: 10.1080/00450618.2017.1304991

[40] Shakibaie F, Walsh LJ. KEY3 laser treatment applications in oral and maxillofacial surgery. Journal of Head Neck & Spine Surgery. 2017;**1**:555574

[41] Shakibaie F, Walsh LJ. Violet and blue light-induced green fluorescence emissions from dental caries. Australian Dental Journal. 2016;**61**:464-468. DOI: 10.1111/adj.12414

[42] Shakibaie F, Walsh LJ. Performance differences in the detection of subgingival calculus by laser fluorescence devices. Lasers in Medical Science. 2015;**30**:2281-2286. DOI: 10.1007/s10103-015-1808-4

[43] Altshuler GB, Yaroslavsky I. Absorption Characteristics of Tissues as a Basis for the Optimal Wavelength Choice in Photodermatology. Burlington, MA: Palomar Medical Technologies; 2004. pp. 1-4

[44] Kane SA. Introduction to Physics in Modern Medicine. London: New York Taylor & Francis; 2003. pp. 67-74

[45] Shakibaie F, Walsh LJ. DIAGNOdent pen versus tactile sense for detection of subgingival calculus: An in vitro study. Clinical and Experimental Dental Research. 2015;**1**:26-31. DOI: 10.1002/cre2.5

[46] Shakibaie F, Walsh LJ. Laser fluorescence detection of subgingival calculus using the DIAGNOdent classic versus periodontal probing. Lasers in Medical Science. 2016;**31**:1621-1626. DOI: 10.1007/s10103-016-2027-3

[47] Hibst R, Gall R. Development of a diode laser-based fluorescence caries detector. Caries Research. 1998;**32**:294

[48] Sainsbury AL, Bird PS, Walsh LJ. DIAGNOdent laser fluorescence assessment of endodontic infection. Journal of Endodontia. 2009;**35**:1404-1407. DOI: 10.1016/j.joen.2009.07.006

[49] Shakibaie F. Detection of subgingival calculus by optical methods [PhD thesis]. University of Queensland, Brisbane, Australia. 2011

[50] Hibst R, Gall R, Klafke M. Device for the recognition of caries, plaque or bacterial infection on teeth. USA Patent 6,024,562. 2000. pp. 1-8

[51] Schwarz F, Sculean A, Georg T, Reich E. Periodontal treatment with an Er:YAG laser compared to scaling and root planing. A controlled clinical study. Journal of Periodontology. 2001;**72**:361-367. DOI: 10.1902/jop.2001.72.3.361

[52] König K, Schneckenburger H. Laser-induced auto-fluorescence for medical diagnosis. Journal of Fluorescence. 1994;**4**:17-40. DOI: 10.1007/BF01876650

[53] Shakibaie F, Walsh LJ. Violet and blue light-induced green fluorescence emissions from dental calculus: A new approach to dental diagnosis. International Dentistry. 2016;**11**:6-13

[54] Hibst R, Paulus R. Molecular basis of red excited caries fluorescence. Caries Research. 2000;**34**:323. DOI: 10.1159/000016607

[55] König K, Hibst R, Meyer H, Flemming G, Schnecken-Burger H. Laser-induced autofluorescence of carious regions of human teeth and caries-involved bacteria. Proceedings of SPIE. 1993;**2080**:170-180. DOI: 10.1117/12.166180

[56] Shakibaie F, Walsh LJ. Effect of oral fluids on dental caries detection by the VistaCam. Clinical and Experimental Dental Research. 2015;**1**:74-79. DOI: 10.1002/cre2.13

[57] Shakibaie F, Walsh LJ. Dental calculus detection using the VistaCam. Clinical and Experimental Dental Research. 2016;**2**:226-229. DOI: 10.1002/cre2.42

[58] Shakibaie F, Walsh LJ. Surface area and volume determination of subgingival calculus using laser fluorescence. Lasers in Medical Science. 2014;**29**:519-524. DOI: 10.1007/s10103-012-1242-9

[59] Buchalla W, Lennon ÁM, Attin T. Comparative fluorescence spectroscopy of root caries lesions. European Journal of Oral Sciences. 2004;**112**:490-496. DOI: 10.1111/j.1600-0722.2004.00173.x

[60] Buchalla W, Lennon ÁM, Attin T. Fluorescence spectroscopy of dental calculus. Journal of Periodontal Research. 2004;**39**:327-332. DOI: 10.1111/j.1600-0765.2004.00747.x

[61] Karlsson L, Johansson E, Tranaeus S. Validity and reliability of laser-induced fluorescence measurements on carious root surfaces in vitro. Caries Research. 2009;**43**:397-404. DOI: 10.1159/000239754

[62] Shakibaie F, Walsh LJ. Differential reflectometry versus tactile sense detection of subgingival calculus in dentistry. Journal of Biomedical Optics. 2012;**17**:106017. DOI: 10.1117/1.JBO.17.10.106017

[63] Walsh LJ, Shakibaie F. Debridement endpoints from subgingival calculus detection. Auxiliary. 2014;**24**:22-23

[64] Schwarz F, Sculean A, Berakdar M, Georg T, Reich E, Becker J. Clinical evaluation of an Er:YAG laser combined with scaling and root planing for non-surgical periodontal treatment. A controlled, prospective clinical study. Journal of Clinical Periodontology. 2003;**30**:26-34

[65] Dommisch H, Peus K, Kneist S, Krause F, Braun A, Hedderich J, Jepsen S, Eberhard J. Fluorescence-controlled Er:YAG laser for caries removal in permanent teeth: A randomized clinical trial. European Journal of Oral Sciences. 2008;**116**:170-176. DOI: 10.1111/j.1600-0722.2008.00521.x

Chapter 11

High-Speed and High-Resolution Photon Counting for Near-Range Lidar

Tatsuo Shiina

Additional information is available at the end of the chapter

http://dx.doi.org/10.5772/intechopen.74350

Abstract

The near-range lidars for hard target and atmosphere detection should follow to the quick motion/activity of the measurement target. The transmitting optical power will be lowered for safety in the human activity space. And the lidar should increase the pulse repetition frequency to get the enough signal-to-noise ratio. The high-speed, high-resolution and high-repetition photon counting is desired for the near-range lidar. It is not a single photon counting at a certain delay time, but a multi-channel scaler with a deep memory for a series of delay times, that is, ranging data acquisition for lidar application. In this chapter, the mini-lidar for near-range observation is discussed. The targets are dust, gas, and atmosphere (aerosols). The activity monitoring of the atmosphere within a few hundred meters is the purpose of this mini-lidar. To follow and visualize the rapid motion of the target, high-speed and high-resolution photon counters (multi-channel scalers) were developed. The observation range can be easily adjustable depending on the lidar setup. The system can visualize the rapid motion of target with the high-resolution of 0.15 m (BIN width of 1 ns) and the minimum summation time of 0.2 s.

Keywords: lidar, dust, atmosphere, high-speed, scaler

1. Introduction

Nowadays, Near-range lidar; LIght Detection And Ranging, is of a particular interest in viewpoint of safety management [1–8]. The word "lidar," however, is used for both of hard target detection and atmospheric/gas detection. The former does not need the high sensitivity in general, but when the target moves quickly, and when the detecting target is a transparent material, the detector sensitivity should be high. The latter needs the high-sensitivity especially for low concentration gas and low altitude atmosphere, which moves rapidly depending on a structure, terrain, and the altitude. The photon counting is one of the choices in such high-sensitive lidar. Our attention is on such a rapid activity measurements of the atmosphere, dust and a certain gas.

The near-range lidars for hard target and atmosphere detection should follow to the quick motion/activity of the target material. The transmitting optical power will be lowered for safety in the human activity space. And the lidar should increase the pulse repetition frequency to get the enough signal-to-noise ratio. The high-speed, high-resolution and high-repetition photon counting is desired for near-range lidar. It is not a single photon counting at a certain delay time, but a multi-channel scaler with a deep memory for a series of delay times, that is, ranging data acquisition for lidar application.

The LED-based mini-lidar for near-range observation is considered in this Chapter. The optical pulsed power of 1nJ is emitted at the pulse width of 10 ns. The activity monitoring of the atmosphere within a few hundred meters is the purpose of this mini-lidar. To follow and visualize the rapid motion of the target, high-speed and high-resolution photon counters (multichannel scalers) were developed. The LED mini-lidar emits the pulsed beam at the pulse repetition frequency of >500 kHz, while the developed photon counter counts the echo signals at the summation time of only 0.2 s at a minimum. The observation range can be easily adjustable depending on the lidar setup. The system can visualize the rapid motion of target with the high-resolution of 0.15 m (BIN width of 1 ns) at the interval of 0.2 s.

At first, the needs of near-range lidar with some fields are explained. Afterwards, The lidar setup will be discussed in the viewpoint of near-range detection. The architecture of the high-speed photon counting is mentioned in the content. The design of the signal processing is considered to optimize the lidar echo detection. Some concrete applications are mentioned as well. At the end, the next step of the high-speed photon counting is finally discussed.

2. Lidar principle and photon counting

2.1. Lidar setup and equations

"Lidar," "LiDAR" or "Laser Radar" come from the word of Radar: Radio Detecting and Ranging. Lidar consists of transmitter part, receiver part and signal processing part. **Figure 1** shows the typical setup of lidar system. The pulsed laser beam was collimated and fired into the atmosphere with almost parallel to the receiver's optical axis. The optical receiver is adjusted its field of view (FOV) to detect the lidar echo from the target with an adequate signal-to-noise ratio. A traditional lidar has individual optics (biaxial optics) for a laser transmitter and a receiver, and the system has a blind area, which the transmitting beam cannot enter within a receiver's FOV, especially in near range. The blind area also causes in the case of coaxial optics that the beam is transmitted within a receiver's aperture. As the receiver's FOV is wider, or the beam divergence is wider, the blind area will be shortened, while the background light enters a lot into the receiver. As a result, the signal-to-noise ratio of the lidar echo will be lower. The lidar echo is detected by high sensitive detectors such as photomultiplier

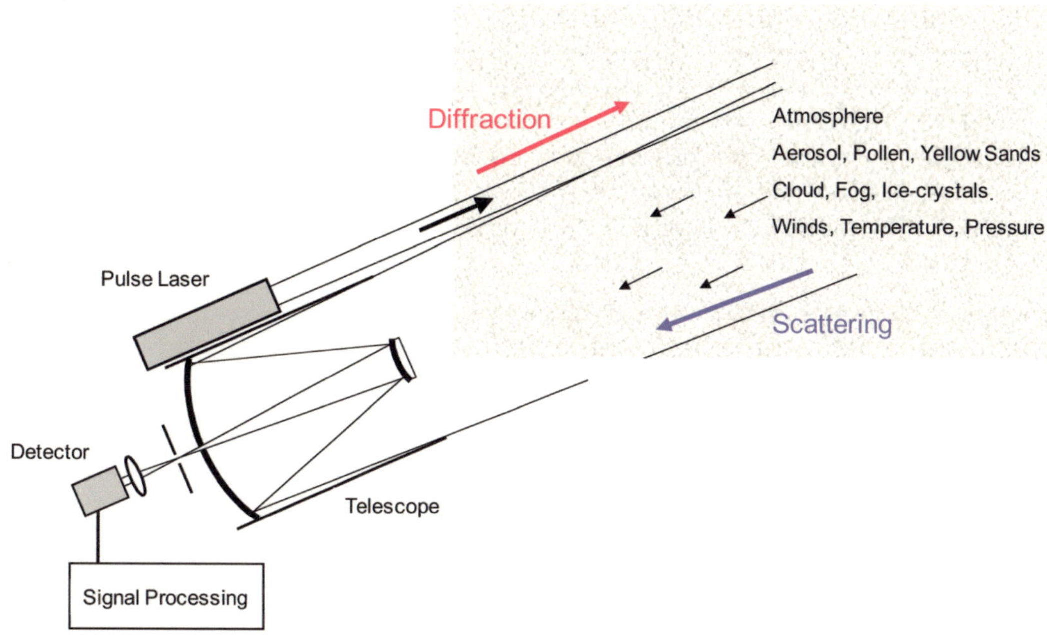

Figure 1. Fundamental setup of lidar system and target samples.

and Avalanche Photo Diode (APD), and stored in PC via a digitizer. When the lidar echo becomes weak and hard to detect an analog signal, photon counting method is selectable. When the collimated beam penetrates the atmosphere, it is attenuated by suspended particles in the atmosphere. As the atmospheric extinction coefficient is represented as α, the transmittance T obeys to the Beer-Lambert's law as follows

$$T(L) = \exp\left[-\int_0^L \alpha(x)dx\right] \tag{1}$$

where L is the propagation distance of the beam. When the beam hits the particles, the scattered lights go back to the same path. The echo, however, is also attenuated by the particles in the return path. Furthermore, the scattered echo spreads in all direction, and its intensity decreases in inversely proportional to the square of the distance because the scattering energy is spread in spherical. The lidar echo power depends on the atmospheric characteristics and also depends on the optical characteristics of the lidar apparatus such as the aperture of the optical receiver, the pulse width of the transmitting beam, the optical efficiency of the lidar system [9–10]. The concrete lidar echo power $P(L)$ is estimated by the lidar equation,

$$P(L) = P_0 K Y(L) A_r \frac{c\tau}{2} \beta(L) T(L)^2 / L^2 \tag{2}$$

where P_0 is the transmitted power [W], K is the system optical efficiency, A_r is the aperture of the optical receiver [m^2], c is the speed of light [m/s], τ is the pulse width of the beam [s], $\beta(L)$ is the backscattering cross section of the target [m^{-1}], *Y(L)* is the geometrical form factor, which is the overlap function between the transmitting beam and the receiver's field of view. It is determined by the telescope specification, the beam divergence, and the size of the field stop aperture; FSA. The geometrical form factor decides the initial receiving characteristics in near range. When you will observe the near-range target, its lidar echo can be corrected by this factor. After the blind area, there is the intermediate distance where the beam partially crosses into the field of view. Here the lidar echo power rises up till the whole of the beam enters in the field of view.

To estimate the signal-to-noise ratio of the lidar echo; SNR, is effective to evaluate the practical lidar echo under the several noises. You should consider the influences of the background light and the noise factor of the detector. The signal-to-noise ratio is written as

$$SNR(L) = \frac{\sqrt{M}\sqrt{\eta \Delta t/h\upsilon P(L)}}{\sqrt{\mu}\sqrt{P(L) + Pb + Pd}} \tag{3}$$

where η is the detector's quantum efficiency, Δt is the sampling interval [s], μ is the detector's noise factor, which is typically equal to 2–3 for APD and 1 for PMT, h is Planck's constant [J s], v is the light frequency [Hz], P_b is the background light power [W], P_d is the equivalent dark current power [W], M is the number of signal summations. That is, the long-term summation will increase the SNR, but its effect does not increase linearly. The numerator of the Eq. (3) improves SNR by the detector's sensitivity and the summation. The denominator weakens SNR by the noises of the atmosphere and the lidar system.

2.2. Analog mode and digital mode

The lidar signal is captured by the photo detector. It is an optical intensity and changes to the electric current. The current is, in general, the flow of many electrons. In the current flow, each electron is not distinguished. This is an analog mode. The lidar echo was come from suspended particles in the air and cause the electric current via a photodetector. The echo intensity as the current is inversely proportional to the square of the distance as shown in lidar Eq. (2). On contrary, the current becomes weak, and the individual electrons behave as discrete flows of pulse responses. In this situation, the lidar echo becomes stochastic scattering process at each distance. The photon signals indicate that the echo intensity is represented by not its signal height, but by its density in a certain time period. It is an photon counting mode (digital mode). **Figure 2** shows such behavior difference on analog mode and photon counting mode of the lidar echo.

2.3. Principle of photon counting

Discrete photon signals are captured by high speed photon detector. Nano-seconds' rise time can follow these discrete signals. The Photo Multiplier Tube (PMT) and Avalanche Photo Diode (APD) are used as photon counting device. In the case of PMT, the device, which the dark current is enough low, is selected as a photon counting device [11]. In the case of APD, the

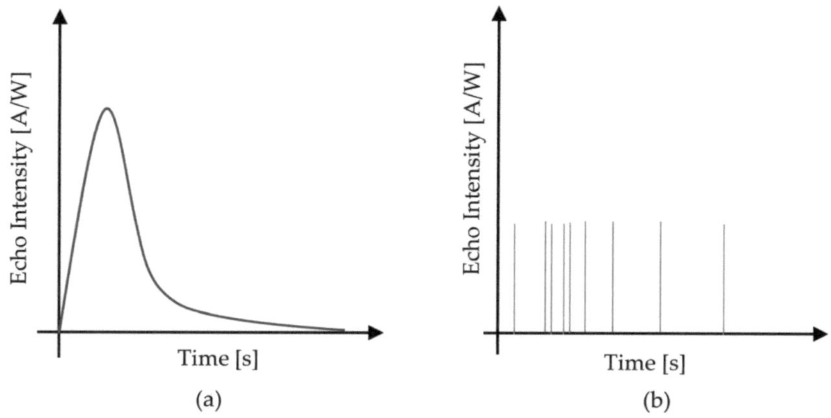

Figure 2. Analog signal (a) and discrete photon signals (b).

high voltage just below its breakdown voltage is implied. APD's photon counting is often called as Geiger mode [12].

Photon counting provides the echo counted numbers within a certain period. Usual photon counting will change the delay time one after another to capture a whole feature of the waveform. It is good for the repeatable echo signal. The sudden change and the intermittent echoes sometimes appear in the atmospheric lidar echo, especially in near range and low altitude echoes. In such a case, the delay time shift should be enough rapid and the echo counts at each delay time will be put in individual counters, which is called as BIN. Series BINs capture the echo counts due to the delay time. Such a device calls "scaler." BIN width indicates the time resolution, which is equivalent with the range resolution. The number of BINs represents the maximum detectable range. The repetition rate or repetition frequency is also the property of the photon counting. It decides the detectable phenomena of the target activity. The balance of the BIN width, its number, and repetition frequency is important for the photon counting device.

In the past, the memory and delay shifter and sequencer was assembled to make rearise the photon counter. Nowadays, however, fruitful functions installed FPGA boards are selectable. Therefore, high-speed and high-precision photon counter can be designed more easily and systematically.

3. Near-range Lidar

3.1. Time scale and spatial scale

The compact lidar is suitable to the near-range observation of less than a few 100 m, while the traditional large lidar is good at the long-range observation from a few kilometers to a 100 km. The compact lidar, however, is not a down-size one of the large lidar. **Table 1** shows the spatial

Phenomena	Spatial scale	Time scale
HP/LP	1000 km	10 h
Typhoon	100 km	3 h
Convection	50 km	2 h
Thunder clouds	10 km	1 h
Cumulus	2 km	10 min
Down burst	600 m	7 min
Tornado	200 m	5 min
Boundary layer	60 m	10 s

Table 1. Spatial and time scales of atmosphere.

and time scales in atmospheric phenomena. The large phenomena such as HP/LP have a large spatial and time scales, while the small phenomena like tornado takes small spatial and time scales. That is, the small phenomena become small structure and quick motion. To detect and visualize such small phenomena, the compact lidar should follow with the quick motion with high resolution. Furthermore, near-range detection is often a sensing in human living space. In that meaning, the transmitting optical power should keep eye-safety. It causes the limit of transmitting power and the selection of optical wavelength.

3.2. Transmitter and receiver requirements

Dust activity in low altitude is rapid, and the lidar should be high-speed and high-resolution detection. The BIN width of the photon counter decides spatial resolution. Pulse width of the transmitting beam is also shortened until the time that at least the detector can response. In the near-range observation, the transmitting beam will be thrown in horizontal direction. the transmitting power should be controlled in the view point of eye-safety. To make a margin to the signal-to-noise ratio under the restriction of the transmitting power, the pulse repetition frequency will be increased.

The lidar is not efficient in the near-range detection because of blind area. To shorten the blind area, the easy way is to give a small tilt to the transmitting beam against the receiver's optical axis. The receiver's field of view (FOV) is narrow of the order of a few milli-radians, and the optical alignment becomes severe. The transmitting beam has its power to spare, its divergence will be wider than the receiver's FOV. The lidar optics is free from misalignment and the system becomes robust. **Figure 3** shows the overlap efficiency of the lidar optics. The main graph (right) indicates the transmitting/receiving ratio under the transmitting beam divergence of 10 mrad. The sub-graph (left) shows the overlap function in the lidar Eq. (2) and its shift to the transmitting/receiving ratio. In the case that the receiver's FOV is 5 and 3 mrad, the transmitting/receiving ratio keeps 30 and 10%, respectively. In other words, when the transmitting beam power has a margin of 3–10 times compared with its usual operation, one can utilize this advantage.

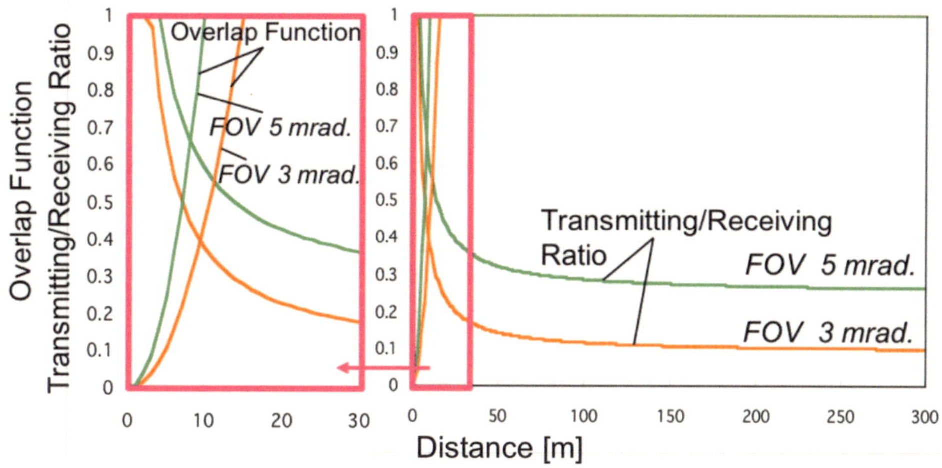

Figure 3. Overlap efficiency of lidar optics.

3.3. Near-range lidar setup

The near-range lidar is better to be compact or mini-size, light weight, and low electric power consumption. The traditional lidar does not have these features because the higher atmosphere has a slow activity, which depends on the altitude in general. The low altitude atmosphere change more quickly due to the ground surface, structure, and weather conditions. Dust flow and a certain gas behavior also follow to the atmosphere activity.

Figure 4 shows one of examples of the LED based mini-lidar optics [13–15]. Its specification is summarized in **Table 2**. The Lamp-type LED of 3 mmϕ is installed for lidar light source. Its transmitting pulse power is 200 mW (=2nJ/10ns). Average power is less than 1 mW. The wavelength is 392 nm and its spectral band width is 10 nm. Beam divergence is controlled to 5 mrad with a Fresnel lens of 50 mmϕ. The receiver is a Cassegrain-type telescope of the

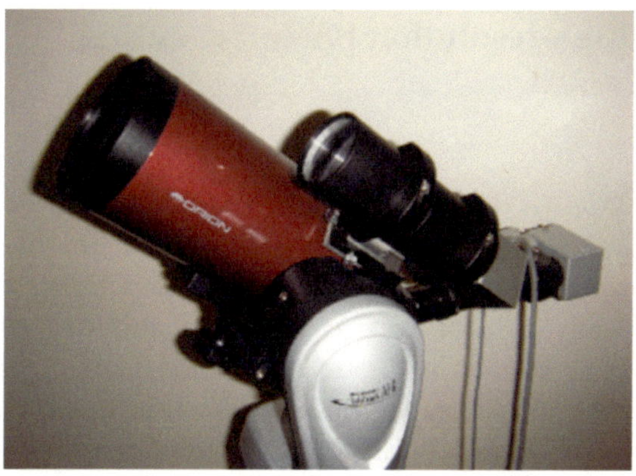

Figure 4. LED-based mini-lidar.

Transmitter	Light source	Lamp-type UV-LED(3mmϕ)
	Wavelength	Center 392 nm, Width 10 nm
	Pulse width	10 ns
	Repetition frequency	380 kHz
	Pulse power	200 mW
	Beam divergence	5 mrad
	Beam size	50 mmϕ
Receiver	Telescope type	Cassegrain
	Aperture	102 mmϕ
	Field of view	3 mrad
	Interference filter	Center 394 nm, Width 10 nm
	Detector	PMT (photon counting purpose)

Table 2. Specification of LED-based mini-lidar.

aperture of 102 mmϕ. The interference filter of the center wavelength 394 nm and its band width 10 nm is installed, too. The PMT for photon counting mode is selected for high sensitive detector. The LED-based mini-lidar is compact (150 mm(W) × 150 mm(H) × 300 mm(D)), light weight (<3 kg), and low electric power consumption (~2 W). The observation range is more than 100 m for atmosphere observation. The electrical power is fixed 12 V, and it can be operated with not only AC adapter but also with small DC buttery.

The pulse power and its repetition frequency are 200 mW and 380 kHz, respectively. The pulse repetition frequency is increased to compensate the low transmitting power. It is hard for the traditional photon counting device to follow to this high repetition frequency. The high-speed and high-resolution photon counter was originally developed.

4. High-speed and high-resolution photon counting

4.1. Architecture

Now several kinds of FPGA (Field-programmable gate array) boards are available. FPGA is one of complex programmable logic device (CPLD). These boards have various kinds of functions such as high-speed signal processing, long memory, plural interfaces and so on. For the lidar application, not only high-speed signal processing but long memory is important. The fundamental architecture is shown in **Figure 5**. At first, lidar signal from the detector is connected into a comparator. It will judge that the signal height exceeds a threshold or not. FPGA device prepares the plural counters (BINs) and de-multiplexer, which shifts the connecting counters due to the delay time. Each discrete echo pulses are entered into the counters due to the delay time. The delay time is generated by the system clock. The trigger

http://dx.doi.org/10.5772/intechopen.74350

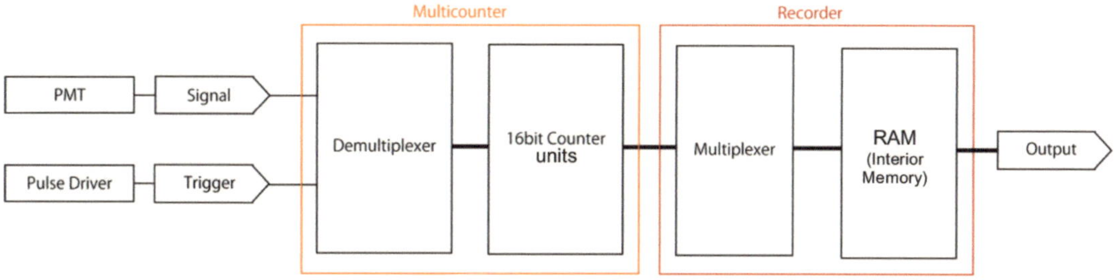

Figure 5. Fundamental architecture of photon counter.

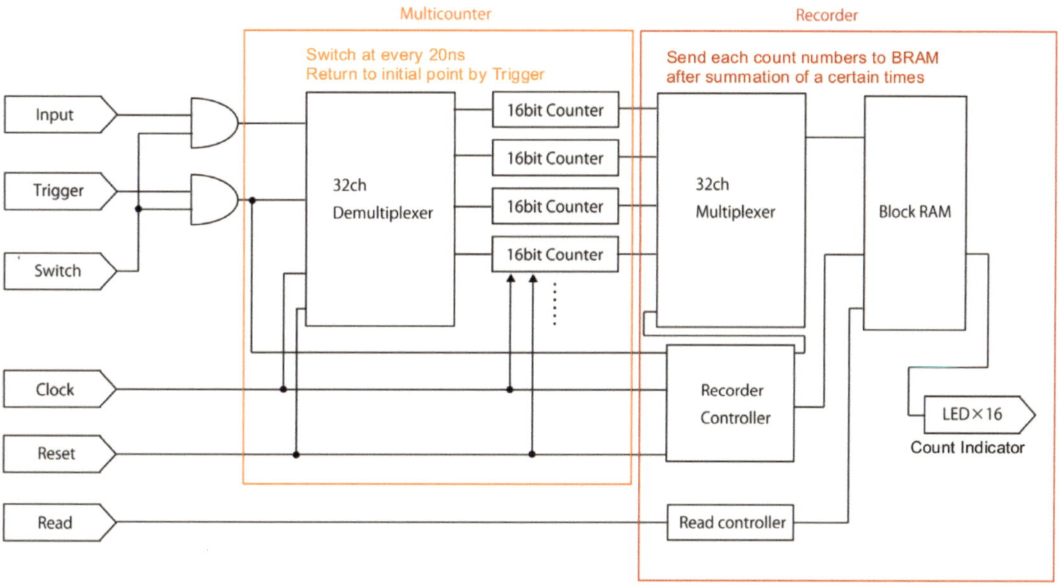

Figure 6. Concrete sketch of signal process on FPGA device.

signal is synchronous to the transmitting pulse beam. This trigger signal links to the repetition frequency of the FPGA process to return the first counter, that is, the nearest distance. After the defined number of times repeats, each counter sends its counted numbers to a memory via a multiplexer. Speed of system clock, FPGA process and memory capacity should have a balance to process the signal on time.

The programming of FPGA device is conducted by Verilog HDL. At first, we fabricated a prototype with a FPGA board (Digilent Basys 2). It is a FPGA trainer board. The FPGA device is Spartan-3E, Xilinx. System clock is 300 MHz. The block RAM is 72 kb. The concrete program design is shown in **Figure 6**. For lidar setup, the 50MHz operation was selected. It is equivalent to BIN width of 20 ns and the range resolution of 3 m. The BIN number is 25, which defines the maximum range of 75 m. The repetition frequency is 500 kHz. The result was successful to detect the sea wave motion with the summation time of 0.2 s.

4.2. FPGA photon counting board

After the trial development, the concrete photon counting board was developed with the help of a company. The schematic diagram and its specification are shown in **Figure 7** and **Table 3**, respectively. The developed photon counting board has 4-channel inputs. They are convenient for multi-channel observation such as orthogonal polarization measurement, certain gas Raman scattering observation, and multi-wavelength detections. The board can be synchronized to the pulse beam oscillation by trigger in port. On contrary, the board can generate the trigger signal to fire the pulse beam.

This photon counting board is commercialized by Trimatiz Co., Ltd., [16]. The FPGA device is Spartan 6, Xilinx. The system clock is 500 MHz. The PC interface to communicate the board setup and to transfer lidar data is PCI express. The board size is 111 × 168 mm. It is the size of expansion board of half size in desktop PC. Power consumption is 7 W, that is, <2 W for each channel. This photon counting board is able to connect parallel till three boards. Synchronous operation is possible up to 12 channels. The board feature is shown in **Figure 8**.

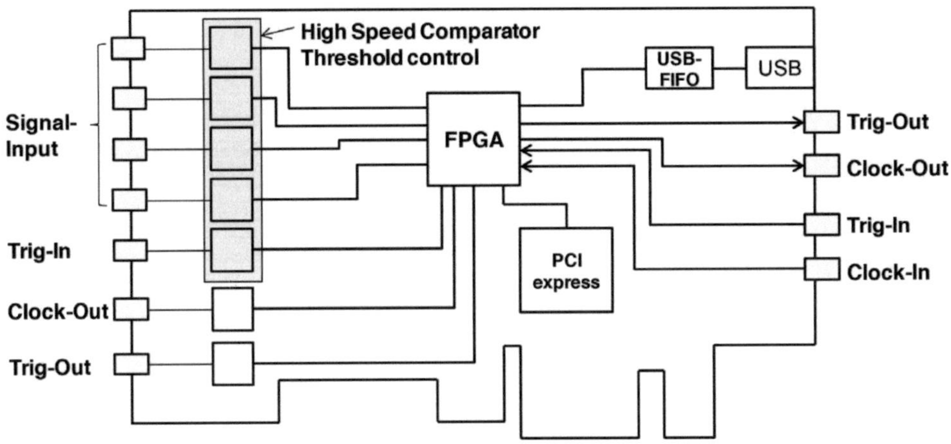

Figure 7. FPGA photon counter board.

FPGA device	Spartan 6
System clock	550 MHz
BIN width	5 ns–10.486 ms [5 ns × 2^n (n = 0–21)]
Number of BINs	32,767
Maximum counts	32,767
Input channel	1–4 (expands to 8–12)
Repetition frequency	>300 kHz
Interface	PCI express

Table 3. Specification of FPGA photon counting board.

Figure 8. FPGA photon counting board (Photon tracker, Trimatiz Co., Ltd.).

The specialized software is bundled to control the FPGA board. The BIN width and number, repetition frequency, threshold and trigger levels, can be adjustable. The highest resolution defined by BIN width is 5 ns, which is equal to the range resolution of 0.75 m. the minimum summation time up to the data transformation to PC is 0.2s.

4.3. Improvement

After the development of the first FPGA photon counting board, minor arrangement was accomplished. The PC interface change to the USB port for convenience via an adapter. As a next generation, the clock signal was integrated by a division program. As a result, the minimum special resolution is accomplished 1 ns, that is, 0.15 m. The repetition frequency is kept the high-speed of 500kHz. Minimum summation time is 0.2s with the enough BIN numbers. In this minimum summation time, the obtained data is stored in a memory within a certain repetition, and sent to PC after the process to reduce the time delay for the PC communication. The channel number is reduced two and PC interface changed to USB port. The signal generation of pseudo random code was installed for future improvement [17–20]. The electrical power consumption is 7W. This second-generation board is produced by Trimatiz Co., Ltd. (**Figure 9(a)**).

Other approach is tried by another company, Shibasaki Co., Ltd. [21]. PMT detector, amplifier and FPGA photon counter board was assembled in one box (**Figure 9(b)**). Single channel for a lidar echo is installed. This assembly contains a microprocessor to control the PC interface of Bluetooth. This assembly is connected to the telescope, and the lidar system does not need to wire to PC. The lidar set on the outdoor and the operator can monitor its observation inside the room. The power consumption is 2W. The LED based mini-lidar with this FPGA photon counting board could operate 2 nighttime observation (continuous 4 hours' operation per night). The minimum time resolution is 4ns, that is equivalent to the distance resolution of 0.6m. The repetition frequency follows to the LED pulse beam oscillation of 500kHz. The functional software is also bundled to visualize and analyze the observation data. The acquired data is visualized in real time.

(a) (b)

Figure 9. Second-generation FPGA photon counter boards. Photon tracker II (Trimatiz Co., Ltd.), and (b) FPGA detection units (Shibasaki Co., Ltd.).

5. Lidar applications

The first trial of LED mini-lidar was conducted with a commercialized photon counter. SR430 produced by Stanford Research Systems was used to detect the near-range lidar echoes. It is a multi-channel scaler. The repetition frequency is 1kHz. **Figure 10** shows some of observation results. The LED mini-lidar of 100mW(=1nJ/10ns) pulse beam received atmosphere echoes in the range of 100m. The inversely proportional decay slopes indicate the atmosphere response and spike like echoes at 55m in rainy day and at 85-115m in clear day came from trees. The notable thing is its summation time. Both of the results took a time of 16 min to gather the enough signal-to-noise ratio. Long summation time is hard to detect the rapid response of the air, dust, and gas flow.

The transmitting pulse power becomes weak, its repetition frequency should be increased to have enough signal-to-noise ratio. When the pulse repetition frequency is increased 100 times,

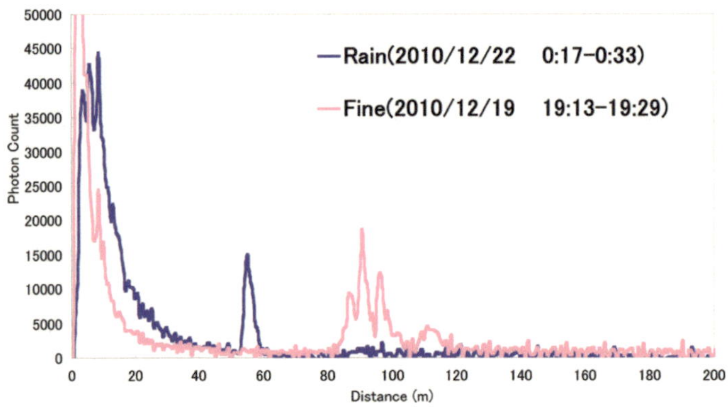

Figure 10. LED-based mini-lidar observation by commercialized photon counter ("SR430" produced by SRS).

the same signal-to-noise ratio will be obtained with the shorten time of 1/10 because of the equation (3). The LED mini-lidar increased its pulse repetition frequency till 500kHz. It is the maximum repetition frequency of the second-generation photon counting boards. In addition to this, the transmitting power also increased by changing the LED device from a Lamp-type (maximum pulse power 250mW) to a power illumination LED (maximum pulse power 1W). As a result, the summation time is shortened to 0.2s for atmosphere observation.

The first-generation photon counting board was used for atmosphere observation as shown in **Figure 11**. The LED mini lidar was improved its pulse power of 200mW (=2nJ/10s), the pulse repetition frequency was 380kHz. The summation time is 10s. Continuous observation of 1 hour was visualized. In this image, horizontal axis is time and vertical axis is distance. Photon counts was range corrected by multiplying the square of distance and taking log-scale, and represent with fake colors. The weather condition was clear. The atmosphere activity was low but one can see the minute change with the image. The atmosphere echoes monitored till 100 m.

When the summation time increases to 30s, the observation range increased till 300m in night time. In the daytime, the observation range was shortened till 100m under the condition that the lidar was fixed on the shade place. Even if in daytime condition, the threshold was adjusted adequately, the atmosphere activity could be monitored.

The other approach was performed by another company. Shibasaki Co., Ltd., tried to develop a high-speed photon counter and small LED lidars with their original architecture. Its structure is mentioned in section 4.3. Their LED mini-lidar transmits the pulsed beam of 250mW (=2.5nJ/10ns) with its repetition frequency of 400kHz. The original FPGA photon counter board has the range resolution of 0.6m, that is, 4ns with the system clock of 250MHz. The repetition frequency can follow the pulse beam oscillation of 400kHz.

Figure 12 shows the one of the observation results. It was the nighttime observation. The summation time was 30s and 50 min continuous lidar echoes was visualized as an image. The echo counts are represented by fake color with range-corrected log-scale. It was a foggy, which came from a near river. Fog activity was clearly captured and its detail can be distinguished.

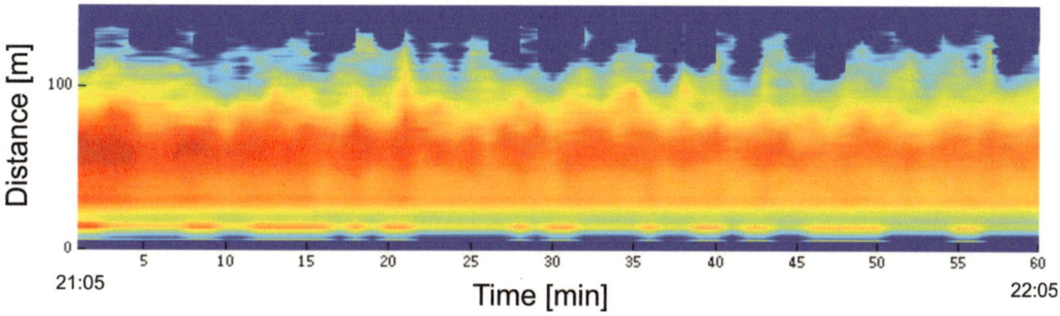

Figure 11. Atmosphere observation on August 14, 2012, by first-generation high-speed photon counter ("Photon Tracker" produced by Trimatiz Co., Ltd.).

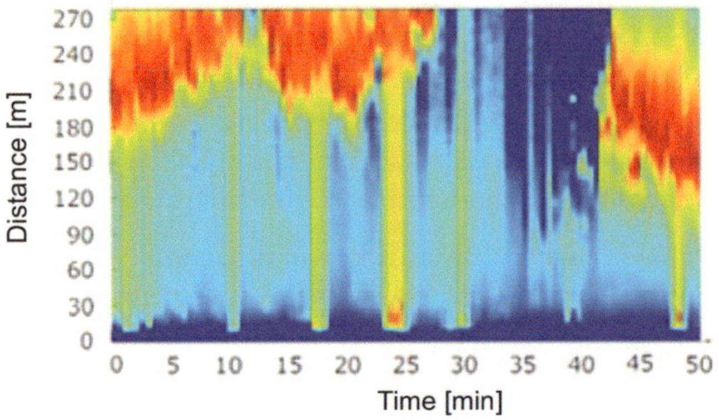

Figure 12. Fog activity observation by a LED-based mini-lidar with a new photon counter, (Shibasaki Co., Ltd.).

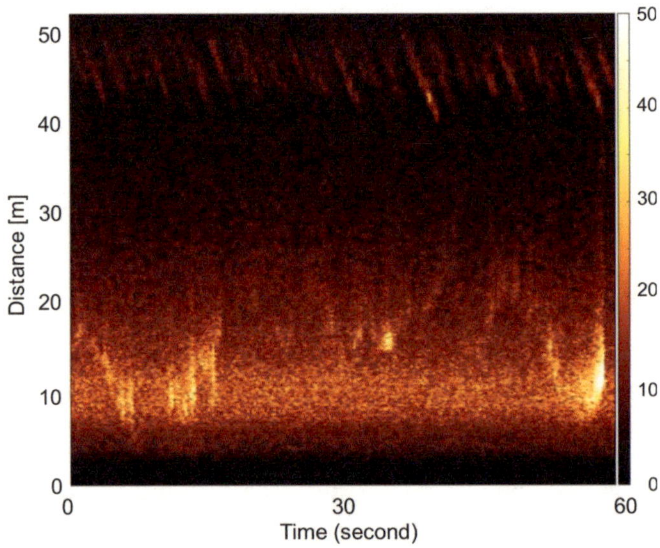

Figure 13. Sea wave observation by second-generation high-speed and high-resolution photon counter ("Photon Tracker II" produced by Trimatiz Co., Ltd.).

Suspended materials activity such as dust, fog, smoke and so on are the suitable targets for near-range lidars.

The second-generation FPGA photon counting board is adapted into the sea wave observation. The power LED beam of 0.75 W(7.5nJ/10 ns) was installed into the lidar setup [22–24]. Its divergence is 10 mrad and the receiver's field of view is 5 mrad. This optical setup of is the same as **Figure 3**. It was a shallow angle measurement. The LED mini-lidar was fixed on a shore and observed the sea wave washing the share with the shallow angle of 2.5°. One of results is shown in **Figure 13**. The summation time was 0.2 s. The range resolution was 0.15 m, that is, the BIN width was 1 ns. The wave motion is represented by fake colors, here with linear scale. It is appeared at 40–50 m. By analyzing the sea wave motion, fundamental profiles of sea wave such as wavelength, period, height and speed are deduced. The atmosphere echoes were also observed in the range of 5–40 m. The sand dusts and splashed waters were captured on the atmosphere, too. The detail of the air flow is obtained vividly. The sea wave frequency is 0.1–1 Hz. The suspended materials flow quickly on the sea surface. The short summation time and high-spatial resolution makes an identical copy of its echo activity. Dust flow on a coast, smoke motion on the chimney, whirlwind between the buildings will be visualized with this high-speed and high-resolution photon counter.

6. Summary

The near-range environmental measurement is major for safety management. In the case of car safety, the target is hard material. Its sensor is not so sensitive and becomes compact. In the near future, such sensors will be improved not only for hard target, but also suspended materials. In a tunnel accident, the smoke density, materials, gases, and their distribution, their sensing devices will give additional information for safety. In a factory, plant, and construction site, the near-range lidar will have advantage for the density measurement, distribution map, and security. Lidar is a remote sensing device, and it has an advantage against traditional contact sensors.

The low altitude atmosphere has rapid dynamics of the order to a few Hz. To follow its activity, the high-speed and high-resolution measurements are required. The traditional photon counting device cannot follow such dynamics. It is because the laser oscillation, especially high-pulse laser power is hard to follow as high repetition frequency as the a few 100 kHz.

The LED light source is one of the promising options for the lidar transmitter. Indeed, laser diode can be used with high pulse oscillation. The fiber laser increases its oscillation frequency, too. Due to this fact, light source selection is possible. With the combination of these light sources, the high-speed and high-resolution photon counting device is capable of detecting near-range atmosphere. To monitor a certain kind of gas, differential absorption method, fluorescent detection method, or Raman scattering signal detecting method will be install into the mini-lidar setup [25–30].

Author details

Tatsuo Shiina

Address all correspondence to: shiina@faculty.chiba-u.jp

Graduate School of Engineering, Chiba University, Chiba-shi, Japan

References

[1] Bennett M, Edner H, Grönlund R, Sjöholm M, Svanberg S, Ferrara R. Joint application of Doppler Lidar and differential absorption lidar to estimate the atomic mercury flux from a chlor-alkali plant. Atmospheric Environment. 2006;**40**:644-673

[2] Edner H, Ragnarson P, Wallinder E. Industrial emission control using Lidar techniques. Environmental Science & Technorogy. 1995;**29**:330-338

[3] Noguchi Y, Miya H, Shiina T, Noguchi K, Fukuchi T, Asahi I, Sugimoto S, Shimamoto Y, Ninomiya H. Compact raman lidar for hydrogen gas leak detection. 25th International Laser Rader Conference; 2010. pp. 178-181

[4] Raj PE, Devara PCS, Maheskumar RS, Pandithurai G, Dani KK. Lidar measurement s of aerosol col umncontent in an urban nocturnal boundary layer. Atmospheric Research. 1997;**45**:201-216

[5] Gong W, Chyba TH, Temple DA. Eye-safe compact scanning LIDAR technology. Optics and Lasers in Engineering. 2007;**45**:898-906

[6] Moorgawa A, Bencherif H, Michaelis MM, Porteneuve J, Malinga S. The Durban atmospheric LIDAR. Optics and Laser Technology. 2007;**39**:306-312

[7] Shiina T, Ninami E, Ito M, Okamura Y. Optical circulator for an in-line-type compact lidar. Applied Optics. 2002;**41**:3900

[8] Shiina T, Ninami E, Ito M, Okamura Y. In-line type micropulse lidar with an annular beam: Theoretical approach. Applied Optics. 2005;**44**:7467

[9] Fujii T, Fukuchi T, editors. Lasar Remote Sensig. Boca Raton: CRC Press; 2005

[10] Weitekamp C, editor. Lidar Range-Resolved Optical Remote Sensing of the Atmosphere. Singapore: Springer; 2005

[11] Available from: https://www.hamamatsu.com/resources/pdf/etd/PMT_handbook_v3aE.pdf

[12] Available from: http://www.bgu.ac.il/~glevi/website/Guides/AvalanchePhotodiodes.pdf

[13] Shiina T, Koyama M. Japanese Patent Application no. 2010-275798, 2010-12-10

[14] Koyama M, Shiina T. Light source module for LED mini-lidar. 2005;**39**(8):617-621

[15] Grishin M, editor. Chapter 8. In: Advances in Solid Stte Lasers Development and Applications. Intech; 2010

[16] Available from: https://www.trimatiz.com/en/

[17] Takeuchi N, Sugimoto N, Baba H, Sakura K. Diode-laser random-modulation cw lidar. Applied Optics. 1983;**22**:1382-1386

[18] Takeuchi N, Sugimoto N, Baba H, Sakurai K. Random modulation cw lidar. Applied Optics. 1986;**25**:64-67

[19] Nagasawa C, Abo M, Yamamoto H, Uchino O. Random modulation cw lidar using new random sequence. Applied Optics. 1990;**29**:1466-1470

[20] Matthey R, Mitev V. Pseudo-random noise-continuous-wave laser radar for surface and cloud measurements. Optics and Lasers in Engineering. 2005;**43**:557-571

[21] Available from: http://www.shibasaki-inc.jp/en-index.html

[22] Lin YC, Chang SJ, Su YK, Chang CS, Shei SC, Ke JC, Lo HM, Chen SC, Kuo CW. High power nitride based light emitting diodes with Ni/ITO p-type contact. Solid-State Electronics. 2003;**47**:1565-1568

[23] Wang JC, Wang RT, Chang TL, Hwang DS. Development of 30 watt high-power LEDs vapor chamber-based plate. International Journal of Heat and Mass Transfer. 2010;**53**: 3990-4001

[24] Lin Y, Chen Y, Chen Z, Ma D, Zhang B, Ye T, Dai Y. Triphenylamine and quinoline-containing polyfluorene for blue light-emitting diodes. European Polymer Journal. 2010; **46**:997-1003

[25] Choi S-C, Ko D-K, Lee J, Kim D-H, Cha H-K. The development of a mobile remote monitoring system by using differntial absorption LIDAR technology. Journal of the Korean Physical Society. 2005;**49**:331-336

[26] Devara PCS, Raj PE, Pandithurai G, Dani KK, Sonbawne SM, Rao YJ. Differential absorption lidar probing of atmospheric ozone over a tropical urban station in India. Measurement Science and Technology. 2007;**18**:639-644

[27] Gondal MA, Mastromarino J. Lidar system for remote environmental studies. Talanta. 2000;**53**:147-154

[28] Cecchi G, Pantani L, Raimondi V, Tomaselli L, Lamenti G, Tiano P, Chiari R. Fluorescence lidar technique for the remote sensing of stone monuments. Journal of Cultural Heritage. 2000;**1**:29-36

[29] Filipčcič A, Horvat M, Veberič D, Zavrtanik D, Zavrtanik M. Scanning lidar based atmospheric monitoring for fluorescence detectors of cosmic showers. Astroparticle Physics. 2003;**18**:501-512

[30] Bengtsson M, Grönlund R, Sjöholm M, Abrahamsson C, Dernfalk AD, Wallström S, Larsson A, Weibring P, Karlsson S, Gubanski SM, Kröll S, Svanberg S. Fluorescence lidar imaging of fungal growth on high-voltage outdoor composite insulators. Optics and Lasers in Engineering. 2005;**43**:624-632

Chapter 12

Detectors for Super-Resolution & Single-Molecule Fluorescence Microscopies

Robert T. Youker

Additional information is available at the end of the chapter

http://dx.doi.org/10.5772/intechopen.71943

Abstract

The resolution of light microscopy was thought to be limited to 250–300 nanometers based on the work of Ernest Abbe. This Abbe diffraction limit was believed to be insurmountable until the invention of Super-resolution microscopic techniques in the late 20th century. These techniques remove this limit and have provided unprecedented detail of cellular structures and dynamics down to several nanometers. An emerging goal in this field is to quantitatively measure individual molecules. Measurement of single-molecule dynamics, such as diffusion coefficients and complex stoichiometries, can be accomplished using fluorescence fluctuation techniques to reveal nanosecond-to-microsecond temporal reactions. These powerful complimentary experimental approaches are made possible by sensitive low-light photodetectors. In this chapter, an overview of the principles of super-resolution and single-molecule microscopies are provided. The different types of photodetectors employed in these techniques are explained. In addition, the advantages and disadvantages for these detectors are discussed, as well as the development of next generation detectors. Finally, example super-resolution and single-molecule cellular studies that take advantage of these detector technologies are presented.

Keywords: biophysical techniques, fluorescence fluctuation, molecular brightness, nanoscopy, palm, protein dynamics, spectroscopy, STED, STORM

1. Introduction

Fluorescence microscopy has been used by biomedical scientists to uncover fundamental processes in cells, tissues, and organisms for ~100 years. The discovery of genetically-encoded fluorescent proteins (e.g. GFP) allowed visualization of molecular dynamics in real-time in biological cells and organisms. A plethora of fluorescent proteins and dyes have been engineered over the last 25 years thus advancing the capabilities of fluorescence microscopy in biomedical research fields. However, a major roadblock in fluorescence microscopy was the diffraction of light that limited conventional light microscopes operating in the visible spectrum to a maximum lateral resolution of approximately 250 nm. Recently, several super-resolution techniques (e.g. PALM, STED, STORM) have been developed to "side-step" this resolution limit and "push" the boundary of optical resolution into the Nano-dimension. Most of these techniques rely on the capture of light emitted from single molecules to obtain increased resolution. Fluorescence fluctuation spectroscopy techniques (e.g. FCS, PCH, N&B) also rely on low-light measurements from few to single molecules. These techniques would not have been possible in cells and organisms without advancements in fluorescent dye chemistry and, importantly, low-light photodetectors.

In this chapter, principles of fluorescence microscopy [Section 2], an overview of super-resolution microscopy (a.k.a. fluorescence nanoscopy), and fluorescence fluctuation techniques (FFTs) will be discussed [sections 3 & 4]. In addition, the different types of detector technologies used in these techniques will be explained [Section 5]. Finally, examples of single-molecule fluorescence experiments employing the discussed photodetectors are presented [Section 6].

2. Principles of fluorescence microscopy

Today, fluorescence microscopy is widely used by scientists all over the world because the high signal-to-noise ratio (SNR) allows for the acquisition of high contrast images of specimens. The high SNR is made possible in part because of bright fluorophore-labeled molecules in the specimen. Labeling can be achieved by chemical attachment of dyes, or genetically-encoded fluorescent proteins such as GFP, or its variants (for a review see [1]). Two commonly employed fluorescence microscope setups are widefield (epifluorescence) and confocal. In-focus and out-of-focus light is collected in widefield fluorescence microscopy, and the out-of-focus light can degrade image quality. In confocal microscopy, out-of-focus light is rejected through an adjustable pinhole placed in front of the detector allowing optical sectioning of the specimen and improved image quality, especially for thick specimens.

The resolution of fluorescence microscopes, or generally all light microscopes, is limited by the diffraction properties and wavelength of light. Resolution of a light microscope is typically quantified by measuring the point spread function (PSF). The PSF is a mathematical description of how blurry a point-like emitter, such as a single fluorescent molecule, would look after being diffracted through a microscope. The full width at half maximum (FWHM) of the PSF is a convenient measurement to characterize resolution. Resolution is the ability to detect two adjacent objects in a sample and is not the ability to detect small structures. However, in biological specimens many subcellular structures are closer together or smaller than the FWHM of the PSF leading to a less than true representation of the specimen being imaged. Ernst Abbe first described the diffraction properties of light in 1873 and determined that resolution is dependent on both the wavelength of light used and the numerical aperture (NA), or light collecting ability, of the objective used in the microscope [2]. A simplified equation based on his work is $d = \lambda/2NA$ where d is the minimum distance to distinguish two point sources. For example, if green light at 500 nm was collected with a 1.0 NA objective then the resolution limit in this case would be d = 500/2 x 1 = 250, or 250 nm. Lord Rayleigh further defined resolution as the distance where the center of the diffraction pattern of image one overlaps with the first minima of the diffraction pattern of image two, and his equation (Rayleigh resolution criterion) is widely used today and is $R = 0.61\lambda/NA$ [2, 3]. For light microscopes, the resolution is roughly half the wavelength of light being used and equates to 150–200 nm. This diffraction limit made it difficult to image small sub-cellular structures such as endocytic vesicles that were about 60–80 nm in size. This resolution "wall" was thought to be insurmountable until new illumination techniques and photo-activatable proteins/dyes led to the development of super-resolution microscopy.

3. Review of super-resolution microscopy

This section provides an overview of several common super-resolution microscopic methods. For an in-depth treatment of the various techniques the reader is directed to these reviews [4–6]. Super-Resolution Microscopy (SRM) refers to microscopic techniques that can circumvent the Abbe diffraction limit and "break" the resolution barrier. SRM can be divided into near-field and far-field optical techniques. In near-field scanning optical microscopy (NSOM, or SNOM) no objective lens is used and instead a glass fiber with an aperture diameter smaller than the excitation light wavelength is used to create an evanescent field. The fiber is scanned over the sample at a short distance and the image resolution is only limited by the aperture dimensions. Lateral resolutions of less than 20 nm and vertical resolutions of 2–5 nm have been reported [7]. The major drawbacks of NSOM are the very short working distance needed, long scan times, and very shallow depth of field, thus limiting NSOM to mostly surface studies [7, 8]. In contrast, far field techniques do not suffer from these limitations and have been more widely adopted in SRM studies of biological specimens.

There are two fundamentally different approaches employed for far-field SRM. One involves manipulation of the PSF geometry, or so-called PSF engineering, by controlling the excitation beam geometry, and the other is probe-based employing Photo-switchable dyes/proteins. PSF engineering approaches include stimulated emission depletion (STED), ground state depletion (GSD), and saturated structural illumination (SSIM) microscopies. Photo-activated localization microscopy (PALM) and stochastic optical reconstruction microscopy (STORM) techniques comprise the probe-based super-resolution techniques. The goal in both techniques is to minimize the signal overlap of the objects and thus improve object discrimination in the image, thereby lowering the resolution limit.

In STED, the excited fluorescent molecules in the sample are exposed to a second-high intensity beam of light that upon saturation de-excites the molecules back to the ground state and thus turns them "off". This second beam of light is annular, or doughnut shaped (generated by phase plate), and the fluorescent molecules in the center remain excited. The resolution of this approach is dependent on the spot size of the remaining excited molecules, and resolutions of 29–60 nm have been achieved with this approach [9, 10]. It is very important to make sure the depletion laser wavelength does not overlap with the excitation wavelength of the dye, or the resolution will be degraded significantly [11]. In addition, the type of dye and specimen mounting medium can both affect image quality [12]. STED microscopy has been used to image protein tangles in Alzheimer's brain slices, electron transport complexes in mitochondria, and myelin sheets in oligodendrocytes [13–15]. GSD microscopy is similar to STED except the excited molecules are not de-excited. Instead they are optically "shelved" by driving the molecules into a long-lived, non-fluorescent dark state (i.e. triple state). The increased resolution is achieved using an annular shaped beam the same as STED. GSD microscopy requires special fluorophores that have a long-lived triplet state; alternatively removal of oxygen from the specimen can be used to modulate the dye's photo-physical properties [16, 17].

Coarse interference patterns, called moiré fringes, occur during illumination of a fluorescent sample. These fringes are not transferred to the image plane during conventional fluorescence microscopy, leading to degraded image resolution. In SSIM, optical gratings are used to create a "bar-code" pattern of high spatial frequency light that is passed over the sample, and images are collected at different angles of illumination. This process allows "Capture" of the moiré fringes and high resolution images can be acquired with the aid of computer algorithms. A two-fold increase in resolution beyond the diffraction limit can be attained and up to eight-fold resolution increase in three dimensions (axial or z-axis) [18–21].

PSF engineering techniques try to limit image overlap of objects through alterations in the geometry of the exciting light. A second approach is to use low levels of excitation light to active a few fluorescence molecules at a time, thereby preventing overlapping signals. The PSF from each molecule can then be fitted mathematically to determine its center. This process is repeated thousands of times to locate nearly-all fluorescent molecules, leading to a point-by-point construction of the super-resolution image. This approach was called pointillism, an analogy to the painting technique, and is used by PALM and STORM techniques [22, 23]. The photoactivatable variant of GFP (pa-GFP) was used in the original PALM experiments, whereas photo-switchable dye-pairs (i.e. Cy3-Cy5), or proteins (i.e. EosFP), were used in STORM. The use of a single photo-switchable dye with long triplet state in place of a dye pair has been developed and is called direct STORM (dSTORM). Images routinely taken with PALM and STORM techniques can have resolutions of 20–30 nm in the lateral axis.

Multi-color (i.e. two or more fluorescent molecules) SRM has been realized for many of the above-mentioned techniques [24–26]. Improved axial resolution has been accomplished for PALM through the implementation of double plane, or biplane (BP) detection, hence the name BP-PALM [27]. The insertion of an astigmatic lens leads to axial resolution improvement for STORM because it allows for precise 3D localization of the molecules, and this technique has been called 3D STORM [28]. Importantly, the performance of pointillism techniques (2D or 3D versions) is highly dependent on the density of fluorophore labeling and the nature of the biological structure. For example, imaging of smaller and filamentous objects tends to work better than densely packed and asymmetric objects [29] .

The use of two or more opposing objectives can be used with conventional or super-resolution microscopes to improve axial resolution. For example, in 4Pi microscopy two precisely aligned objectives are used to illuminate and detect the sample form opposite sides. This allows for coherent superposition of light onto the detector and thus constructive interference that reduces the axial resolution to 100 nm or less [18]. This type of interferometric microscope configuration has been applied to STED (called isoSTED) [30], SSIM (called I^5S) [31], and PALM (called iPALM) [32]. These technically demanding techniques require precise piezoelectric controlled alignment, careful specimen preparation, and as such are employed usually by microscopy specialists.

The type of SRM employed depends on proper matching of the experimental requirements with the technical capabilities of the SR technique. For example, PALM and STORM may not be suitable for live cell imaging of embryos during development due to photo-toxicity from the large number of images required, but SSIM, or 4Pi microscopy, is more suitable. Both SSIM and STED techniques do not require extensive image processing (unlike PALM and STORM), and what you see is what you get. Currently, there is great interest in development of dyes/proteins that will allow long-term (i.e. hours) SRM experiments in live samples.

4. Review of fluorescence fluctuation techniques (FFTs)

Fluorescence fluctuation techniques (FFTs) are a group of spectroscopic/microscopic approaches that extract information about the dynamics (e.g. size, diffusion rate, binding partners) of fluorescently-labeled molecules from the small variations encoded in the emitted fluorescence signal (See **Figure 1**). This section provides an overview of FFTs, but for an in-depth explanation of the theory of the techniques the reader is directed to these reviews [33, 34].

The original FFT called fluorescence correlation spectroscopy was developed by Magde, Elson, and Webb in the 1970s, and measurements were initially performed in cuvettes *in vitro* [35, 36]. Cellular measurements were not possible until the invention of the confocal microscope in the 1990s. The reason for this is explained below. FCS measurements work on the principle that fluorescent molecules diffuse by Brownian or directed motion through the microscope illumination volume leading to changes, or fluctuation in the signal being recorded. Plotting the autocorrelation function $G(\tau)$ of this signal versus time leads to the FCS auto-correlation curve. The time at which there is half-maximal decay of the FCS curve is the diffusion time of the molecule. The diffusion rate can be calculated once the waist dimension of the illuminated volume is determined [33]. The inverse of the y-axis ($G(\tau)$) on the FCS curve corresponds to the average number of molecules diffusing through the microscope illuminated volume. The fluorescent molecule concentration must be nanomolar to micromolar to prevent averaging out of fluctuations; otherwise the FCS technique will not work. The detection volume of wide-field fluorescence microscopes is incompatible with FCS because the light from hundreds of molecules is registered, but a confocal microscope has a very small detection volume (~ 1 fL) due to the adjustable pinhole, thus allowing the measurement of small numbers of molecules. Very sensitive detectors such as avalanche photodiodes (APDs, discussed below) are needed to record the signals from these few molecules.

The FCS technique can be applied to two different fluorescently-labeled molecules and this is called fluorescence cross correlation spectroscopy (FCCS). In FCCS, the auto-correlation curves from two different fluorescently labeled molecules are compared to each other, cross-correlation

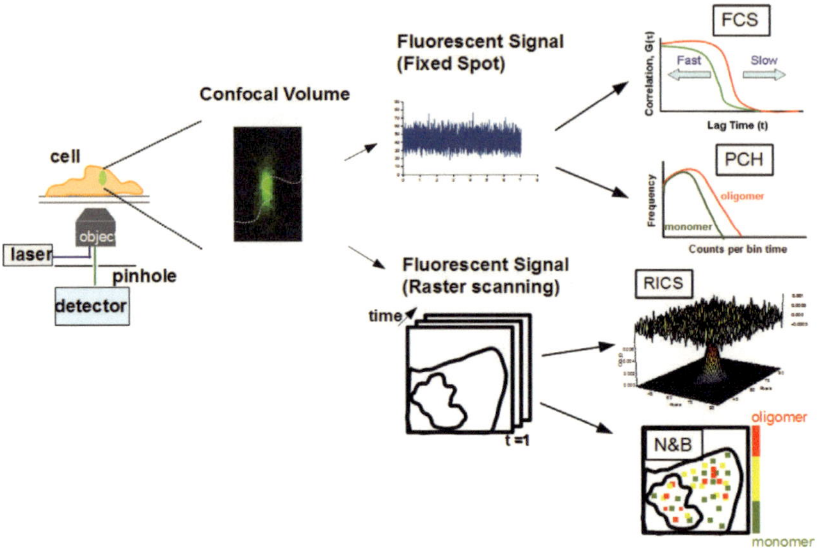

Figure 1. Schematic illustrating the principles of FCS, PCH, RICS and N&B analyses. *Left:* Cartoon of confocal microscope setup and image of XZ profile of one-photon PSF with dotted line representing diffusion of molecule. *Middle:* Signal trace from fixed point, or raster scan images of cell specimen. *Right:* Fixed point data can be used to generate FCS and PCH curves. Raster scanned data can be used to calculate diffusion (RICS) or molecular brightness (N&B) values on a pixel-by-pixel basis. Figure reproduced with permission from *J. Biomed. Opt.* 19(9), 090801 (Sep 26, 2014) doi: 10.1117/1.JBO.19.9.090801.

curve, to determine if the molecules are diffusing in a complex [37, 38]. FCS and FCCS techniques perform well if the molecules investigated undergo isotropic diffusion (diffusion rate is constant), but they are not well suited for molecules with anisotropic behavior (diffusion varies in time and space). The approach of raster image correlation spectroscopy (RICS) is better suited for the measurement of anisotropic dynamics and can measure large regions of a cell compared to spot measurements (FCS and FCCS) that are fixed in one location [39, 40].

In RICS, the light source is scanned across the sample allowing the measurement of multiple adjacent and nonadjacent volumes thus providing a more complete description of the diffusion dynamics of the molecules [39]. The appropriate scan speed and pixel size are critical for RICS to be successful [41]. If the scan speed is too fast compared to the molecule then the molecule will "appear" to be immobile. If the scan speed is too slow then the opposite problem emerges and the molecule will diffuse away before being registered by the detector. For example, a scan speed of 25 μs per pixel and pixel size of ~0.05 μm are usually sufficient to measure the diffusion of a 25 kDa protein by RICS [42].

As stated earlier, the average number of molecules can be determined from the amplitude of the FCS curve ($G_o \approx 1/N$), and changes in the amplitude of the curve have been used to infer protein-ligand binding and protein–protein dimerization [43]. For example, as the protein dimerizes the concentration of individual monomers decreases, leading to a doubling of the amplitude. This relationship holds true if the concentrations of the different species are constant throughout the measurement. This type of analysis where intensity (1st moment) and variance (2nd moment) are used to determine protein size and number from FCS data is called moment analysis [44, 45]. This type of analysis breaks down in complex systems where there are multiple species of varying concentrations that are common in biological specimens [46, 47]. However, photon counting histogram (PCH) analysis does not suffer the same drawbacks as moment analysis. In PCH analysis, the fluorescence signal, recorded as photon counts, is separated into different bins and plotted as a function of frequency to generate a histogram plot. The amplitude and shape of the histogram plot is influenced by the microscope PSF, detector noise, and fluctuation in molecule numbers. Careful fitting of the histogram returns the first and second moments for the entire photon distribution. This allows measurements of complex samples and determination of the counts per second molecule (cpsm) or molecular brightness. The molecular brightness can provide information on the size of molecules with proper calibration [46].

An imaging analogue to PCH analysis is called number and brightness (N&B) analysis. In N&B analysis the average molecular brightness and number of fluorescent molecules can be calculated from individual pixels of a raster-scanned image using moment analysis [48]. N&B analysis has been used to study the assembly/disassembly of focal adhesion complexes in cells, the formation of protein oligomers during vesicular transport, and many other cellular processes [48–52]. Like RICS, an appropriate scan speed (pixel dwell time) is required to "capture" the fluctuations of the molecules. Analogue detectors can be used for N&B acquisition but photon counting detectors are highly recommended because no adjustment is needed for digital-to-analogue conversion.

Importantly, the brightness values obtained by these FFTs are dependent on the illumination intensity used and the detector settings. For example, doubling the laser power used to excite the molecules will lead to a doubling of molecular brightness, assuming all other factors are equal. Therefore, it is imperative that all measurements be performed with the same settings for proper comparison of specimens. The molecular stoichiometries obtained by FFTs are many times minimum estimates because unlabelled molecules are mixed with labeled molecules in the biological specimen. Ideally, all molecules under investigation should be labeled in the specimen. New genetic approaches such as CRISPR make it possible to replace endogenous proteins with fluorescently-tagged versions in the genome of the cell or organism [53]. Finally, complete characterization of the detectors used for data acquisition is required to account for detector artifacts, such as after-pulsing and dead-time, during FFT analyses.

5. Detector technologies used in SRM & FFTs

5.1. Brief history of photodetectors

The first detectors for microscopes were the human eye as seen in the 17th century hand-written drawings of van Leeuwenhoek's animalcules [54]. In the 1800s, light sensitive silver halide emulsions on either copper plates or nitrate film were used to record images.

Photographic film dominated many aspects of microscopic imaging until the mid-20th century. Photodetectors are devices that record changes in light intensities (photons) and then create an electrical or optical output. Modern photodetector technologies began with the invention of photomultiplier tubes (PMTs) in the 1930s [55–57] followed by the invention of avalanche photodiodes (APDs) in the 1960s [58]. Advances in semiconductor materials and integrated circuit technologies led to creation of focal plane array (FPA) detectors, such as charge-coupled devices (CCDs) in the 1970s, and to the present day scientific grade complementary metal oxide semiconductors (sCMOS) [59–62]. Different detector technologies have advantages/disadvantages, and there are several criteria used to evaluate them such as quantum efficiency (QE), readout noise, dark current levels, and SNR (**Table 1**).

This section provides an overview of basic photodetector technologies with an emphasis on detectors commonly used in SRM & FFTs (e.g. APDs, CCD cameras, sCMOS). For an in-depth technical explanation of photodetectors the reader is referred to the textbook Optical Systems Engineering by Kasunic [63].

5.2. Photomultiplier tubes (PMTs)

The working principle of PMT is based on the photoelectric effect. PMTs are composed of vacuum tubes consisting of a cathode, multiple dynodes, and an anode (**Figure 2**). Incident photons are absorbed by the cathode ejecting primary electrons (~3 eV) that are accelerated by an electrostatic field toward a series of dynodes. Additional numbers of electrons are ejected (5–10 electrons) because each subsequent dynode is held at a more positive voltage potential (~100 eV), leading to an amplification of the signal. The electron current is then detected by an external electrical circuit. PMTs usually have 10–14 dynodes with a cathode-to-anode voltage gap of ~1 kV and current gain of 10^6 to 10^8. The composition of the photocathode determines the PMTs spectral response, quantum efficiency (QE), and dark current levels. Cathodes made of multi-alkali semiconductor mate-

Detector Type	QE @ 600 nm (%)	Dark Count (e–/pixel/sec)	Readout Noise (e- pixels/rms)	Detector Gain
R10699 PMT	20	2[#]	0	1.3 x 10^7
silicon APD (Geiger mode)	75	3.0 to 9.2	5 to 20	10^5 to 10^6
Hybrid detector	10 to 40	0–500	0	100,000
CCD	40 to 95	0.0002 to 0.001	8 to 12	4
EMCCD	>90	0.0001 to 0.07	40 to 65	10,000
sCMOS	55 to 80	0.01 to 2	5 to 25	1
H33D Gen I*	4.5	<1**	0	10,000,000
H33D Gen IIB*	30.9	<15**	0	1,000,000

*not commercially available.

**kHz across total surface.

[#]after 30 minutes' storage in dark & supply voltage 1 x 10^6.

Table 1. Typical characteristics of detectors used in single molecule fluorescence experiments.

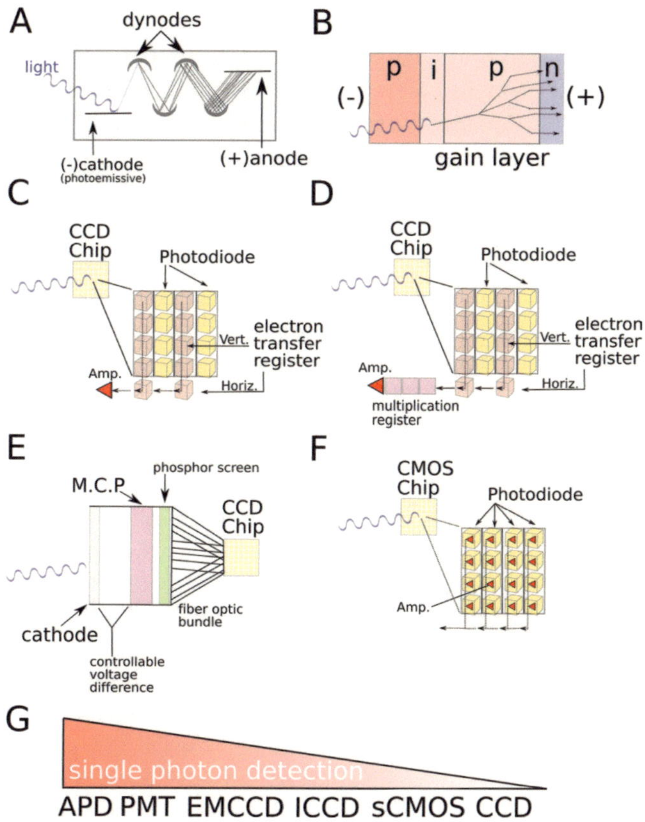

Figure 2. Schematic diagram of Photodetectors. A) Schematic for PMT illustrating amplification of incident light by dynodes. B) Schematic for APD where p = p-layer, i = i-layer, and n = n-layer are indicated. Impact ionization and electron multiplication is illustrated. C) Schematic for CCD sensor chip where amp. (triangle) = amplifier, Vert. = vertical registration, and Horiz. = horizontal registration D) schematic for EMCCD sensor chip E) schematic for ICCD sensor chip where M.C.P = micro-channel plate F) schematic diagram for CMOS sensor chip where the amplifier (triangle) and associated digital processing is incorporated into each photodiode G) relative scale for single photon counting suitability for the various photodetectors with APD being the most suited and CCD the least. Absolute suitability will depend not only on detector type and characteristics but also instrument design.

rial (e.g. GaAs, or GaAsP) have maximal sensitivity at 400 nm, providing good spectral response in the UV and blue regions but a rapid drop-off in response in the green and red region of the visible spectrum. PMTS have large gain and low noise but suffer from low QE (20–40%) compared to other technologies (e.g. APDs). Importantly, single photons can be detected with PMTs, but discrimination of single versus multiple photons is difficult. Avalanche photodiodes have better QE and better sensitivity in the green and red region of the visible spectrum compared to PMTs [64].

5.3. Avalanche photodiodes (APDs)

Avalanche photodiodes (APDs) can be thought of as solid-state versions of PMTs with the exception that there is no photocathode and thus utilize primary photoelectrons more efficiently than PMTs. APDs are composed of three semiconductor layers called p-layer, i-layer, and n-layer (a.k.a *p-i-n*, **Figure 2**). The n-layer has extra electrons: whereas the p-layer is electron poor and has "holes". APDs are like *p-i-n* photodetectors except there is an extra p-layer between the i-layer and n-layer to create a *p-i-p-n* orientation where the second p-layer is much thicker than the other layers. A negative voltage (reverse bias) is applied across the junction, and absorption of a photon creates an electron–hole pair that is accelerated through the thicker p-layer (acts as gain layer) where conductive electrons impact non-conductive electrons thus making them conductive. This impact ionization can cascade as occurs with PMTs, leading to a large amplification of electrons called an avalanche. Average gain for APDs is 100 times, which is too low for single-photon detection, but operation of the APD in Geiger mode above breakdown voltage allows for single photon avalanche detection (SPAD) [65, 66]. The i-layer in APDs allows for better photon absorption, shorter photoelectron diffusion time, lower capacitance, and faster response. Researchers have wanted the sensitivity of APDs combined with the gain and dynamic range of PMTs. This motivation led to the development of hybrid photodetectors.

5.4. Hybrid photodetectors

Hybrid photodetectors are a combination of PMT-APD technologies and were developed in the 1990s for high energy physics experiments [67, 68]. Hybrid detectors have a photocathode, electron multiplication component, and output terminal housed inside a vacuum tube. The difference from PMTs is the electron multiplication is performed by a silicon avalanche diode (AD), instead of dynodes. The silicon diode contains a thin p-layer facing the photocathode followed by a thicker middle silicon layer and finally a p-n junction that is attached to the bias terminal. Photoelectrons ejected from the cathode are accelerated toward the AD by a very large voltage difference compared to PMTs (~8 kV). The electrons are multiplied in the AD first through electron bombardment and then by avalanche gain. Total gain can be greater than 100 times that is considerably lower than PMTs, but hybrid detectors have other benefits that make-up for the low gain. Hybrid detectors have better SNR compared to PMTs because the first gain step can be up to 1000 times (one photoelectron yields 1000 secondary electrons). This higher SNR allows for discrimination between one photon and multiple photons. The response time is faster for hybrid detectors compared to PMTs, and there is virtually no after-pulsing (false detection of photon). Hybrid detectors are well suited for fluorescence applications where after-pulsing can cause artifacts such as in time correlated single photon counting (TCSPC) and FCS. Hybrid detectors outperform SPADs and PMTs for FCS and other single-molecule fluorescence experiments [69]. Unfortunately, these point-like detectors discussed above are inefficient for imaging large regions. However, arrays of photodetectors are inherently suited to large field imaging. One popular array photodetector is the charged coupled device (CCD).

5.5. Charged coupled devices (CCDs)

Charge-coupled device (CCD) cameras have completely replaced photographic film cameras for scientific experiments, and are routinely used for microscopic imaging. A CCD is an array of photosensitive elements attached to a silica semiconductor substrate [62, 70]. Each element is composed of a metal oxide semiconductor (MOS) capacitor consisting of surface gate electrode, aluminum or polysilicon, deposited on top of charge carrying channels that are insulated by silicon dioxide (**Figure 2**). Incident photons "strike" individual MOS elements generating electron–hole pairs, leading to an accumulation of charge in the potential well below the MOS. Application of an external voltage controls the movement and release of the built-up charge in each photosensitive element or pixel. The architecture of the electrodes required for the charge readout acts as a shift register for charge transfer, and there are several different register types employed in CCDs [63, 71]. The three main CCD types are full-frame, frame-transfer, and interline transfer architectures. Full-frame CCDs utilize the entire sensor array for light collection providing maximum efficiency. However, a mechanical shutter is needed to stop exposure and allow transfer of the built-up charge limiting image acquisition rates. Frame-transfer CCDs have half of the photosensitive surface covered by an opaque mask that acts as a photoelectron storage space. Charge is transferred to the masked area allowing the next exposure to commence while the first is being processed. This setup allows for faster frame rates but half the sensor is not available for image acquisition meaning a larger chip is needed compared to a full-frame, thus adding to cost.

Interline CCDs have alternating rows of pixels that are masked with an opaque material (e.g. aluminum) thus allowing acquisition and charge transfer to occur simultaneously. The charge in the unmasked rows is transferred to the masked row allowing for a second round of exposure during readout of the previous first exposure. This dramatically speeds up acquisition rates at the expense of reduced sensor surface. The addition of micro-lenses to the interline CCD design focuses more light onto the pixels increasing efficiency, from 50–75%, of collected light [72, 73]. A second added benefit of using micro-lenses is that it extends the spectral sensitivity of the CCD into the blue and UV light range that is ideal for imaging with GFP. One of the most effective strategies to increase QE is back illumination of the sensor where the wiring is behind the photocathode layer leading to less light scattering and up to 90% QE [74].

For all three architectures (full-frame, frame-rate, interline) the charge readout is fed into a CCD output amplifier, and then an analogue-to-digital converter (ADC). The stored charge in each sensor pixel is linearly proportional to the light absorbed up to the full well capacity (FWC). The FWC determines the maximum signal a pixel can record and is a major factor affecting a CCDs dynamic range. Traditionally, CCDs were composed of square sensor pixels arranged in a rectangular pattern with a 4:3 aspect ratio. Common CCD image sensors range in size from 6 to 16 mm (diagonal measurement). Many current scientific grade CCD cameras employ square sensor arrays to better match the microscope field of view (**Table 2**). Addition of an electron multiplication register between the shift register and output amplifier can increase the signal from the image sensor, and are called electron multiplication charge-coupled device (EMCCD). This modification of the CCD improves the SNR and increases the QE to 95%, or greater in most cases (**Figure 2**). EMCCD cameras have replaced CCD cameras for many imaging applications, including SRM (**Table 3**). Drawbacks to the electron multiplication are gain instability, performance decay with age, and potentially increased dark current [75]. Finally, an intensifying screen can be put in front of the CCD sensor (ICCD) to increase sensitivity to single-photon detection [76]. The intensifying screen is composed of a photocathode, micro-channel plate (MCP), and a phosphor screen (**Figure 2**). The photons strike the photocathode generating photoelectrons that are amplified by the MCP, a plate with angled tubes that creates a "shower" of electrons like dynodes in PMT. Secondary electrons from the MCP strike the phosphor screen creating photons that are read by the CCD sensor. Importantly, ICCD sensors provide gate-ability (100's of picosecond temporal resolution) in addition to enhanced sensitivity over EMCCDs [77]. A drawback to ICCD and EMCCD cameras are their cost (~$30,000–40,000), but advances in semi-conductor fabrication have led to smaller and more cost-effective photodetectors, such as complementary metal oxide semiconductors (CMOS).

5.6. Complementary metal oxide semiconductors (CMOS)

CMOS sensors, like CCDs, are arrays of photosensitive pixels, but are smaller in size due to advanced manufacturing techniques [78]. Specifically, the readout and processing circuitry are miniaturized, and incorporated into each pixel creating an "active-pixel sensor" (APS). The miniaturized transistors are fabricated using a complementary MOS (CMOS) technology (**Figure 2**). This miniaturization comes with a price, most notably higher noise compared to CCDs and the APS takes up pixel area affecting light absorption. An advantage of CMOS is that each pixel can be read out randomly and no electrons are lost by charge transfer across a row. CMOS chips consume less power and are more suitable for low-price products like cell phone cameras. The lower performance of early generation CMOS sensors, compared to CCDs, prohibited their use for scientific applications. Recently, higher performance CMOS sensors have been fabricated and are called scientific grade CMOS (sCMOS) [79]. These new sCMOS sensors were introduced in 2010 and were marketed as low noise, high QE (55–70%), and fast frame rate (>100 fps) cameras (**Table 4**). There is improved image resolution due to the smaller pixel size of sCMOS sensors compared to EMCCDs. However, distortions can occur in images due to rolling shutters that are used (i.e. each row captured at different time). The sCMOS camera noise is not Gaussian distributed and the improved resolution is at the expense of decreased sensitivity compared to EMCCDs [80]. Which camera technology (EMCCD, or sCMOS) is best suited for low-light microscopy experiments?

Model	Manufacturer	QE @ 600 nm (%)	Dark Current (e–/pixel/sec)	Readout Noise (e- pixels/rms)	Resolution (um-pixels)	Imaging Array
Cool-SNAP DYNO	Photometrics	75	0.0006	5.2	4.54 x 4.54	1940 x 1460
Retiga R1	Qimaging	75	0.001	<5.5	6.45 x 6.45	1360 x 1024
SOPHIA 2048B	Princeton Sci.	>95	0.00025	22 @ 4 MHz	15 x 15	2048 x 2048
Clara	Andor	>40	0.0003	5 @ 10 MHz*	6.45 x 6.45	1392 x 1040

Table 2. Examples of commercially available CCD cameras.

Model	Manufacturer	QE @ 600 nm (%)	Dark Current (e–/pixel/sec)	Readout Noise (e- pixels/rms)	Resolution (um-pixels)	Imaging array
Evolve 128	Photometrics	>92	0.0069	46 @ 10 MHz	24 x 24	128 x 128
Evolve 512	Photometrics	>95	0.003	65 @ 10 MHz*	16 x 16	512 x 512
ProEM-HS:1KBX3	Princeton Sci.	~95	0.002	26 @ 10 MHz*	10 x 10	1024 x 1024
iXON Ultra 888	Andor	>95	0.00011**	40 @ 10 MHz*	13 x 13	1024 x 1024

*with EM gain <1.
**at max cooling.

Table 3. Examples of commerically available EMCCD cameras.

A 2012 application note by the camera manufacturer ANDOR found no performance difference between the tested EMCCD (ANDOR iXon3) and sCMOS (ANDOR Neo) cameras using a spinning disk confocal setup where the emission light from the sample was equally split between the two cameras [81]. However, the authors are quite right to raise the important caveat that the samples imaged contained near-perfect labeling, and were fluorescently very bright and stable. These artificial conditions are near ideal and do not reflect the typical 3D spinning disk imaging experiments. Typical signals are much lower and on the order of 0–20 photons/pixel [82]. The authors agree with other researchers in the field that there exists a SNR cross-over point where the sCMOS will outperform the EMCCD [83]. In this experiment, the cross-over point was somewhere between 40 and 100 photons/pixel. A second study found the cross-over point to be greater than 4 photons/pixel or 180 photons/pixel when comparing the ANDOR iXon DU897BV to the Hamamatsu ORCA-Flash 4.0, or ORCA-Flash 2.8 sCMOS cameras, respectively [83]. Image artifacts (streaking lines) were seen at specific illumination intensities for the sCMOS but not the EMCCD cameras.

Model	Manufacturer	QE @ 600 nm (%)	Dark Current (e–/pixel/sec)	Readout Noise (e- pixels/rms)	Resolution (um-pixels)	Imaging array
Prime sCMOS	Photometrics	72	0.01*	1.3	6.5 x 6.5	2048 x 2048
Optimos sCMOS	Qimaging	55	0.5	1.9	6.5 x 6.5	1920 x 1080
KURO sCMOS	Princeton Sci.	95	1.9 (−10 C)	1.5	11 x 11	1200 x 1200
Zyla 4.2 Plus	Andor	>80	0.019*	1.1 @ 216 MHz	6.5 x 6.5	2048 x 2048

*water cooled.

Table 4. Examples of commercially available sCMOS cameras.

The brightness of the fluorescent sample is highly dependent on the microscope setup. Therefore, the SNR cross-over point must be determined empirically for each experimental situation (typical ranges observed are ~4–200 photon/pixel). EMCCD cameras have inherent excess noise due to the amplification process that contributes to about 50% of the total noise. This prevents manufacture of a shot-noise limited EMCCD detector [83]. In contrast, sCMOS camera performance (increased QE and reduced noise) can be improved through hardware and software optimization [78]. It is predicted that sCMOS would be the camera of choice when greater than 50 photons/pixel is reached [84]. Currently, EMCCDs are better suited to measure small fluorescence changes in live cells with high spatial resolution compared to sCMOS [85]. In addition, EMCCDs superior imaging capability for low light samples outweigh the benefits of sCMOS for spinning disk confocal microscopy at this time [81]. This hotly debated comparison between EMCCD and sCMOS camera performance is not expected to slow down soon. In fact, implementation of in-camera signal-processing algorithms are being introduced to enhance both EMCCD and sCMOS camera technologies (see Section 5.7), and could re-ignite the debate.

5.7. Next generation photodetectors

Technological advances have brought the performance of EMCCD and CMOS cameras closer to point-like detectors such as APDs and Hybrid detectors. The sensitivity and readout noise are still generally better for point-like detectors, but these types of detectors cannot "count" photon numbers unless external electronics and software are used to bin the photons (~1 nano-sec to 10 msec). Importantly, there are inherent throughput limitations for point-like detectors in contrast to array detectors. One solution around this problem is parallelization of the point-like detector (i.e. an array of point-like detectors). Factors that must be considered for parallelization include: parallel excitation, parallel detection, excitation and detection alignment, and data processing [86]. Each light source must be sufficiently separate to prevent crosstalk during multiple-spot excitation and the spot separation must be a few diameters apart as a general rule of thumb [87]. An eight pixel custom linear SPAD array and a 32 x 32 CMOS SPAD array were used recently to perform parallel FCS measurements on a fluorescent dye in solution with quasi-diffraction limited spots [88–90]. Custom liquid crystal light modulators, or micro-lens are required to direct and separate the multiple PSFs. These results with highly-parallel arrays are encouraging but these parallelized detectors have lower sensitivity and larger dark counts thus leading to higher background counts for FCS measurements compared to individual detectors [90]. Recently, a frame summing/filtering scheme called "smart-aggregation" was introduced to increase SPAD array performance [91]. This approach promises to "push" SPAD camera performance beyond EMCCD and CMOS cameras, but significant technological advances are still required.

Bright photo-stable dyes that emit in the red and near-infrared range of the visible spectrum are commonly used in imaging studies involving animals. This is due to their favorable excitation using two-photon light sources and reduced scattering of emitted light in thick animal tissues [92]. The wide use of these dyes has led to the development of red enhanced SPAD (RE-SPAD). Traditional SPADs have a QE of 15% at 800 nm wavelengths but newer RE-SPADs have a QE of 40–60% [93, 94]. Recently, a parallelization of RE-SPADs has been fabricated and tested [94]. A new type of SPAD architecture has been implemented to enhance the electrical isolation between individual SPAD elements to reduce crosstalk [94]. These RE-SPAD array detectors are still in their infancy and further characterization of detector parameters (temporal resolution, dark count rate, after-pulsing, etc.) is necessary before commercialization and mass production are a reality.

SPADs have sub-nanosecond time resolution but are inefficient imaging detectors because scanning is required for image formation. In contrast, wide-field detectors that have picosecond response, such as time-gated ICCDs, are photon-inefficient due to the lens coupling of the intensifier screen to the CCD sensor. Recently, several research groups have developed single-photon wide-field detectors with high temporal and spatial resolutions thus attempting to combine the best attributes of both SPADs and ICCDs [86, 90, 95]. One such detector called H33D (pronounced "HEED") is composed of S20 multi-alkali cathode, triple MCP stack, and cross-delay line anode [95]. A front-end field programmable gate array (FGPA) is used for time-stamping and synchronization of photon events. The H33D detector has 18% QE at 400 nm diminishing to 3% at 630 nm. Temporal resolution was 100 picosecond FWHM using a red diode laser. Fluorescence lifetimes of dyes in solution, colloidal quantum dots and quantum dot labeled receptors on the surface of cells have been measured using this high temporal and spatial single photon detector [95–97]. Several other groups have fabricated large-area photon counting detectors with a similar architecture [98].

There are multiple sources of noise (dark, read, shot) that affect the SNR collected by the detector. Cooling of the detector sensor can reduce dark noise and read noise can be minimized through thoughtful electronic design and sensor performance optimization. Photon shot noise is an intrinsic property of light and has a Poisson statistical distribution. Shot noise varies with the square root of the signal (shot noise = $\sqrt{\text{signal}}$) and thus increasing light levels leads to improving SNR. However, the low light levels encountered with SRM and FFTs measurements can lead to shot noise dominating the signal. Increased exposure time, frame averaging, and increased excitation light intensity have been used to circumvent this low SNR problem. A fourth method that is more compatible with SRM is the use of de-noising algorithms to dynamically analysis the acquired image and filter out the noise. Spatial and temporal filtering of images and video has been extensively studied [99–104]. One very popular spatial filter is the averaging-based non-local means (NLM) filter [99]. It is important to choose an appropriate algorithm and parameter settings that do not introduce artifacts (e.g. aliasing, blurring or ringing).

Manufacturers are starting to add real-time de-noising algorithms to the firmware of their cameras for enhanced SNR and improved performance. The Prime™ family of cameras from Photometrics employs an algorithm developed at INRIA (NLM with patch based refinement) and optimized at the Curie Institute [101]. This algorithm called Prime-Enhance allows up to an 8-fold decreased exposure time while retaining a high SNR (due to reduction of photon shot noise effects) thus leading to reduced photo-toxicity in live cell experiments [101, 105]. The Prime-Enhance algorithm is also purported to not introduce common processing artifacts such as aliasing, blurring or ringing.

6. Examples of photodetector used in SRM & FFTs

A popular photodetector for probe-based SRM is EMCCD cameras. These camera-types have been used to study membrane protein dynamics in plant cells, assembly of HIV virus particles, and viral protein receptor interactions just to name a few applications [106–109]. New sCMOS cameras are inherently faster than EMCCDs, allow high-throughput capabilities, and are starting to be used for some SRM applications [83, 98]. Currently, EMCCD cameras are employed for rapid high resolution live cell imaging [85]. While, sCMOS cameras are employed for slower cellular dynamic studies, or fixed cell super-resolution imaging [98]. For fluorescence fluctuation studies, APD and hybrid detectors are commonly employed because of their sensitivity, efficiency, and faster response compared to PMTs. Usually APDs are used for FCS and PCH measurements of fluorescently-labeled molecules in tissue culture cells. For example,

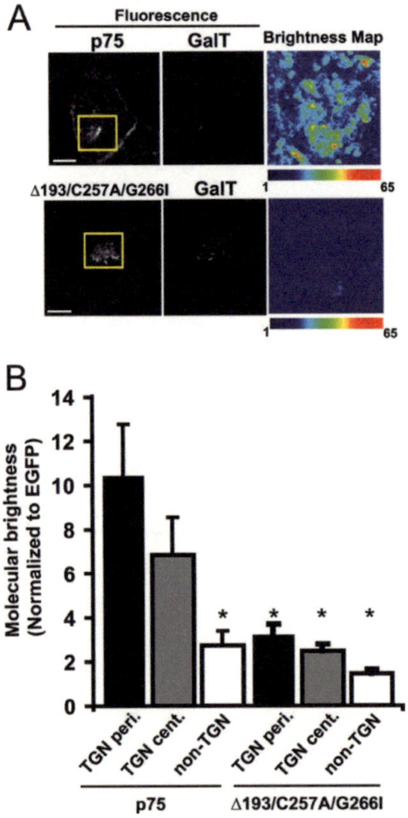

Figure 3. N&B analysis reveals spatially heterogeneous clustering of the p75 receptor at the trans-Golgi network. A) *Left:* Fluorescence images of MDCK cells expressing wildtype p75 receptor and apical sorting mutant (Δ193/C257A/G266I) at the trans-Golgi network (TGN). Cells co-expressing TGN marker GalT-mcherry. *Right:* Molecular brightness maps of inserts with scale equal to normalized brightness (× EGFP per pixel). B) Normalized brightness values (B values) for wildtype and mutant p75 in non-TGN, peripheral-TGN (TGN peri.), and central-TGN (TGB cent.) asterisk, $p < 0.05$ unpaired T-test. Figure reproduced with permission from *Mol. Biol. Cell* 24(12), (Jun 15, 2013) doi: 10.1091/mbc.E13-02-0078.

oligomerization (protein–protein association) was shown to be important for the trafficking of a membrane receptor (p75) to the apical surface of epithelial cells [49]. These studies used APDs to make FCS, PCH, and N&B measurements of wildtype and mutant receptors. The wildtype receptor formed higher-order oligomers in the Golgi membrane and dimers at the plasma membrane (**Figure 3**). In contrast, the mutant proteins that could not traffic to the apical membrane did not form higher-order oligomers at the Golgi (**Figure 3**).

Hybrid detectors have been employed to measure the formation of lipid rafts in a cell model of Fabry's disease [110]. In Fabry's disease, lysosomal function is disrupted due to reduced activity of a specific enzyme (alpha-galactosidase A) that leads to accumulation of neutral glycosphingolipids such as globotriaosylceramide (Gb3). N&B analysis was performed on wildtype and alpha-galactosidase deficient cells, which act as a model for Fabry's disease. Antibody-induced clustering of a model lipid raft protein was increased in the mutant compared to control cells [110]. These results suggested that accumulated Gb3 may alter lipid raft protein interactions in membranes of alpha-galactosidase deficient cells. These two examples are just a snippet of the many experiments that have employed a variety of low-light photodetectors.

7. Summary and concluding remarks

In the past seventy-five years, advances in micro-circuitry and semiconductor materials have led to dramatic advances in photodetector QE, sensitivity, and resolution. EMCCD and sCMOS cameras are the detectors of choice for probe-based SRM. In contrast, APD and hybrid detectors are becoming more common for use in PSF-engineered SRM and FFTs. Manufacturing attempts to combine the best attributes of point-like detectors with array detectors (CMOS and SPAD arrays) has been met with moderate success. These parallelized photodetectors are still in their early stages and are not routinely used. Shot noise and read noise are a problem especially as pixel densities have increased in array detectors. De-noising algorithms are being used to combat shot noise and increase SNR for low-light applications. Finally, A detector suitable for SRM and single molecule fluorescence experiments must have high sensitivity, high temporal resolution, and low readout noise. The researcher should compare the SNR, dark count, read noise, and frame rate to determine which detector type best fits their experimental needs. No single photodetector technology is suitable for all techniques and some researchers choose to outfit their microscopes with multiple camera technologies (e.g. EMCCD and sCMOS) to allow greatest flexibility when imaging. Advances in hardware and software promises to enhance detector technologies and push the boundaries of single-molecule detection even further in the coming years.

Acknowledgements

I apologize to authors whose research I was unable to cite due to page constraints. My sincere thanks to Dr. Greg Adkison and Dr. Haibing Teng for proof-reading the manuscript. The author is grateful for start-up funds from the Department of Biology and the College of Arts & Sciences Dean's Office at Western Carolina University.

Author details

Robert T. Youker

Address all correspondence to: rtyouker@wcu.edu

Department of Biology, Western Carolina University, Cullowhee, North Carolina, USA

References

[1] Chudakov DM, Matz MV, Lukyanov S, et al. Fluorescent proteins and their applications in imaging living cells and tissues. Physiological Reviews. 2010;**90**:1103-1163

[2] Rayleigh L. On the theory of optical images, with special reference to the microscope. Journal of the Royal Microscopical Society. 1903;**23**:474-482

[3] Rayleigh L XII. On the manufacture and theory of diffraction-gratings. Philosophical Magazine. 1874;**47**:81-93

[4] Schermelleh L, Heintzmann R, Leonhardt H. A guide to super-resolution fluorescence microscopy. The Journal of Cell Biology. 2010;**190**:165-175

[5] Lauterbach MA. Finding, defining and breaking the diffraction barrier in microscopy – A historical perspective. Optical Nanoscopy. 2012;**1**:8

[6] Huang B, Bates M, Zhuang X. Super-Resolution Fluorescence Microscopy. Annual Review of Biochemistry. 2009;**78**:993-1016

[7] Betzig E, Trautman JK. Near-field optics: Microscopy, spectroscopy, and surface modification beyond the diffraction limit. Science. 1992;**257**:189-195

[8] Hwang J, Tamm LK, Böhm C, et al. Nanoscale complexity of phospholipid monolayers investigated by near-field scanning optical microscopy. Science. 1995;**270**:610-614

[9] Willig KI, Harke B, Medda R, et al. STED microscopy with continuous wave beams. Nature Methods. 2007;**4**:915-918

[10] Hell SW, Wichmann J. Breaking the diffraction resolution limit by stimulated emission: Stimulated-emission-depletion fluorescence microscopy. Optics Letters. 1994;**19**:780-782

[11] Thorley JA, Pike J, Rappoport JZ. Super-resolution Microscopy. In: Fluorescence Microscopy. Elsevier, pp. 199-212

[12] Beater S, Holzmeister P, Pibiri E, et al. Choosing dyes for cw-STED nanoscopy using self-assembled nanorulers. Physical Chemistry Chemical Physics. 2014;**16**:6990-6996

[13] Steshenko O, Andrade DM, Honigmann A, et al. Reorganization of lipid diffusion by myelin basic protein as revealed by STED Nanoscopy. Biophysical Journal. 2016;**110**:2441-2450

[14] Benda A, Aitken H, Davies DS, et al. STED imaging of tau filaments in Alzheimer's disease cortical grey matter. Journal of Structural Biology. 2016;**195**:345-352

[15] Ishigaki M, Iketani M, Sugaya M, et al. STED super-resolution imaging of mitochondria labeled with TMRM in living cells. Mitochondrion. 2016;**28**:79-87

[16] Hell SW, Kroug M. Ground-state-depletion fluorscence microscopy: A concept for breaking the diffraction resolution limit. Applied Physics B: Lasers and Optics. 1995;**60**:495-497

[17] Halpern AR, Howard MD, Vaughan JC. Point by point: An introductory guide to sample preparation for single-molecule, super-resolution fluorescence microscopy: Sample preparation for single-molecule super-resolution fluorescence microscopy. In: Mahal L, Romesberg F, Shah K, et al., editors. Current Protocols in Chemical Biology. Hoboken, NJ, USA: John Wiley & Sons, Inc. pp. 103-120

[18] Hell SW, Stelzer EH, Lindek S, et al. Confocal microscopy with an increased detection aperture: Type-B 4Pi confocal microscopy. Optics Letters. 1994;**19**:222

[19] Heintzmann R, Cremer CG. Laterally modulated excitation microscopy: improvement of resolution by using a diffraction grating. In: Bigio IJ, Schneckenburger H, Slavik J, et al. (eds), pp. 185-196

[20] Gustafsson MGL, Agard DA, Sedat JW. Sevenfold improvement of axial resolution in 3D wide-field microscopy using two objective lenses. In: Wilson T, Cogswell CJ (eds), pp. 147-156

[21] Ward EN, Pal R. Image scanning microscopy: An overview. Journal of Microscopy. 2017;**266**: 221-228

[22] Rust MJ, Bates M, Zhuang X. Sub-diffraction-limit imaging by stochastic optical reconstruction microscopy (STORM). Nature Methods. 2006;**3**:793-796

[23] Betzig E, Patterson GH, Sougrat R, et al. Imaging intracellular fluorescent proteins at Nanometer resolution. Science. 2006;**313**:1642-1645

[24] Dempsey GT, Vaughan JC, Chen KH, et al. Evaluation of fluorophores for optimal performance in localization-based super-resolution imaging. Nature Methods. 2011;**8**: 1027-1036

[25] Winter FR, Loidolt M, Westphal V, et al. Multicolour nanoscopy of fixed and living cells with a single STED beam and hyperspectral detection. Scientific Reports. 2017;**7**:46492

[26] Shroff H, Galbraith CG, Galbraith JA, et al. Dual-color superresolution imaging of genetically expressed probes within individual adhesion complexes. Proceedings of the National Academy of Sciences. 2007;**104**:20308-20313

[27] Juette MF, Gould TJ, Lessard MD, et al. Three-dimensional sub–100 nm resolution fluorescence microscopy of thick samples. Nature Methods. 2008;**5**:527-529

[28] Huang B, Wang W, Bates M, et al. Three-dimensional super-resolution imaging by stochastic optical reconstruction microscopy. Science. 2008;**319**:810-813

[29] Wegel E, Göhler A, Lagerholm BC, et al. Imaging cellular structures in super-resolution with SIM, STED and Localisation Microscopy: A practical comparison. Scientific Reports; 6. Epub ahead of print July 2016. DOI: 10.1038/srep27290

[30] Schmidt R, Wurm CA, Jakobs S, et al. Spherical nanosized focal spot unravels the interior of cells. Nature Methods. 2008;**5**:539-544

[31] Shao L, Isaac B, Uzawa S, et al. I5S: Wide-field light microscopy with 100-nm-scale resolution in three dimensions. Biophysical Journal. 2008;**94**:4971-4983

[32] Shtengel G, Galbraith JA, Galbraith CG, et al. Interferometric fluorescent super-resolution microscopy resolves 3D cellular ultrastructure. Proceedings of the National Academy of Sciences. 2009;**106**:3125-3130

[33] Youker RT, Teng H. Measuring protein dynamics in live cells: Protocols and practical considerations for fluorescence fluctuation microscopy. Journal of Biomedical Optics. 2014;**19**:90801

[34] Weidemann T, Mücksch J, Schwille P. Fluorescence fluctuation microscopy: A diversified arsenal of methods to investigate molecular dynamics inside cells. Current Opinion in Structural Biology. 2014;**28**:69-76

[35] Elson EL, Webb WW. Concentration correlation spectroscopy: A new biophysical probe based on occupation number fluctuations. Annual Review of Biophysics and Bioengineering. 1975;**4**:311-334

[36] Magde D, Elson EL, Webb WW. Fluorescence correlation spectroscopy. II. An experimental realization. Biopolymers. 1974;**13**:29-61

[37] Eigen M, Rigler R. Sorting single molecules: Application to diagnostics and evolutionary biotechnology. Proceedings of the National Academy of Sciences. 1994;**91**:5740-5747

[38] Schwille P, Meyer-Almes FJ, Rigler R. Dual-color fluorescence cross-correlation spectroscopy for multicomponent diffusional analysis in solution. Biophysical Journal. 1997; **72**:1878-1886

[39] Digman MA, Sengupta P, Wiseman PW, et al. Fluctuation correlation spectroscopy with a laser-scanning microscope: Exploiting the hidden time structure. Biophysical Journal. 2005;**88**:L33-L36

[40] Digman MA, Gratton E. Analysis of diffusion and binding in cells using the RICS approach. Microscopy Research and Technique. 2009;**72**:323-332

[41] Rossow MJ, Sasaki JM, Digman MA, et al. Raster image correlation spectroscopy in live cells. Nature Protocols. 2010;**5**:1761-1774

[42] Digman MA, Stakic M, Gratton E. Raster image correlation spectroscopy and number and brightness analysis. In: Methods in Enzymology. Elsevier; 2013. pp. 121-144

[43] Schwille P, Haupts U, Maiti S, et al. Molecular dynamics in living cells observed by fluorescence correlation spectroscopy with one- and two-photon excitation. Biophysical Journal. 1999;**77**:2251-2265

[44] Hillesheim LN, Müller JD. The dual-color photon counting histogram with non-ideal Photodetectors. Biophysical Journal. 2005;**89**:3491-3507

[45] Chen Y, Müller JD, Ruan Q, et al. Molecular brightness characterization of EGFP in vivo by fluorescence fluctuation spectroscopy. Biophysical Journal. 2002;**82**:133-144

[46] Macdonald P, Johnson J, Smith E, et al. Brightness analysis. In: Methods in Enzymology. Elsevier; 2013. pp. 71-98

[47] Dross N, Spriet C, Zwerger M, et al. Mapping eGFP oligomer mobility in living cell nuclei. PLoS One. 2009;**4**:e5041

[48] Digman MA, Dalal R, Horwitz AF, et al. Mapping the number of molecules and brightness in the laser scanning microscope. Biophysical Journal. 2008;**94**:2320-2332

[49] Youker RT, Bruns JR, Costa SA, et al. Multiple motifs regulate apical sorting of p75 via a mechanism that involves dimerization and higher-order oligomerization. Molecular Biology of the Cell. 2013;**24**:1996-2007

[50] James NG, Digman MA, Gratton E, et al. Number and brightness analysis of LRRK2 Oligomerization in live cells. Biophysical Journal. 2012;**102**:L41-L43

[51] Ross JA, Digman MA, Wang L, et al. Oligomerization state of Dynamin 2 in cell membranes using TIRF and number and brightness analysis. Biophysical Journal. 2011;**100**:L15-L17

[52] Adu-Gyamfi E, Digman MA, Gratton E, et al. Investigation of Ebola VP40 assembly and Oligomerization in live cells using number and brightness analysis. Biophysical Journal. 2012;**102**:2517-2525

[53] Jiang F, Doudna JA. CRISPR–Cas9 structures and mechanisms. Annual Review of Biophysics. 2017;**46**:505-529

[54] Dobell C. A Collection of Writings by the Father of Protozoology and Bacteriology, Antony Van Leeuwenhoek and his 'Little Animals'. New York: Dover Publications; 1960

[55] Lubsandorzhiev BK. On the history of photomultiplier tube invention. Nuclear Instruments and Methods in Physics Research Section A: Accelerators, Spectrometers, Detectors and Associated Equipment. 2006;**567**:236-238

[56] Bay Z. Electron multiplier as an electron counting device. Nature. 1938;**141**:1011-1011

[57] Zworykin VK, Morton GA, Malter L. The secondary emission multiplier-a new electronic device. Proceedings of the IRE. 1936;**24**:351-375

[58] Goetzberger A, McDonald B, Haitz RH, et al. Avalanche effects in silicon p—n junctions. II. Structurally perfect junctions. Journal of Applied Physics. 1963;**34**:1591-1600

[59] Boyle WS, Smith GE. Charge coupled semiconductor devices. Bell System Technical Journal. 1970;**49**:587-593

[60] Amelio GF, Tompsett MF, Smith GE. Experimental verification of the charge coupled device concept. Bell System Technical Journal. 1970;**49**:593-600

[61] Wanlass F, Sah C. Nanowatt logic using field-effect metal-oxide semiconductor triodes. Institute of Electrical and Electronics Engineers. pp. 32-33

[62] Mendis S, Kemeny SE, Fossum ER. CMOS active pixel image sensor. IEEE Transactions on Electron Devices. 1994;**41**:452-453

[63] Kasunic K. Optical systems engineering. New York: McGraw-Hill Professional, 2011

[64] Lawrence WG, Varadi G, Entine G, et al. A comparison of avalanche photodiode and photomultiplier tube detectors for flow cytometry. In: Farkas DL, Nicolau DV, Leif RC (eds), p. 68590M

[65] Michalet X, Ingargiola A, Colyer RA, et al. Silicon photon-counting avalanche diodes for single-molecule fluorescence spectroscopy. IEEE Journal of Selected Topics in Quantum Electronics. 2014;**20**:248-267

[66] Dautet H, Deschamps P, Dion B, et al. Photon counting techniques with silicon avalanche photodiodes. Applied Optics. 1993;**32**:3894

[67] Suyama M, Kawai Y, Kimura S, et al. A compact hybrid photodetector (HPD). IEEE Transactions on Nuclear Science. 1997;**44**:985-989

[68] Suyama M, Hirano K, Kawai Y, et al. A hybrid photodetector (HPD) with a III-V photocathode. IEEE Transactions on Nuclear Science. 1998;**45**:572-575

[69] Michalet X, Cheng A, Antelman J, et al. Hybrid photodetector for single-molecule spectroscopy and microscopy. In: Enderlein J, Gryczynski ZK, Erdmann R (eds), p. 68620F

[70] Nakamura J, editor. Image Sensors and Signal Processing for Digital Still Cameras. Boca Raton, FL: Taylor & Francis; 2006

[71] Burke B, Jorden P, Vu P. CCD technology. Experimental Astronomy. 2006;**19**:69-102

[72] Spring KR, Fellers TJ, Davidson MW. Introduction to Charge-Coupled Devices (CCDs) https://www.microscopyu.com/digital-imaging/introduction-to-charge-coupled-devices-ccds (accessed 10 September 2017)

[73] Gale MT. Active alignment of replicated microlens arrays on a charge-coupled device imager. Optical Engineering. 1997;**36**:1510

[74] Swain PK, Cheskis D. Back-illuminated image sensors come to the forefront. Photonics Spectra. 2008 https://www.photonics.com/Article.aspx?AID=34685 (2008, accessed 10 September 2017)

[75] Zhang L, Neves L, Lundeen JS, Walmsley IA. A characterization of the single-photon sensitivity of an electron multiplying charge-coupled device. Journal of Physics B: Atomic, Molecular and Optical Physics. 2009;**42**:114011

[76] Schühle UDO. Intensified solid state sensor cameras (chapter 25). In: Observing Photons in Space. Noordwijk, the Netherlands: ISSI; 2010

[77] Dussault D, Hoess P. Noise performance comparison of ICCD with CCD and EMCCD cameras. In: Dereniak EL, Sampson RE, Johnson CB (eds), p. 195

[78] Bigas M, Cabruja E, Forest J, et al. Review of CMOS image sensors. Microelectronics Journal. 2006;**37**:433-451

[79] Fowler B, Liu C, Mims S, et al. A 5.5Mpixel 100 frames/sec wide dynamic range low noise CMOS image sensor for scientific applications. In: Bodegom E, Nguyen V (eds), p. 753607

[80] Jonkman J, Brown CM. Any way you slice it: A comparison of confocal microscopy techniques. Journal of Biomolecular Techniques. 2015 jbt.15-2602-003

[81] EMCCD vs sCMOS - Cameras For Spinning Disk Confocal Microscopy. Application Note, ANDOR Technology http://www.andor.com/learning-academy/emccd-vs-scmos-cameras-for-spinning-disk-confocal-microscopy-application-note (accessed 15 August 2017)

[82] Pawley JB. Handbook of Biological Confocal Microscopy. Springer-Verlag US: Boston, MA; 2006

[83] Long F, Zeng S, Huang Z-L. Localization-based super-resolution microscopy with an sCMOS camera part II: Experimental methodology for comparing sCMOS with EMCCD cameras. Optics Express. 2012;**20**:17741

[84] Joubert JR, Sharma DK. Emccd vs. scmos for microscopic imaging. Photonics Spectra. 2011:46-50

[85] Beier HT, Ibey BL. Experimental comparison of the high-speed imaging performance of an EM-CCD and sCMOS camera in a dynamic live-cell imaging test case. PLoS One. 2014;**9**:e84614

[86] Michalet X, Colyer RA, Scalia G, et al. Development of new photon-counting detectors for single-molecule fluorescence microscopy. Philosophical Transactions of the Royal Society B. 2012;**368**:20120035-20120035

[87] Michalet X, Colyer RA, Scalia G, et al. High-throughput single-molecule fluorescence spectroscopy using parallel detection. In: Razeghi M, Sudharsanan R, Brown GJ (eds), p. 76082D

[88] Colyer RA, Scalia G, Rech I, et al. High-throughput FCS using an LCOS spatial light modulator and an 8 × 1 SPAD array. Biomedical Optics Express. 2010;**1**:1408

[89] Colyer RA, Scalia G, Villa FA, et al. Ultra high-throughput single molecule spectroscopy with a 1024 pixel SPAD. In: Enderlein J, Gryczynski ZK, Erdmann R (eds), p. 790503

[90] Michalet X, Colyer RA, Scalia G, et al. New photon-counting detectors for single-molecule fluorescence spectroscopy and imaging. In: Itzler MA, Campbell JC (eds), p. 803316

[91] Gyongy I, Davies A, Dutton NAW, et al. Smart-aggregation imaging for single molecule localisation with SPAD cameras. Scientific Reports; 6. Epub ahead of print December 2016. DOI: 10.1038/srep37349

[92] Perry SW, Burke RM, Brown EB. Two-photon and second harmonic microscopy in clinical and translational cancer research. Annals of Biomedical Engineering. 2012;**40**:277-291

[93] Gulinatti A, Rech I, Panzeri F, et al. New silicon SPAD technology for enhanced red-sensitivity, high-resolution timing and system integration. Journal of Modern Optics. 2012;**59**:1489-1499

[94] Gulinatti A, Ceccarelli F, Rech I, et al. Silicon technologies for arrays of Single Photon Avalanche Diodes. In: Itzler MA, Campbell JC (eds), p. 98580A

[95] Michalet X, Siegmund OHW, Vallerga JV, et al. Photon-counting H33D detector for biological fluorescence imaging. Nuclear Instruments and Methods in Physics Research Section A: Accelerators, Spectrometers, Detectors and Associated Equipment. 2006;**567**: 133

[96] Pinaud F, King D, Moore H-P, et al. Bioactivation and cell targeting of semiconductor CdSe/ZnS Nanocrystals with Phytochelatin-related peptides. Journal of the American Chemical Society. 2004;**126**:6115-6123

[97] Dahan M, Laurence T, Pinaud F, et al. Time-gated biological imaging by use of colloidal quantum dots. Optics Letters. 2001;**26**:825-827

[98] Ma H, Fu R, Xu J, et al. A simple and cost-effective setup for super-resolution localization microscopy. Scientific Reports; 7. Epub ahead of print December 2017. DOI: 10.1038/s41598-017-01606-6

[99] Buades A, Coll B, Morel JM. A review of image Denoising algorithms, with a new one. Multiscale Modeling and Simulation. 2005;**4**:490-530

[100] Park SW. Image denoising filter based on patch-based difference refinement. Optical Engineering. 2012;**51**:67007

[101] Boulanger J, Kervrann C, Bouthemy P, et al. Patch-based nonlocal functional for Denoising fluorescence microscopy image sequences. IEEE Transactions on Medical Imaging. 2010;**29**: 442-454

[102] Boulanger J, Kervrann C, Bouthemy P. Space-time adaptation for patch-based image sequence restoration. IEEE Transactions on Pattern Analysis and Machine Intelligence. 2007;**29**:1096-1102

[103] Buades A, Coll B, Morel J-M. Nonlocal image and movie Denoising. International Journal of Computer Vision. 2008;**76**:123-139

[104] Dabov K, Foi A, Katkovnik V, et al. Image Denoising by sparse 3-D transform-domain collaborative filtering. IEEE Transactions on Image Processing. 2007;**16**:2080-2095

[105] PrimeEnhance: 2D Active Image Denoising. Technical Note Rev A4-02032017, Photometrics. https://www.photometrics.com/resources/technotes/pdfs/PrimeEnhance-TechNote.pdf (2017, accessed 28 August 2017)

[106] Dirk BS, Heit B, Dikeakos JD. Visualizing interactions between HIV-1 Nef and host cellular proteins using ground-state depletion microscopy. AIDS Research and Human Retroviruses. 2015;**31**:671-672

[107] Muranyi W, Malkusch S, Müller B, et al. Super-resolution microscopy reveals specific recruitment of HIV-1 envelope proteins to viral assembly sites dependent on the envelope C-terminal tail. PLoS Pathogens. 2013;**9**:e1003198

[108] Wang X, Li X, Deng X, et al. Single-molecule fluorescence imaging to quantify membrane protein dynamics and oligomerization in living plant cells. Nature Protocols. 2015;**10**:2054-2063

[109] Dirk B, Van Nynatten L, Dikeakos J. Where in the cell are you? Probing HIV-1 host interactions through advanced imaging techniques. Virus. 2016;**8**:288

[110] Labilloy A, Youker RT, Bruns JR, et al. Altered dynamics of a lipid raft associated protein in a kidney model of Fabry disease. Molecular Genetics and Metabolism. 2014;**111**: 184-192